Total Productive Maintenance

Total Productive Maintenance

Steven Borris

McGraw-Hill

New York Chicago San Francisco Lisbon London Madrid
Mexico City Milan New Delhi San Juan Seoul
Singapore Sydney Toronto

Cataloging-in-Publication Data is on file with the Library of Congress

Copyright © 2006 by The McGraw-Hill Companies, Inc. All rights reserved. Printed in the United States of America. Except as permitted under the United States Copyright Act of 1976, no part of this publication may be reproduced or distributed in any form or by any means, or stored in a data base or retrieval system, without the prior written permission of the publisher.

1 2 3 4 5 6 7 8 9 0   DOC/DOC   0 1 0 9 8 7 6 5

ISBN 0-07-146733-5

The sponsoring editor for this book was Kenneth P. McCombs and the production supervisor was Richard Ruzycka. It was set in Century Schoolbook by TechBooks. The art director for the cover was Anthony Landi.

Printed and bound by RR Donnelley.

This book was printed on recycled, acid-free paper containing a minimum of 50% recycled, de-inked fiber.

McGraw-Hill books are available at special quantity discounts to use as premiums and sales promotions, or for use in corporate training programs. For more information, please write to the Director of Special Sales, McGraw-Hill Professional, Two Penn Plaza, New York, NY 10121-2298. Or contact your local bookstore.

Information contained in this work has been obtained by The McGraw-Hill Companies, Inc. ("McGraw-Hill") from sources believed to be reliable. However, neither McGraw-Hill nor its authors guarantee the accuracy or completeness of any information published herein, and neither McGraw-Hill nor its authors shall be responsible for any errors, omissions, or damages arising out of use of this information. This work is published with the understanding that McGraw-Hill and its authors are supplying information but are not attempting to render engineering or other professional services. If such services are required, the assistance of an appropriate professional should be sought.

Contents

List of Figures, Formulas, and Tables   ix
Preface   xxi
Acknowledgments   xxiii

Introduction   1

The Development of Maintenance Systems	1
The Writing Technique and the Contents of the Book	3
The Pillars of TPM	7
The Toyota Production System (Also Known As Lean Manufacturing)	12
Finally, Advice for Using the Techniques	12

Chapter 1. TPM—Basic, Use, and Ideal Conditions   15

Fault Development	15
The Basic Condition	21
Technical Standards	24
Overall Equipment Efficiency	26
The availability of the equipment	29
The performance of the equipment	31
The quality of the product	31
Natural and Forced Deterioration	32
Use Conditions	34
The Ideal Condition	37
Improvement Methodology	37
How Do We Restore the Basic Condition?	41

Chapter 2. TPM Jishu-Hozen—Autonomous Maintenance   43

The TPM Initial Clean and Inspect and F-Tagging	45
The Cleaning Map: What and Where to Clean	50
F-Tags: How to Record Fuguai	52
Discovery of a Serious Fault during the Cleaning	58
Tracking the Progress of the Initial Cleans	58

Chapter 3. TPM—Analyzing and Categorizing the Failure Data — 61

F-Tags, The Machine History Log, and Minor Stops or Unrecorded Losses
   Categorizing — 63
Finding Out the TPM Causes for the F-Tags to Help Find the Cure — 71
Pareto Charts — 80
The Defect Map — 83

Chapter 4. TPM—Creating Standards and Preparation for Autonomous Maintenance — 85

Task Transfer: Red to White F-Tags or PM to AM Tasks — 89
Explanation of the Embedding and Responsibility Spreadsheet — 95
PM Teams (Kobetsu Kaizen) — 97

Chapter 5. TPM: The Education & Training and Safety Pillars — 103

The TPM Education & Training Pillar — 106
   Equipment training — 113
   A sequence for training equipment — 116
   Competency: How does TPM assess the skill level of the team members? — 120
The TPM Safety Pillar — 126
   The area map — 127
   The hazard map — 129
   Risk assessment — 131
   Safe working procedures: Using as standards — 146

Chapter 6. 5S: Organization and Improvements by Default — 153

5S: SSSSS—The Meaning — 154
The Benefits of 5S — 154
The Decision to Implement 5S — 156
Initial Management Implementation — 157
   Audit sheets — 159
   The red tag holding area — 160
Step 1: Seiri—Sort — 162
   Red tag details — 162
   Step 2: Seiton—Set in Order — 164
Step 3: Seiso—Shine — 174
   The 5S cleaning map or assignment map — 175
Step 4: Seiketsu—Standardization — 179
Step 5: Shitsuke—Self-Discipline — 181

Chapter 7. SMED—Single Minute Exchange of Die — 183

Where Did SMED Originate? — 183
Step 1: Creating the SMED Team — 186
   The team members and their responsibilities — 186
Step 2: Select the Tool — 187
Step 3: Document Every Step of the Changeover — 188
Step 4: Viewing the Changeover as a Bar Graph — 196
Step 5: Define the Target Time for the Changeover — 197

Step 6: Analysis of the Elements	198
The SMED Analysis	204
Implementing ideas	208
Create the new procedure	208
Step 7: Repeating the Exercise	211
Applying SMED to Maintenance and the Use of Turnaround Parts	211

Chapter 8. Deciding on a Maintenance Strategy   213

The TPM PM Analysis	214
The malfunction and PM maps	215
Interpreting PM maps	220
Scheduled maintenance or scheduled restoration	223
Scheduled replacement or scheduled discard	226
The RCM PM Analysis	227
The RCM decision diagram	228
Failure is unacceptable: Redesign the system	243
Recording the process on the decision worksheet	249
Failure finding and calculating acceptable risk	252

Chapter 9. RCM—Reliability Centered Maintenance   257

The First Stage in an RCM Analysis: The Operating Context	257
Example of a Furnace Boatloader Operating Context: Tool Analysis Level	261
Equipment Defined as Functions	267
Identifying Functions and Labeling	280
Functional Failures to Failure Effects	284
Failure modes	285
Failure effects	289
Where Did RCM Come From?	292
Non-Time-Based Failures	297
Infant mortality	298

Chapter 10. Time- and Condition-Based Maintenance   299

Introduction to On-Condition Maintenance	302
Friction between Maintenance and Production	304
What if we were starting from scratch?	311
In Summary	313

Chapter 11. Fault Analysis: A Few Ways to Help Find Root Causes   315

The 5 Why's	317
Fishbone Diagrams	320
Fault Tree Diagrams	323
OCAPs: Out-of-Control Action Plans	324

Chapter 12. Team Objectives and Activity Boards   327

Activity Boards	328
Team Goals	329
Monitoring Progress	331

What do we monitor?	331
How do we calculate the failure rate and the target improvement?	331
Authority for Working in Specific Machine Areas	334
What Do the Results of a Real RCM Analysis Look Like?	335
Summary of the boatloader analysis	337
Lean Manufacturing	340
Defects	342
Overproduction	343
Waiting	344
Transporting	344
Overprocessing	347
Unnecessary inventory	348
Unnecessary operator movement	349
Value	350
Equipment	353
Pull	355

Chapter 13. Six Sigma: A High-Level Appreciation — 359

Graphs and Their Use in Six Sigma	361
Average and standard deviation	362
Standard deviation and z scores	364
The x–y graph	365
The Main Terms of Six Sigma	367
The customer	367
The teams and the leaders	368
The Champion or Sponsor	368
Six Sigma Controller	368
The Rules and Expectations	368
The Six Sigma Charter	368
The Technical Stuff	369
The sigma value	369
Defects per opportunity	369
Defects per million opportunities	369
The Stages of a Six Sigma Analysis	371
Considerations or Limitations in Using Six Sigma	373
Faultfinding the cause of a lamp failure	374
Possible Limitations with Using Statistics	376

Index — 379

List of Figures, Formulas, and Tables

Figure I.1   The pillars of TPM. *There are eight pillars in TPM. They are intended to cover every department and function in the company. There are special pillars for Education and Training and for Safety, as TPM recognizes these two areas as being a major cause of poor performance.*

Formula 1.1   Availability formula. *The availability referred to in OEE is a cause of many disputes. There are industry standards, but people disagree on their definitions. Most people want the downtimes to be "blamed" on other groups. To keep it simple, here it is based only on available time and downtime.*

Formula 1.2   Performance formula. *As with the Availability Formula, this is simple. It is based entirely on how many units are produced and how many could have been produced.*

Formula 1.3   Quality formula. *This formula is based on the number of units produced but it takes into consideration the number of units lost as a result of defects. There is often debate as to whether compensation should be made to allow for downtimes.*

Formula 1.4   OEE formula. *This is the critical measure used by TPM to evaluate the capability of a piece of equipment within a production system. It has only three main components, performance, availability, and quality, but these values encompass all of the issues that can affect how much usable product the equipment and operator system can make.*

Table 1.1   Availability, performance, and quality as a percentage and a probability. *OEE is a sensitive formula designed to pick up problems early. This table shows how rapidly the OEE values fall in response to any changes. It also shows that, even with 100% availability, there can still be problems with the equipment or the setup.*

Table 1.2   A comparison of the vacuum system's exhaust pipework. *This is a nice table that illustrates a point about use conditions and how simple it can sometimes be to avoid several years of problems, downtime, and secondary damage. In such instances, reading the manuals or talking to the vendor could have prevented the bad layout of a system.*

Figure 1.1   Example of a failure analysis sheet. *The failure analysis sheet is designed to ensure the correct data about a problem is recorded. If the fault is repaired and the issue returns, the original data is available for reference. It also provides the necessary information for evaluating equipment performance data.*

Figure 2.1   The process flowchart. *This chart summarizes the steps required to set up and run an Autonomous Maintenance team. The procedure is virtually identical to that which would be applied to the autonomous maintenance part of a Zero Fails team. It includes the training, safety, and repeat scheduling following the completion of the initial clean.*

Figure 2.2   Illustration of a cleaning map of a part of a tool. *This is a simple photograph showing two views of a tool and where it should be cleaned. The image was imported to PowerPoint and the dots were added. The data is for illustration only. Notice there is also text on the map to give guidance to risk assessments.*

Figure 2.3   An example of an F-tag log sheet. *The log sheet is used to record the type of tag that has been issued, what task it is for, where it is and to link it to the relevant risk assessments and safe working procedures. It also establishes if the problem is a recurring fault.*

Figure 2.4   Task certification sheet example. *The log sheet is used to identify the type of tag that has been issued, what it is for, where it is and to link it to the relevant risk assessments and safe working procedures. It also certifies who all are permitted to work on the task and the date they were approved to do so.*

Figure 2.5   Example of the drop-off of F-tags. *As the initial clean progresses, improvements will be made. This chart is one way to see the successes and monitor the progress.*

Figure 3.1   A product handling system showing the initial failure area. *This is a complex handling system that is prone to drift and set up issues. The design was around 15 years old. It was only specified to move 1000 units without assistance. It was replaced by a robot system in later tools. In one situation the robot lasted 230,000 movements without an assist.*

Table 3.1   Note of initial fault symptoms illustrated in Fig. 3.1. *The image gives an immediate visual image of the magnitude and areas of the problems. Although the image has dots on it, there is a limited amount of precision. The table permits more detail and the capability to simply put a check mark any time the fault reappears. It also makes analysis simpler.*

Figure 3.2   A product handling system showing the 6-month failure areas. *This is a representation of the changes seen in the dot (fault) distribution after a period of 3 months. This one clearly shows areas of recurring issues and not simply a couple of early, single fault issues.*

Figure 3.3   Example of a possible F-tag category spreadsheet. *The categories are the groupings that TPM uses to identify root causes. They include unchecked deterioration, inadequate skill level, basic condition neglect, operating standards not followed, and design weakness. This purpose links the F-tag, the fault description, and the possible (machine) root cause with the category. It also includes the anticipated fix and W3 (What, Where, and Who).*

List of Figures, Formulas, and Tables   xi

Figure 3.4   A standard defect chart—to visually display the JIPM categories. *The categories recorded in Fig. 3.3 above are analyzed and plotted as a simple histogram to get a visual display of the spread of the faults. This can show a weakness in the overall maintenance system.*

Figure 3.5   Modified defect chart showing the historical fails—sorted as number of fails. *This is a bar chart like Fig. 3.4, except it has a second axis added. The new axis is the actual downtime hours for each category; this shows a real amount of lost production time in addition to the number of faults.*

Figure 3.6   Historical fails data sorted as repair time. *A modified defect chart, again with two axes. This time the data has been sorted as downtime. The reason is to highlight how the distribution of the categories actually changed when the different perspective was used.*

Table 3.2   Sample data as number of fails and hours of repair time. *This is a simple table showing the layout needed for Excel to offer the double axis graph. It is sorted in fails.*

Table 3.3   Data as percentage of fails and percentage hours of repair time. *This is the same data as in Table 3.2, but we have started the conversion to percentages. The next chart shows the final stage in preparing for the Pareto chart.*

Table 3.4   Data as a Pareto chart—sorted against fails. *The percentages in Table 3.3 have been changed into cumulative values for the Pareto to be plotted.*

Figure 3.7   Data as a Pareto chart—sorted as fails. *This is another modified Pareto. The data in the table format in Table 3.4 will offer the combined bars and line graph when Excel is used to chart the data. It does no harm to keep all the data as displayed as it makes evaluating the individual contributions easy.*

Table 3.5   A simpler method of explaining the Pareto data calculation. *This is a simple table but it is so much easier to understand than using just the words above.*

Figure 3.8   Defect map example. *The defect map is a diagram of the whole tool, showing the positions of the red and white F-tags. One of its advantages is highlighting areas where faults accumulate. This can be a guide to a skill shortage or a lack of standards.*

Figure 4.1   Sources of faults through to final working standards. *This is a complex flowchart. It summarizes the sources of the faults and how to divide them. It breaks the tasks into two groups: one for the AM groups and one for the PM groups. It also details the procedures to follow for correcting the problems and organizing the meetings.*

List 4.1   Pre-AM safety checks. *This list is a confirmation that the teams are competent to work without risk of causing injury to themselves or any others.*

Figure 4.2   F-tag embedding and responsibility spreadsheet. *This is a summary of all of the tasks taken (or to be taken) to ensure that an F-tag becomes part of the system. It will become a part either in a scheduled AM clean or in a maintenance PM inspection. The frequencies for the task and the responsibilities are also included.*

List 4.2 Suggested AM team responsibilities. *The team is also autonomous and has a range of responsibilities. The responsibilities are defined by the managers. They are based on the competence of the teams. This list contains suggestions for each team's responsibilities.*

Figure 4.3 Improving "hard-to-access areas" by modifying a panel. *This is an illustration of a simple cutout in a panel to permit easy viewing of a gauge. Previously the main panel had to be removed to gain access.*

List 4.3 Suggested PM team responsibilities. *The team is also autonomous and has a range of responsibilities. The responsibilities are similar to an AM team and are defined by the managers. They are based on the competence of the teams, but cover an AM and a Zero Fails component that have interrelated responsibilities. This list contains suggestions for each team's responsibilities.*

Figure 4.4 Example of a master fails list and weekly chart. *This is a stacked histogram designed to show the number of failures on a weekly basis and the number of times each fault occurred.*

List 5.1 Prerequisite training for TPM teams. *TPM has a range of skills it needs to carry out the specific tasks, but there are also a number of foundation skills also required. These tend to be more general, but are still essential for working safely and efficiently in a team.*

Figure 5.1 Sample training record as would be used on the "activity board". *This is the common training toolkit that is provided to all team members and includes training on TPM.*

List 5.2 Equipment specific training required. *This is linked to Fig. 5.2 and is more technical and tool-specific training.*

Figure 5.2 Example of tool-specific training records. *Associated with List 5.2.*

Figure 5.3 Example of a one-point lesson. *The one-point lesson is the favored TPM way. It is simple and makes only one or two points. It works on the same principles as an advertising poster.*

Figure 5.4 Zero Fails team composition—membership.

Figure 5.5 Zero Fails team composition—overlapping management. *This is a simple diagram based on the triangular, overlapping management structure used by the Japanese in forming teams. The structure ensures the information is disseminated both upwards and downwards.*

Figure 5.6 A layout drawing of an ion implanter (NV10-160 high current implanter) mentioning the main modules. *This is a diagram you will see often, except the labels will be different. It is one way of displaying the basic layout of a tool. In this case it is a simple plan diagram as would be provided by the manufacturer. It has all of the major components listed.*

Figure 5.7 Information on the process gases used. *This is simple data as supplied by the manufacturer.*

Figure 5.8 Emergency off switches (EMOs). *This a variation of Fig. 5.6. It has been modified to show detail missing from the first one. In this case it is the positions of the EMOs.*

Figure 5.9 Competency levels for four tasks. *It is easy to give a speech and call it training. It is harder to confirm whether people understand. TPM, as does most of industry, has a variation on five levels from knowing nothing to being able to teach the subject. It involves supervision and training levels.*

Figure 5.10 Example of a skill log that can be used for transferring tasks to operators. *I like this spreadsheet, it shows the fuel gauge level as defined in Fig. 5.9 but also has information on training and approval dates.*

Figure 5.11 A spin track system, drawn using PowerPoint. *This is a PowerPoint drawing. It is highly simplified. It is planned to show the main path the product follows and the steps the process computer looks for. It is not intended to be a work of art, just to pass on the essential information.*

Figure 5.12 The area map for a Nova implanter. *This is another variation of the layout. This time it is being used to illustrate how a tool can be broken down into functional areas that can be used by all areas of TPM and safety. The areas are particularly handy for modular risk assessments.*

Figure 5.13 Sample hazard map. *The hazard map is a table, like a shop floor plan that tells which floors sell which goods. The difference being that the floors are the areas (from the area map) and the goods are the hazards on each floor.*

Table 5.1 Three natural levels of a risk assessment.

Figure 5.14 The sequence of steps for developing modular risk assessments and safe working procedures. *This is a flow diagram that outlines the sequence of steps and evaluation of risk for all three levels.*

Table 5.2 How risk varies for the initial assessment. *There are different levels of risk. This one defines three levels from operator to task.*

Figure 5.15 Example showing part of a Level 1 risk assessment. *This is a spreadsheet format. There will be many number of formats to choose from. This format is not suitable for sorting.*

Formula 5.1 Risk assessment formula.

Formula 5.2 Example of a risk calculation. *This calculation is to illustrate why we need to find a numerical method to define risk levels.*

Formula 5.3 A numerical risk calculation. *This time the calculation in Formula 5.2 has been substituted for numbers and gives us a way of comparing risk.*

Figure 5.16 Example showing part of a Level 2 risk assessment.

Figure 5.17 Example showing part of a Level 3 risk assessment.

Figure 5.18 Example #1 of a "step-by-step" safe working procedure. *This is a method for writing procedures that has proved to be effective over a number of years. It is based on a table format with absolute instructions on the left and general information on the right. It aims to teach understanding as well as following instructions.*

Figure 5.19 Example #2 of a "step-by-step" safe working procedure.

Table 6.1 A table of currently used "S" equivalents. *When 5S was exported, there was a mad rush for equivalent words that had the same* meanings as the Japanese names. This table is a collection of the names used. There are probably more, but these ones make the point.

Figure 6.1 A schematic illustration of the 5S process. *This is a graph showing the improvements toward the ultimate goal gained by each step of 5S.*

Figure 6.2 A summary of the stages of implementation. *There are a number of points to be considered before implementing 5S. This is a simplified flowchart showing the main considerations.*

Figure 6.3 An example of an implementation plan. This one is for 5S. *This is a project plan drawn out on a spreadsheet. The times are arbitrary but give a rough idea of the sequence.*

Figure 6.4 An example of a red tag. *The red tag can be detailed or can be simply a number. It can be a dot or a number written on electrical tape. This tag is a complex printed example.*

Figure 6.5 Storage labeling. *This is a PowerPoint drawing. It is limited in detail. The shelving should be "open" to enable seeing straight in and they should be sloped for the same purpose. The diagram does show the clear labeling that makes it possible for the operator to establish the destination (The shelf location) without the need to hunt.*

Figure 6.6 Preferred storage locations. *This is a bit of a misnomer. It concerns distances from the operator and not types of storage. It is the distance where the parts should be located. It works on the basic principle that the parts used most often should be the closest to the operator.*

Figure 6.7 The best storage should display the contents. *The shelving or boxes should enable viewing of the contents without having to look inside from above. This box is tilted and it has a cutaway front for clear viewing.*

Figure 6.8 The shape of the parts can be drawn on the shelf so that it is visible when the part is removed. *This is a way to see what is missing from the shelf. It should ideally also give ordering and minimum stock levels too.*

Figure 6.9 Spaghetti map—before. *This spaghetti map is a representation of a workshop layout. The lines are the path followed by the workers when they carry out the tasks. The complexity of the pattern is a clue to the need for a rearrangement of the tools. See Fig. 6.10.*

Figure 6.10 Spaghetti map—after. *This is the new layout map following Fig. 6.9.*

Figure 6.11 Example of a floor layout and photo locator. *Large items have to be stored on the floor, so they can be given "parking bays" to live in. However, there has to be a method to see what is missing, when the part is in use. This diagram explains a simple method.*

Figure 6.12 Standard safety warnings that can be used with different colors. *An example of color as used on warning signs. Floor markings should not contradict the official coding. Details can be found from The Health and Safety Executive or their equivalents in different countries.*

Figure 6.13 Plan-Do-Check Act cycle. *Deming's cycle for ensuring a plan works through all four phases.*

List of Figures, Formulas, and Tables   xv

Figure 6.14   Alternative style cleaning map. *There are cleaning maps that look like photographs, as we saw in the chapter on Autonomous Maintenance. But they do not have to be so complex. This one is a simple PowerPoint layout that defines the tool as a block and who is responsible for cleaning it.*

Figure 6.15   Clean and Inspect Checklist. *This sheet defines F-tags (tasks) with the person responsible for carrying it out. It also defines the frequency for the checks. If the task becomes embedded on to a computer system, the sheet will become redundant.*

Figure 6.16   5S area audit sheet. *This is a simple sheet that allocates points to areas depending on how well they are scored against the 5S steps. It enables areas to be totaled and so they can be compared.*

Figure 7.1   Four production days, each divided into 24 h. *This is exactly what it says. There are 96 squares divided into four "days" of 24 h.*

Figure 7.2   Four production days, two products, 4-h change time. This time the changeover is after 24 h . *This is the same layout as Fig. 7.1, except after each 24 h, there are four "red" hours that represent the time taken to carry out a changeover. In total, we can see three changes totaling 12 h—or half a day's production.*

Table 7.1   The pros and cons of large batch production. *A table listing the benefits associated with batch and queue production and the benefits of smaller runs.*

Figure 7.3   The analysis sequence for a changeover. *This is the seven-step changeover sequence to be followed for carrying out an SMED analysis.*

Table 7.2   The elements of making a cup of tea. *This table has four degrees of analysis from the simple command (element) through some smaller elements to microelements. The intention is to highlight the steps that are normally not considered when evaluating an operation. Color is also used for tracking the elements.*

Figure 7.4   Parallel task allocation in SMED—shown as a project plan. *Many changeovers use only one man. It is realized, I believe, that two men might make a saving. This is a simple project-style chart to show a number of steps where time can be saved if parallel tasks are carried out. This figure illustrates the concept, the chapter explains the cost benefits.*

Figure 7.5   The Observation Sheet. *A form that has a horizontal bar chart added. It is used to calculate and display the element times for a changeover.*

Figure 7.6   Example of a reduction plan: A standard bar graph with start and end points.

Figure 7.7   The SMED analysis chart. *This is the chart that is used to record and calculate the data. In this example, a large chart is used and Post-it Notes are attached to identify the various entries. Post-it Notes allow simple changes to be made.*

Figure 7.8   How the elements are changed by an SMED analysis. *This is a project-style chart that shows a before and after flow of tasks.*

Table 7.3   Flip chart analysis example. *A similar table to that of Table 7.2. It is not as elegant as the analysis chart system.*

Figure 7.9 The cost vs. improvement impact chart. *Every solution has a cost. This chart illustrates the argument for deciding in which order to carry out the solutions.*

List 7.1 Element conditions. *These are the questions to ask of every microelement to find out whether it is necessary or whether it can be altered.*

Figure 7.10 The SMED analysis chart with improvements and ideas. *This is a chart similar to that of Fig. 7.7. It has the addition of some Post-it Notes for suggesting changes.*

Figure 8.1 Malfunction map—recognize the base diagram? *The malfunction shows the location of all of the different types of F-tags including the minor stops. It shows areas of the tool that the teams have to concentrate impovements on.*

Table 8.1 Green Dot Allocation Table. *This is a table that records the failures and the PM that should have avoided the failure. A green dot is allocated to each failure.*

Figure 8.2 A PM map of a mechanical "electrode" assembly. *This is similar to a photographed cleaning map except it is a photograph of the part that is maintained. The blue dots (which appear as gray in the figure) and the green dots (which appear as white in the figure) annotate where the work has been done and where failures have occurred.*

Table 8.2 PM analysis sheet. *A similar table to Table 8.1. This one identifies each step of the PM (as elements) and allocates each a number. A blue dot is also numbered according to each step.*

Figure 8.3 The flowchart is a simplified set of steps for analyzing the PM map. *This chart analyzes the different dot patterns.*

Figure 8.4 Sample PM map of a moving assembly that maintains a vacuum as it moves. *This is a drawing of an assembly that is maintained. The blue and green dots (gray and white dots resp. in the figure) still annotate where the work has been done and where failures have occurred.*

Figure 8.5 A visual guide of the main TPM steps that brought us here. *This is a complex flowchart that tracks from fault or F-tag to PM map or malfunction map.*

Figure 8.6 The statistical pattern of failures of a part over time. *This is a binomial distribution that represents the pattern followed by parts that wear out according to the rules of natural deterioration.*

Figure 8.7 The decision diagram column sequence in tabular format. *A simple group of four columns representing hidden, safety, operational, and nonoperational failures.*

Figure 8.8 On-condition monitoring decision blocks for hidden failures. *This is the first decision block on the chart: the first row of the first column. It is looking for parts whose failure can be detected using warning signals from the failing part.*

Figure 8.9 The decision diagram—left side. *This diagram and Fig. 8.10 are the two halves of the decision diagram—a flowchart that will analyze any failure mode and turn it into a PM task.*

Figure 8.10 The decision diagram—right side. *See Fig. 8.9.*

List of Figures, Formulas, and Tables   xvii

Figure 8.11   The scheduled restoration and scheduled discard decision blocks for hidden failures. *These two decision blocks determine whether the failure can be maintained using stripping and rebuilding or if it needs to be replaced as a complete module.*

Figure 8.12   The failure-finding decision blocks. *This decision block checks to find out if a part can have its function tested to confirm it is operating reliably. This applies to hidden failures only.*

Figure 8.13   Decision diagram. *This diagram is too small to read, but can be used in conjunction with Table 8.3 to get a better understanding of its function. They are drawn to the same scale.*

Table 8.3   The decision diagram sequence in tabular format. *This explains the functions of the equivalent areas of Fig. 8.13.*

Figure 8.14   Three rows of operational decision blocks. *These are basically the same questions that are asked in the corresponding "Hidden" column except the consequences are financial or product-related.*

Figure 8.15   Decision diagram—multiple failure frequencies. *Some parts have components that fail at different times and in different ways. This part will ask if this is the case here.*

Figure 8.16   The missing sign. *This is a worked example of how to use the decision diagram to define a solution to a failure.*

Figure 8.17   Failure of a thermocouple measuring the temperature of a heating plate. *This is a worked example of how to use the decision diagram to define a solution to a failure.*

Figure 8.18   Vacuum pump exhaust line blocks with dust. *This is a worked example of how to use the decision diagram to define a solution to a failure.*

Figure 8.19   Decision worksheet layout. *This is a decision worksheet overlaid to explain the purpose of each section.*

Figure 8.20   Completed decision worksheet using previous examples. *This demonstrates the simplicity of the decision worksheet.*

Figure 8.21   Proving the testing frequency formula. *This is the only real proof in the book. I had to be certain I understood what is happening. This is the simplest proof I could find.*

Table 8.4   The most common FFI values as a percentage of MTBF. *This is a table calculated from the formula.*

Table 9.1   The ideal team composition.

Figure 9.1   The RCM process flowchart.

Figure 9.2   Examples of the contents of an operating context. *This is really a table with different details. If any are in the wrong columns, they can be changed if necessary.*

Figure 9.3   A TMX furnace schematic. *This is a 3D drawing I put together using PowerPoint. It shows a very simple layout.*

Figure 9.4   Wafers, boats, and a paddle.

Figure 9.5   The paddle drive and positioning. *Another PowerPoint schematic showing the mechanics and the main drive electronics.*

Figure 9.6   Use of overlapping diagrams to show linkage. *This is viewed in conjunction with Fig. 9.5 to get a better understanding of functionality.*

Table 9.2 Features versus functions of a pen. *The main difference between a feature and function is the degree of detail and accuracy of the description. This table gives a set of examples.*

Table 9.3 Secondary functions. *These are a few examples of general Secondary Functions.*

Figure 9.7 The ESCAPES as a memory prompter. *The ESCAPES were defined by John Moubray and there is probably no better memory prompter.*

List 9.1 Hidden function examples. *About 24 different "hidden" items: most can be found in safety catalogs.*

Figure 9.8 The Information Worksheet. *The Information Worksheet is virtually an FMEA form. This one is the original style and asks only the basic information. It requires the user to understand all the concepts fully.*

Figure 9.9 Duplicate function elimination flowchart. *A duplicate function is one that has an effect on another function when it fails. Having duplicates only causes extra work, but this flowchart will help you identify them.*

Figure 9.10 Modified Information Worksheet. *This one has been modified to expand the failure effects and ensure that more of the details are recorded. I believe it is more suitable than the original but it still does not always get completed properly.*

Figure 9.11 The extra "numbers" columns on the Information Worksheet. *This sheet turns the form into a working spreadsheet. Just enter the numbers and the costs are calculated.*

Figure 9.12 Defining the functions—flowchart.

Figure 9.13 Simplifying finding functions by subdividing the areas. *When there are a large number of functions, the task of defining them all can be simplified if the area is broken down into smaller areas.*

Figure 9.14 Example of functions, functional failures, and failure modes. *This is a partial information worksheet assembled to give the reader an understanding of how to use it.*

Figure 9.15 Information Worksheet Example 1.

Figure 9.16 Information Worksheet Example 2.

Figure 9.17 Information Worksheet Example 3.

Figure 9.18 Time-based causes occurred in only 11 percent of the failures analyzed. *The bathtub and the half-bathtub curve. Two different graphs with three failure patterns are shown. These patterns are identifiable with components that follow time-based failure patterns.*

Figure 9.19 Failures that are not time-related occurred in 89 percent of the failures. *Probability-based failures and the added effect of infant mortality. Manmade problems?*

Figure 10.1 Time-based maintenance example. *This is part of a PM in a format I devised to support a customer developing a maintenance schedule.*

Figure 10.2 The P-F curve used for on-condition maintenance. *The P-F curve is the pattern often followed by components that do not have a normal or a predictable period for failure. It is a tendency toward giving a failure warning before the system collapses to total failure.*

Figure 10.3 Downtime saved by replacing an "electron shower" module compared to removing, servicing, and reinstalling. *This is a table showing the large time savings that can be made using turnaround parts.*

Figure 11.1 Example of a "Why-Why" analysis sheet. *This is only a sample sheet, which is why it only has five data entry points. There can be any number of "Why's" in an analysis.*

Figure 11.2 The general fishbone diagram. *This one is the generic for starting from cold.*

Figure 11.3 The alarm fishbone. *The alarm example was a simple analogy made up to cover a few examples.*

Figure 11.4 The fault tree diagram. *This also uses the alarm theme.*

Figure 12.1 A sample layout of an activity board. *This is just a quick layout on PowerPoint to give an illustration of the groupings and contents.*

Figure 12.2 Example of an initial improvement sheet. *This is a slightly complicated spreadsheet.*

Figure 12.3 Responsibility map. *This is like an area map except it coordinates with Table 12.1 and confirms certification to work.*

Table 12.1 Example of an Area Responsibility and Certification Table. *Links to the area map and confirms certification to work.*

Figure 12.4 A boatloader assembly.

Figure 12.5 The 7-wastes as defined by Taiichi Ohno. *A pie chart.*

Figure 12.6 Production villages compared to production cells. *Two layouts on one diagram showing the options of a four-machine "village" or four single production "cells."*

Figure 12.7 Balancing costs. *A vector diagram illustrating the power of the wastes.*

Figure 12.8 A simple value stream flow. *In the example a car service is used. Taking the car to the garage versus having the car collected. The objective is to highlight the areas of waste.*

Figure 13.1 The binomial distribution. *The statistician's dream.*

Table 13.1 The data and histogram of the wage distribution in Profit & Co. *An example designed to demonstrate the difference between the mean and the mode.*

Figure 13.2 A simple x–y graph.

Figure 13.3 A slightly less obvious graph.

Figure 13.4 Percentage failures plotted against Six-Sigma value.

Preface

If you are planning to learn or apply continuous improvement techniques, this is very likely the book for you.

This book is written as simply as possible, in plain English, by a Scottish engineer/trainer. It initially explains what each method or individual topic is about and then goes on to list the steps you should follow to implement what you have just learned. Then you learn the next bit and follow the steps—it encourages the readers to adapt some parts and tailor them to their own needs. This approach works on the basis of a tried and tested method.

The scope of the book is not limited to a specific topic; it covers important continuous improvement techniques such as Total Productive Maintenance (TPM), 5S, Quick Changeover (SMED), and Reliability Centered Maintenance (RCM). Additionally, this book provides a more than liberal sprinkling of Lean Manufacturing, Risk Assessment, Safe Working Procedures, Problem-Solving Techniques and also takes a critical look at Six Sigma.

Since this book is intended to be a practical volume, it taps into the best of 30-year-old methodologies. It does not rigidly adhere to the pure techniques; but like good food, it tends to blend the various flavors together where the author, in his experience, believes it to be most beneficial to the reader. For example, TPM promotes the idea that every problem on a tool must be fixed. This is an excellent idea but it does not consider the cost involved. The author uses RCM cost versus consequence arguments to help prioritize the repairs. This method works better on a limited budget and ensures real returns on investment.

If you could deconstruct a consultant and use the bits to make a book, this is probably what you would get .... I hope you enjoy the read.

Steven Borris

Acknowledgments

There are a number of people I want to thank for their help and support in writing this book. Some of them are legends, like Taiichi Ohno, Shigeo Shingo, W. Edwards Deming, John Moubray, and Henry Ford. Others have become anonymous in the mists of time. These are the believers who have taught and applied the techniques over thirty years as consultants, college and university lecturers, or even as trainers, relaying the knowledge throughout their companies. Most of the techniques covered in this book have been taught to millions of people all over the world. They have been repeatedly modified and adapted to encompass new technologies as they diffused outwards over the years. It is a testament to them all that not only have the methods lasted for so long, but they have flourished and will survive well into the future.

There are people closer to me that I would like to thank, too. Friends and colleagues who gave me the opportunity to develop my writing and training techniques: Mike Bray and Kelvin Smith from the early days, when I worked with an equipment vendor called Eaton Semiconductor UK (now known as Axcelis). All of the implanter diagrams used in the book are Eaton equipment. Then, from my latter days within the semiconductor industry, I was supported and encouraged by Mick O'Brien, Marshall North, and Neil Martin.

Over the last year or so, Dave Hale has also been a major influence. His different perspectives and years of experience have been a sounding board for ideas. He has taught me many things, including how to handle an "unwilling or suspicious" audience and why it is very important they understand the reasoning for any *change*. He also taught me the reasons continuous improvement techniques can fail and the need for managers to be trained and take an active role.

Writing this book as you see it today took me about a year of tapping away at my PC during the day, in the evenings, and even in the wee small hours of the morning. While I was fighting with every word and sentence, my wife, Carol, has shown an extraordinary amount of patience and tolerance. Having said that, I am just a tad grateful that she never learned the art of knife throwing....

My final thanks are to Kenneth McCombs of McGraw-Hill. I believe his guidance and advice has made this a much better book than it was when I first wrote it.

My thanks and gratitude to you all.

Reader Response Request:

If you have any comments to make that might improve the book, please write to me via McGraw-Hill Scientific & Technical, New York or e-mail me at steven.borris@ntlworld.com. I would also like to hear of any successes or failures you have while implementing the procedures. Perhaps we can pass on your advice to the next generation.

Thank you for taking the time to read this book. I hope you enjoyed it and found it useful.

Total Productive Maintenance

Introduction

Two manufacturing techniques have fought their corner against a series of new rivals for more than thirty years and they still come out on top: one is *Total Productive Maintenance* (TPM) and the other is *Reliability Centered Maintenance* (RCM). Both methods started their lives in America, although one of them, TPM, emigrated to Japan where it took on a life if its own. In the United States, it was known as PM (no "T") and it stood for *Preventive Maintenance*. PM was arguably the industry standard for production and provided a solid enough foundation, but it was purely an engineering tool intended to make equipment more reliable. After all, that is what a manufacturing technique is all about, isn't it?

The Development of Maintenance Systems

It is difficult to imagine a time when equipment was not maintained. Remarkably enough, maintenance and productivity have not always been the Holy Grail of industry that they have become in today's most successful companies. Yet it surprises me to discover that even in the twenty-first century, there are still a large number of companies who appear to be oblivious of the potential gains that await them. They will probably never consider improvement techniques unless they find themselves in difficulty, at which point they will seek help from everyone: professional and government organizations like the Manufacturing Institute, the Department of Trade and Industry (DTI), and Scottish Enterprise. It does make me wonder, however, how many companies actually are aware that there are better ways out there but simply will not take steps to do anything about them.

Reactive maintenance ruled the roost in the early days of manufacturing. If we look back, we discover that there was no real need to be efficient. There was a huge surplus of workers and cheap labor. Such a pool of labor and the capability to produce all of the goods that everyone wanted was enough to satisfy industry. When production halted (not *if*) the problems would be fixed and production would restart . . . whenever . . . . The goods would simply be delivered late. There was no *need* to avoid breakdowns. Any preventive maintenance was limited to a tap with a hammer, oil, or a grease gun. Besides, the equipment was very solid, robust, and built to last. Why should they have given a second thought to efficiency?

As we entered the twentieth century, mass production was seen as the way to reduced prices. The more product that was made, the cheaper the selling price could be. So some experts started thinking about ways of increasing output, but not really from today's engineering perspective. Henry Ford's efficiency expert (Frederick Taylor) basically just wanted to make employees work harder. Taylor's approach was to break down manufacturing and assembly into the smallest practical steps possible, to use as many men as it took to carry out all of the steps, and to minimize any need for skilled employees. To be fair, he did seem to have some ideas on factory layout and Ford did have production cells in his first factory. In any case, in 1913, Henry Ford hit on the monumental idea of the moving production line, which did speed up production.

Then along came two World Wars. The overflowing labor pools were drained as they poured their contents in the direction of the war effort. Supporting the war and, of course, supplying soldiers knocked production off balance. In order to keep industry moving, it was necessary to use women to top up the pools. Women were trained to carry out tasks previously regarded as suitable only for men. As it turned out, many of the women did the jobs even better than the men. My mother became a welder—a very good one, I have been told. She built ships on the "Clyde," the river in Central Scotland that was once famous for shipbuilding. The "Queens" were built on the Clyde.

Having the labor was only part of the answer. The goods were still not shipping fast enough. With the extra demand on the machines came an increase in breakdowns. It rapidly became apparent that the previous expectations of industry and the inefficiency of reactive maintenance were just not good enough. Breakdowns had to be reduced if there was to be any way to increase output. Not only that, the shortage of raw materials now focused attention on the levels of waste. Cutting the cost of losses became a serious consideration. As if all this was not enough, the manufacturing equipment designs were increasing in complexity and required an even higher level of skilled support to operate and maintain them. The engineering answer of the day was Time-Based Maintenance, which did help a bit. This was the era of PM.

By the time we get to the seventies and onwards, the bogeyman was profits. Now we see the development of TPM and RCM. With customers looking for even more reduced *cost of ownership* of equipment—the cost of running a piece of equipment had to be reduced, despite its complexity still increasing. Then another requirement crept into play: quality.

In the sixties and seventies, the car and the electronics industry were not too reliable. To be fair, at one point, both the British and the Japanese products were equally bad, but the Japanese goods began to get better. And better. Today, Japanese car manufacturers are always in the top five of the car reliability tables and their equipment level and price position them at the top of the sales tree. Why? Japanese industry believes the needs of the customer come first. This, fortunately, leads to increased sales, greater customer satisfaction, fewer after-sales problems, and finally even more repeat sales. This leads to more profit.

The last impetus for manufacturing improvement was safety. This was driven by legislation. Today, it is possible to be imprisoned if an injury can be proven to be the result of negligence by a manager. Similar requirements have been placed on environmental pollution, but the laws tend to be a tad more complicated.

How was it possible for the Japanese to transform their industry so effectively? One treatment was the development and successful application of TPM.

Industry in Europe and America believed that only technical groups had any practical input on improving the performance of equipment. This was true to a point. If the maintenance situation was very poor, then the benefits in repairing that situation could seem out of proportion. The Japanese knew maintenance was not the complete solution. In fact they held the view that production had a major input. In retrospect, this view was so obviously correct that it seems ridiculous that no one else realized it. Nevertheless, it was pretty innovative to see companies making improvements on the basis of the needs of all of the equipment users as well as the engineers. True optimization of equipment could only be achieved with the active input of the production groups, particularly the operators. Lucky for us, though, the benefits of TPM are achievable by anyone who is prepared to make the commitment and follow a few simple disciplines, which, funnily enough, are described in this book.

RCM is also equipment-based, different from TPM and a bit harder to understand. It too relies on input from operators, production, and process engineers. Its longevity is based on a fundamental premise that any maintenance that *could* be carried out should be considered and evaluated against the cost and the consequences of failure. It requires a different way of thinking but it is well worth the effort. It is my view that TPM and RCM complement each other. As the book progresses I will explain further.

The Writing Technique and the Contents of the Book

There are parts of this book that are not original. I make no claims that the ideas are all mine. After all, many of the procedures we will discuss have been in use for over thirty years. This makes it very likely you will have seen the odd similar diagram or spreadsheet before. In fact, most of the subjects are taught at degree level in universities and colleges. However, what I think what you will find different is the practical, step-by-step guide used for actually applying the techniques and the use of bullets to separate out points of importance. For technical issues, I much prefer to use bullets than to have a train of points in a paragraph format. I think bullets are easier to follow. I also find that when reading bullet points, each new point seems to prompt the brain to get ready to absorb it and that, when looking for a specific point, it is easier to find.

My writing technique developed over a period of many years, improving and adapting as the need arose. I started writing procedures during my early field service work and improved them from there on. Eventually, they evolved again while working in the semiconductor industry; there was a different need. I tackled a TPM workbook and developed a new style of Safe Working Procedures. This book uses many aspects of this writing technique.

Just as the three primary colors blend into a rainbow, in the book, I hope to illustrate how a selection of continuous improvement methods, particularly TPM and RCM, overlap and mesh together to create a complete improvement program that works. Once you bite the bullet and start, you will discover that the same basic training, problem-solving techniques, and cross-functional project teams become virtually interchangeable, making each new procedure easier to implement. The cards stay the same, only the rules of each game are different. Each of the procedures covered here have been modified and improved by their users over the years. Even so, their development will never be over. The foundations will always stay much the same, but every day someone will find a new problem, find a better way to adapt the method to suit *their* needs, or find a simpler way to apply them.

I started using TPM techniques before I even knew TPM existed. This is because much of TPM is simply good engineering practice. I have worked in two fields that required root-cause solutions: the hospital environment and equipment field service. Both had to be certain that faults did not return; one would have seriously affected patients and the other, remarkably the more stressful of the two, would have affected customers and profit. The version of TPM explained here was developed to suit modern industry and be adaptable for any type of equipment—not just heavy industry.

The "formal" method, on which my training was based, was developed by the Japanese Institute of Plant Maintenance (JIPM). It was quite difficult to apply because of the illustrations used. For example, it would discuss "cutting blade" accuracy and how the blades deteriorate over a single shift and their lack of accuracy following replacement, until they have warmed up. In this book, I will discuss the bell curve and the P-F curve, preconditioning and improving setups. All of which can *also* be applied to cutting blades. The JIPM method was very good, despite its heavy industry bias. This, coupled with the way the book was translated, made it quite difficult to follow. In this book, to help you grasp the techniques, I will refer to two quite complex tools and some really simple stuff: cars, making tea, and refrigerators; things that you will recognize as obvious, but will illustrate the points I am trying to make.

Just like TPM, I learned very early in my career that most equipment problems revolved around people. Either through lack of skills, lack of training, poor or no procedures. Often, the solution boiled down to the writing of a proper, standard procedure and some training. I mentioned above that as my part of a TPM pilot team, I used my documentation and training skills to write the workbook that came to be the guide used by all of the company's TPM teams—it was also used as a reference by other sites in the United States. Like this book, the workbook was also highly visual—much to the dislike of the IS department—but computer storage was much smaller in those days and networks were less efficient. (This sounds like a history lesson, but it was less than ten years ago!) The addition of the images to the documents greatly increased file sizes, which, naturally, caused the issues with the computer department. It was decided to retain the method because the document users wanted the benefits the document style had to offer. The end result was that I usually ended up working at home, where my own PC setup was better. The workbook proved to be very popular. To save time and reduce the demands on the network, I would print the copies overnight and sometimes they would simply disappear....

One shortcoming of technical books I hope to avoid is that they only promote the good bits. I have tried to bring a degree of real life to the subjects. As each topic is covered, there are a series of actions to follow. If I don't think a step is practical, I will tell you what the technique recommends, what I would do in its place, and I will explain why. There are not many areas of contention. Although TPM and RCM are separate entities, there are areas where they naturally overlap and complement each other. This relationship makes them both better to use.

RCM is the technique developed by the aircraft industry to generate maintenance procedures. By considering "the tool" as functions and how they might fail, then comparing the maintenance cost against the cost (and the consequences) of a failure, and using a special *decision diagram*, the most suitable PM method is identified. It is also possible, under specific circumstances, to conclude that the cost of maintenance is more than the cost of a failure and so no maintenance would be recommended for that part.

The use of RCM is flexible; it can be applied to equipment at any point in its life cycle. I like to apply RCM to a tool after it has been restored to its *basic condition* using TPM. It is at this time when the tool has no obvious "manmade" deterioration issues and the application of RCM can be applied to specific areas of the tool that present issues. The improved technical infrastructure, documentation of standard procedures, technical training, and skill levels developed through using TPM makes for a better RCM analysis and might lead to previously unanticipated causes of the problems.

The book also covers 5S, a foundation organizational tool that is used to declutter work areas, improve production support, consider minor production area layout, and solve productivity issues. We also look at quick changeovers, also known as Single Minute Exchange of Die or SMED for short. SMED teaches teams a methodology to speed up sequential tasks with a view to minimizing the length of time a production tool is off-line. Like 5S, it includes how to identify areas of waste, poor preparation, lack of organization, and lack of standards. Both 5S and SMED were developed from the Toyota Production System (also known as lean manufacturing) but, because of the benefits they provide, they have been "spun off" and promoted as stand-alone procedures for many years.

More generally, the book also includes a group of support techniques, including problem solving and modular risk assessment, which will help to simplify the use of all of the techniques. An explanation of basic graphing techniques including histograms and Pareto charts will aid technicians and operators in the analysis of equipment performance and productivity. There is even a short overview of Six Sigma, the problem-solving tool. It was written to give the reader an insight to its usefulness.

Even this book has evolved in the time it has been written. Originally the chapters were arranged in the exact order the techniques would have been applied. This version is better. The order of the chapters is easier to follow. It has not lost the original flow, however, as you will find prerequisites clearly identified. For example, before allowing a *Zero Fails* (ZF) or an *Autonomous Maintenance* (AM) team to work on equipment, the reader will be directed to the essential training, how to understand competency, safety, and any other areas that must be understood. There are also detailed flowcharts peppered throughout the book that summarize and guide the reader through the methodology.

I try to demonstrate that many companies exercise cost control but they are less aware of the financial impact of losses. Just as in the real world, costs and cost reduction are a constant priority. The application of the techniques in this book will identify areas where losses are often made and direct the reader how to find and eliminate them.

The book will also demonstrate how to optimize a maintenance system; make technicians, engineers, and operators more effective; and will standardize the way all tasks are carried out. I have found that the system works.

Even though the reader's company will have its own training and safety departments, please take the time to consider the methods explained in this book. TPM has a Health & Safety Pillar with a target of zero accidents and an Education & Training Pillar. The safety and training material might be useful for the reader to adapt and use in conjunction with their own systems. There are also a range of suggested spreadsheet formats that can be used to develop record systems.

All of the above procedures are designed to improve both maintenance and productivity, with lean manufacturing being the cement that bonds everything together. Although there is a section that discusses the basic concepts of lean manufacturing, two of its components, value and waste, are referred to often as they are as valuable to most techniques as salt is to flavoring food.

The Pillars of TPM

TPM now comprises of eight different sections which have come to be known as pillars. Each pillar has its own areas of responsibility, but they also have areas where they overlap. This book does not consider all of the eight pillars in depth. It concentrates on those that are most related to maintenance and productivity, although the information will provide enough detail to give the reader a sound understanding of the others.

The pillars, identified in Fig. I.1, are as follows:

1. *Health & Safety*
   This is crucial as it sets the goal of zero accidents. Its importance is emphasized by the need to protect operators, who will be trained, initially, to carry out simple technical tasks. Bear in mind that most of the operators that will be participating in AM were not employed with maintenance in mind, no matter how simple. To this end, we must cover risk assessments, hazard maps, and some other safety concepts in detail. To build confidence in the operators, they should be trained in how to carry out risk assessments. They are also encouraged to help with the development of the safe working procedures.
2. *Education & Training*
   In many companies, training is not given the importance it deserves. Procedures are often passed on informally *on the job*, and the trainee is required to make his own shorthand notes in his log book. These

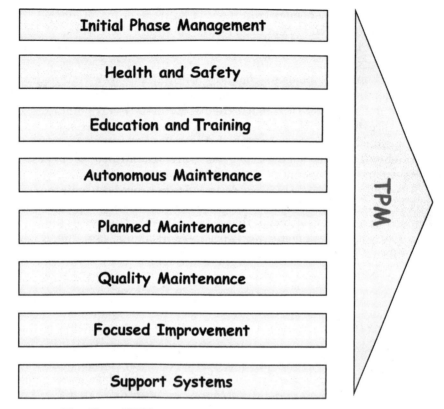

Figure I.1 The pillars of TPM.

are the instructions he is expected to use in the future when he carries out the tasks by himself. This is highly ineffective as a training technique, as it assumes

- ➤ The trainer—the *qualified* technician—actually knows the correct method;
- ➤ That the trainer can, without using a proper procedure, recall all of the steps and relevant facts in the correct order;
- ➤ That he has the ability to explain what he is doing;
- ➤ That the trainee is capable of understanding the topic;
- ➤ That the trainee is capable of accurate note-taking;
- ➤ That the trainee can draw proper, accurate diagrams;
- ➤ That the trainee can learn at the same time as taking notes and following instructions.

Reliance on this method of training will cost the company lots of money in the long run.

Without proper training, TPM, and maintenance in general, will simply not work. This pillar explains what knowledge is necessary, how to teach it, and how to confirm it has been absorbed and has been understood. It is important that the competency of the operator is confirmed, not simply that they have attended a course. (Take a look at some vendor certificates; they often say that a person "has attended" a course. There is rarely any reference to understanding or performance.) All details of training must be recorded.

TPM also recognizes that the absence of proper training methods is not limited to industry, which is why it promotes the use of standard operating procedures. As a short example to reinforce the point, let me relate the story of a conversation with a trainee theater nurse. He was annoyed because he failed his test on an anaesthetic machine. It turned out his *experienced* instructor had missed out a few steps. The qualified nurse is expected to memorize the procedure and, once that has been done, there is often no further need to refer to the instructions. Which brings us neatly back to the reason for the poor training: the instructor should have used the instructions. It is the only realistic way to ensure every, essential step is covered and is correct.

3. *Autonomous Maintenance (AM)*
Using highly skilled technicians or engineers to carry out very simple maintenance tasks is not cost-effective. If operators could be trained to carry out these basic tasks, it gives them an opportunity to increase their skill level, makes them more responsible for the operation of the tool, increases their job prospects, and frees up the technicians to work on more complex tasks including TPM teams. It also has the benefit that the $cost to do the job is reduced.

Just as operators wash their own cars and check for damage, this pillar is intended to increase operators' skill to a level where they are able to carry out the basic maintenance on their own equipment. By adopting "clean and inspect" procedures, they are taught to recognize abnormal operation and identify problems that are developing. Through time, as the operators' skill improves, the AM teams will progress to more complex maintenance. They might even be capable of transferring to a technician grade.

4. *Planned Maintenance (PM)*
Planned Maintenance looks for the underlying causes of equipment problems and identifies and implements root-cause solutions.

In many organizations maintenance is rarely managed, with the engineers choosing the jobs they want to tackle and using their "own experience" to carry out the work. Most technicians I know dislike routine maintenance as it is too repetitive and is not a challenge.

Besides, the best wages go to the firefighters: the ones who come to the rescue when the tool crashes. There are even technicians who are the company's experts on the problems that happen every week! They are so good at resolving the issues that they still happen every week. Some people do not appreciate that recurring faults are unresolved faults. The technical term for firefighting is Reactive Maintenance or it would be if the problem was actually repaired on failure.

I dislike the term firefighting. It is an inappropriate name designed to glamorize a bad maintenance practice. When firemen put out fires, the fires don't usually return. Equipment breakdowns frequently do. Time after time, we see the same issues! If a fireman does not put out the fire properly, there is nothing left to return to. So, does this mean that putting out the fire is the root-cause fix for a fireman? Not a chance! When the fire is out, the fire department will search the ashes and continue to investigate until it discovers the root-cause of the fire. Even at this point the task is not over. Action is taken to learn from the findings and to circulate what has been learned to all the other fire departments. The fire department even passes on its knowledge to government offices and visits companies to teach them what it has learned about fires. In short, the firefighter strives to prevent fires from happening in the first place. This is the purpose of the PM pillar: to prevent breakdowns.

For most readers of this book, it will be no surprise to discover that good maintenance (and productivity) follows the same practices and high standards as the fire department. How, then, can we use the same principles to improve maintenance standards? Simple, we adopt TPM. The PM pillar covers all aspects of equipment analysis and improvement in a nice, simple way.

Where the core of an AM team is operators that have a dedicated technical supervision and a support network, Planned Maintenance teams are cross-functional and are known as *Zero Fails* (ZF) teams. ZF teams include operators and technicians. In addition to the basic problems covered by AM teams, the ZF teams also tackle the more complex issues. These include the effectiveness of current maintenance, eliminating recurring problems, and improving equipment efficiency.

Overall Equipment Efficiency (OEE) is the measure used by TPM to attain the best equipment performance.

5. *Quality Maintenance*

Even what is regarded as a perfect tool will not produce perfect product. There will always be some kind of variation in the quality or the physical attributes of the product. The cause of the variation is the limitations in the equipment design and the choice of the components used. This pillar utilizes cross-functional teams to analyze areas of equipment performance where the product variation should be reduced.

Once a cause has been found, the team would investigate if a modification or an upgrade might be implemented to increase yield. Alternatively, it could search for a different manufacturing process that might not exhibit the same limitations.

6. *Focused Improvement*

   There will be outstanding issues with equipment or processes that have been difficult to identify in the past. Cross-functional teams are used to investigate the issues and to find permanent solutions.

   The problems under consideration have to be evaluated to justify if a fix would provide a positive, cost-effective benefit.

7. *Support Systems*

   Every department within an organization has an impact on production: stores, purchasing, facilities, quality control, scheduling, goods in, office staff, and sales. Have I missed any? This pillar uses TPM techniques to identify and resolve problems.

   The problems might manifest as problems like a lack of spares, incorrect spares, excessive lead times, poor quality materials, lack of standardization of dimensions of materials, parts shipped with the wrong specification, parts not arriving on time or arriving in Goods In but no one passing on the information....

   There are a huge number of issues. Basically, we look for problems and then apply the TPM procedures to analyze and then eliminate them.

8. *Initial Phase Management*

   This is the organizational or planning pillar. Teams are set up to consider every stage of production.

   The methodology follows a kind of Value Flow Analysis. How does the company get the ideas for new products? How does it make the selection of and design of new products? How can the customers' needs and wants be better served? When the customer approaches the company, is the call handled efficiently? What about the stages between the call and the product being shipped? Is the documentation necessary and effective? Is the billing correct? Does the customer get the goods as promised and when promised?

   Another area covered is intended to improve the manufacturability of the product. Is it easy to make? Can it be assembled the wrong way or are the parts made Poke Yoke and so only fit together in one way? Is it reliable? Is it easy to maintain? Is it easy to operate? Is the machine efficient from an energy and efficiency perspective? Does it have a low cost of ownership?

The team must investigate the complete system from start to end and look for ways to make improvements.

The Toyota Production System (Also Known As Lean Manufacturing)

Much of the content of this book is based, somewhere in its history, on the Toyota Production System. It has been officially recognized by governmental organizations that lean manufacturing can be a very positive asset in virtually any company and so they have started promoting its merits. Lean sprung from American techniques. One American in particular, W. Edwards Deming, who had a system known as *Total Quality Manufacturing* (TQM) became very popular in Japan. Many of the ideas he promoted became the basis of lean, except lean was adapted to a great extent to conform with the Taiichi Ohno's belief that productivity improvement was not limited to engineering groups. In Chap. 12 lean is discussed more fully. It is included with the activity boards and team goals only because it is not the main theme of the book although it is certainly the spice that gives the book its flavor.

Lean has been included to provide an explanation of the importance of many of the components of this book including the search for waste, 5S and SMED, Value, Value Stream Analysis, and, of course, Pull.

Finally, Advice for Using the Techniques

As I mentioned at the beginning of the introduction, even before I knew what TPM and RCM were, I was using many of the same techniques because they are simply good engineering practices. The intention of this book is to help you get a good understanding of all the methods, to help you use Autonomous Maintenance and Zero Fails, and to guide you through the analyses leading to the use of the RCM decision diagram.

If you are reading this, then I would hope that you know TPM stands up on its own merits. I have used it and I know it works. From its PM base, it took twenty years before Total Preventive Maintenance was born. Even then, it was still subject to changes until it evolved into today's factory-wide improvement tool.

It was necessary to adapt TPM to enable it to be used in an electronics company. The same methods could be applied to any light industry to make them easier to understand. In my case, a pilot team analyzed and applied the JIPM method to see how it worked. The disadvantage of the pilot team, especially a large one, is a need to make a lot of compromises. In the main, compromise can be a good thing: excessive compromise can take the edge of the blade. As part of the team, I included a fair degree of generalization in the original workbook. This gave the workbook a broader audience, not just electronics equipment. The method used in this book is still based on JIPM, my own experience as engineer, and that of introducing the modified version of TPM in a modern company.

Do you remember the one major omission of PM? It did not use non-maintenance personnel. Well, TPM does and this means extra special consideration must be taken to ensure the safety of the operators. Before jumping in head first to any of the procedures, there are a few obvious guidelines to consider.

> *Make sure you know what you are doing.*
> If you need training first—arrange it.
> *Make sure you do everything safely.*
> Take advice from where you can.
> *You don't need to be alone.*
> Go to your colleagues for advice and support.
> *The timescales are at your own discretion.*
> There will always be managers who want more and faster. Having tight or tough targets is not a bad thing—unless they cause you make mistakes or cut corners.
> *Spend the time needed for creating proper procedures and training.*
> This is a major benefit. It will be a drain on manpower, but it will be well worth it in the long term—and the short term for that matter! They might prevent you from ending up in jail.

For further information, there are excellent books by the Japanese Institute of Planned Maintenance, edited by Kunio Shirose (it is a bit hard to read), and *TPM: The Western Way* by Peter Willmott.

Chapter 1

TPM—Basic, Use, and Ideal Conditions

Why does equipment suffer from faults and how do these faults change from minor issues to major problems?

The question is not as simple as it seems, or to be more precise, the answer isn't. As we start to investigate and apply *Total Productive Maintenance* (TPM) methodology, we will quickly find that faults have a range of causes. Some are due to real unpredictable, "sporadic," equipment failures, but we will also discover that a very large number of failures, if not most of them, are due to what TPM rightly calls "deterioration from the *basic condition*." This definition covers a lot of options as we have to define both deterioration (its types and the different causes) and the basic condition. If we begin by considering the root causes of deterioration, we will discover that it subdivides into two different kinds: *natural* and *forced*. Their names are a bit of a giveaway as to what they mean but an understanding of the concepts of natural and forced deterioration will help immensely in appreciating what the basic condition is and why equipment becomes unreliable. Did you notice I used "becomes" unreliable and not "is"?

Fault Development

Deterioration from the basic condition is the first step on the way to developing faults. The further away we get, the greater the likelihood of a failure. To illustrate a failure mechanism, we will look at two examples and two fault types: *total* and *partial* failures, where a partial failure is one that does not immediately stop the equipment from running. Partial failures often start off their lives as small irritations. Unfortunately, if left to their own devices, they can grow into significant issues. It is only when the complete problem is analyzed over its lifetime that the true causes can be identified. It might be surprising to discover that the analysis can highlight other problem areas, many of them systematic and which do not involve the operation of the equipment. It might also be a surprise to discover that the same system problems, or at least a variation of them, occur in many companies. We should take a quick look at this *alternative* fault development route before going into depth about technical faults, although I guess even these failure causes would still be categorized by TPM as a source of "forced deterioration," which pretty much means we caused them ourselves. When we start our analysis, we know that the problem has to have a beginning. There will be a first time the fault happens, which might not be the same as the first time it was recognized as a failure or a problem. It is necessary to delve deeper. We need to find out whether it was an obvious issue that took only a few minutes to resolve or whether it was a temporary fix that was never followed up or was it just overlooked?

Once we have established the source of the infection, we can move on to how the issue developed and spread. To do this, we must look at the *big picture* and follow the progress of the mechanism and the flow to the end. Did the initial fault return on the same part? Did it gradually increase, becoming more disruptive until the problem reached a point where it changed category from a partial to a total failure? Was the cause such that it allowed the fault to spread to other parts of the tool? For example, in the case of a tight-fitting part, a metal tool is used to hammer it into position, and in so doing, damages the location pins. Then, rather than repairing the pin and reverting to correct fitting, the same *hammer* is used on other, less difficult parts, damaging their posts. The damage does not stop at the posts. The burrs on the posts scrape and gouge the holes in the changeover parts, which makes them even harder to fit and, as a consequence, need more *assistance*. This could be a difficult concept to grasp, so I will illustrate it with two examples.

Production equipment that has the potential to make multiple products must use different parts to facilitate the different processes. They are known as changeover parts. Let's consider an imaginary toolmaker, F.I. Hackem and Sons, who sells two ranges of cutting tools, which need two styles of base plate on which the cutting and grinding tools are mounted. Plate 1 is for the smaller tools and Plate 2 is for the larger, high-power tools. The base plate supports both the precision cutting tools and the clamping systems for the workpieces. When it is time to make the larger base plates, one stage of the setup change involves a fitter who must increase the bit size to drill bigger holes. The larger plate (Plate 2) needs the bigger holes to fit the 6-mm bolts, which are essential to support the heavier workpieces. If the toolmaker changes to the 6-mm drill bit and does not notice that there is something wrong with the drill, every hole drilled will be damaged by the defective cutting edge. The burring and scoring in the holes will cause problems that might not appear immediately. It will take time for all the damaged parts to interact during use. The mounting bolts might need to be inserted, removed, and replaced a few times to stir up and aggravate the problems. Eventually what we, the users, will see is a small problem. This would be the first time we have a clue to any issues. Then the next time we make a change, we might see the same problem, but where the bolt is a bit harder to fit into the same hole. If more than one hole was affected, we might see two small problems in different holes. The next time we could see three or more problems and so on. The burrs in the hole could even damage the bolts.

Now, we should consider the same sort of problem but on a larger scale. This time there is not only 1 but 20 or 30 standard, preformed, *changeover* parts that need to be replaced. They are normally supplied as optional "kits" with the tool—one set for each size of product to be made. The kits are color-coded to tell them apart, one color for each set of parts. The equipment manufacturer is aware how important a fast changeover from one size to another is, because it affects productivity. Hence, the use of color-coded parts. Now let us consider the possibility of a poor fit of some of the parts. This will slow down the changeover. So each time a badly fitting part is encountered, a *persuading* tool of some sort is employed to help with the positioning of the part. Often the persuader that makes the parts fit together comes in the shape of a stronger operator, or tapping the part with a (progressively) heavier tool. Sometimes a new tool, a mallet, is added to the changeover kit, and, of course, because they are harder to fit together, they are also harder to take apart. Now we have to introduce the use of levers or screwdrivers to help remove the parts that took so long to fit.

Where we started with one small issue that added a few minutes to the changeover, after a time we have a task that has become heavy, stressful, manual labor and where several of the parts must be forced to fit. The damage spreads through the use of the hammers and the acceptance by the operators that force will be required. They no longer expect a proper, push-on fit. This damage infection has now doubled the changeover time.

The tooling problem is not the only failure that is having an influence here. Analyzing the fault has brought to light a few other possible "systems" failings. The first few times the fault appeared, they could possibly have been overlooked because they were so minor. However, this is only acceptable if it could be reasonably attributed to skill level issues making the change awkward to carry out. As soon as it is recognized that there is a problem, a series of systems should have automatically kicked in. In the event the fault continued to be unrecognized by the operator, other systems should have identified a problem, if not what the problem was.

The information listed below was gathered by discussing an issue with all of the operators for the area and a couple of trapped engineers. The operators are closest to equipment problems and should always be considered as one of the best sources of data. What are the underlying problems they raised? Take a look at the following list:

- *The user has realized there is a fitting problem but has not initiated a repair.*
  There must be a formal fault reporting structure that leads to a repair, but somehow it was not used or, unacceptably, was used but was not responded to. "Of course there is a system!" "We all have that kind of system!" "What do you think we are: stupid?" Hmmm....

  Remarkable as it might seem, there really are companies where internal politics builds barriers. I know that this does not happen in YOUR workplace, but let's imagine it could be happening to someone else's.

- *Perhaps it was decided during the performance of a continuous improvement exercise that the previous system, which has been followed since the beginning of time, was due to be reviewed.*
  With the introduction of TPM on this tool set and a task analysis, it was decided that this particular changeover was suitable, simple, and safe enough to be carried out by an *Autonomous Maintenance* (AM) team. Although currently carried out by engineers, they will no longer be directly involved in the changeover. The responsibility for the change now rests in the hands of the trained operators. The engineers, who were never excluded, now (wrongly) feel they have been rejected.

  This is not how AM works. The aim is for engineers to have time released for them to carry out more complex and appropriate work suitable to their superior technical skills. Returning to our example, we now have the perception of bad feeling and lack of willingness to support the changeover, or worse, to support the line. "Right then, do it yourself. Let's see how well you get on...."

  On the basis of this bad feeling, we appear to have a situation where faults are no longer reported unless they are total failures and the line is dead. If the operators are not competent to make their own minor repairs, which will be the case in the early days, we have a doomed situation that allows the fault to grow and become malignant.

The AM team should have had a dedicated engineering support team member, and a supervisor/manager who should have recognized and resolved the fault before it became an issue.
- ⊗ *Equally, we can have bad feelings between engineers and production, which lead to a reduced support service.*
The effect is similar to a husband and wife in a huff. Not pleasant, but likely where personalities are involved.

- *There might not be a formal fault reporting system.*
Many systems rely on an operator paging or finding an engineer, and, if available, he will respond.

- *There is also the possibility that the Engineering Manager or Lead Hand follows a different set of priorities.*
Perhaps, the task in this example, since it has not stopped the line, has been recorded, but has not reached the threshold for priority. (Please refer to the section in Chap. 9 that discusses the RCM "Functional Failure" modes, Total and Partial failures, and the difference between specified limits and the limits that are routinely "accepted" by production.)

There is also, amazingly, a historical argument that the number of hours "this tool" works does not allow time for any maintenance. Thank God these guys do not have responsibility for aircraft, trains, ships, ambulances, or medical equipment!

- *The production system fails to detect the trend that the tool off-line time is increasing.* Even if the system is paper-based and relies on a manual analysis and a spreadsheet program like Excel for the graphs, it should still be possible to see as little as a quarter to half an hour's extra downtime in a day. This time translates to a performance loss in the range of 1 to 2 percent.

Depending on the number of changeovers, the time will be multiplied over a week or a month, as will the £cost equivalent of lost production. The major difference when looking at the cost is that the fault will be expressed in terms of tens of thousands of pounds.

- *There are more general causes that apply across all aspects of production and maintenance.*
  - ⊗ *There are no standard procedures.*
    - The operator should know the proper sequence of the changeover steps and how it should flow, making any abnormal procedures obvious.
    - The operator should also know, and be able to recognize, the most common failures. I suspect that this is one of them and it should have a standard response.

- A standard procedure should be available for times when we need a reminder for workers who have not carried out the task for a long time and might have forgotten a few steps.
  They might even have forgotten the order of the steps, causing them to take extra time to remove wrongly fitted parts and reinstall them later in the correct order.
- The documented procedure should specify
  - How long the task should take.
    There will be a range of time that allows for the slowest and the fastest operator.
  - The actions to take in the event of the task time being outwith the acceptable window.

☹ *There are no competency checks.*
Refer to Chap. 5.
Suffice to say, inexperience breeds variability in performance and a tendency not to recognize errors.
Keep a record of the training—a "heads up, on-the-job explanation" is not good enough.

☹ *There are no opportunities for operators to talk to engineers on issues affecting their work.*
Interdiscipline interaction in the form of production or line meetings should be encouraged. Apart from the benefits gained by discussing problems, barriers are torn down and friendships are built. They also allow different groups to see issues through each other's eyes and help them understand that they do not work for a part of the company (a production department, an engineering department, etc.), but for a complete company.

Many companies have simply never considered performance analysis. I am not even talking about complex themes, just the basics. I am unsure if any see it as a waste of resources. The uptime of a tool should be the responsibility of the engineers and should include all sources of downtime. If the engineers create the figures, they will become more involved. It is impossible to say how many companies suffer from the problems in the above examples. In my own personal experience, it is more than 50 percent, although many of them were forced by the demands of competition to make changes. In fact where one issue is seen, there is often a strong likelihood that many of the others will also be seen. The problems are cultural and developed within the organization. Even where fresh blood is introduced into areas of the company, it can be stifled by the rear guard, who tends to resist change, and the new employee can be prevented from making positive changes. Let me quickly add, I am not an ageist. Far from it, I am old(er). I believe that the people who resist simply do not have the capability to understand new ideas and could be afraid to move away from their comfort zones. Their age is irrelevant; their nature is the overriding factor.

We have reviewed examples of how machine problems, or at least their magnitude and impact, can be affected by issues other than the equipment itself. We have also seen that ineffective technical management can also seriously affect the performance of a tool. TPM realizes this fact too. It is time to take a more detailed look at how faults develop on equipment. To explain failures and deterioration in TPM, we must first have a defined starting or reference point: this point is known as the basic condition.

If I show you a length of wooden rod, you will be unable to tell me how much is missing until I tell you its original length. Similarly, if we consider a square, sandstone block that has been exposed to the elements, before we can evaluate how much it has been damaged, we must define the original reference. The reference has to be the initial state of the block or the average block, immediately following its manufacture. It must be assumed that there are no flaws and that the block was constructed to the stonemason's standard. This state is called the basic condition. Knowing the initial dimensions, we can now define the deterioration in units: "...it has worn by 0.4 inches in its length, and 0.3 inches in its width...." The original flatness was 0.1 ± 0.05 mm over the full area of the surface." Now we can say, "The deterioration has changed the flatness to a range of 0 to 2.5 mm. The worst damage to the block is 43 mm down and 34 mm right from the top right-hand corner, where it was most exposed to the elements."

The deterioration is the amount of change from the basic condition caused, in this case, by weathering and any damage caused to it as it performs its function. This deterioration manifests itself in a range of ways, such as loss of material properties like strength, erosion, and wear of the surfaces, changes in dimensions, chips, internal and external cracks, and visual discoloration. We must also define a similar basic condition for equipment, too. It can be a single definition for the whole tool, but this tends to be a bit impractical. It is more likely to be redefined to cover each specific area under analysis.

The Basic Condition

This is the starting point: the zero line of the rule. Any variation is measured from here, so it has to be accurate. The simplest description of the basic condition is the condition the tool would be expected to be in when it was first manufactured and operating to its original specification. So the basic condition would contain all of the machine's original specifications and that of any internal modules, power sources, etc. This means that all aspects of every module within the tool must be working and any software should be bug-free. The tool must be capable of producing the maximum amount of product (capacity) for which it was designed. Where variations in capacity are conditional, provided the conditions are met, the individual specifications should be correct. This final point is a tricky one, and might have a *get-out clause* written into a specification: the equipment should not have to be overrun to maintain the maximum production rates.

To explain the overrun limitation, we need to appreciate that some tools can just barely perform at their maximum specification and can only sustain it for a short time. For some it might be a maximum speed, but it can only be achieved if everything is set up perfectly and when the motor or pump is driven at full power. This is not an issue, provided running at full power for any length of time does not cause the part to overheat or fail. Some power supplies, for example, are also often unhappy to run at their full load for a prolonged time, so if failure is not an option, the supply should be uprated to allow the unit to sit at a comfortable point in its performance range. Another good example is the *standard* music system in a car. Often the speakers are not capable of running for prolonged periods at full volume: the speaker coils can blow the wires, like a fuse, when running at high power or single frequencies. I had a car with speakers that were very prone to failing. That being said possibly Emerson, Lake & Palmer, and Led Zeppelin no doubt had something to do with it. But the point is, the speakers should be able to cope with anything the system does without failing.

There is also a visual component of the basic condition: the tool should be clean inside and out, like a new car. It should have no leaks, all doors and covers should be in place and fit, any interlocks must be functional, all cables and connectors should be supported, all modules should be properly mounted, and so on. The machine should not look secondhand. The owner of any equipment that is not properly maintained from the onset will find that, in about a year, the cumulative deterioration takes over and it becomes extremely difficult for uptime to remain at the basic condition levels. The end result is unreliability, underperforming modules, drift, bypassed interlocks (who me!), dirty floor panels, oil and air leaks, and an invitation to a lifetime of unscheduled breakdowns and quick fixes.

One of the early goals of TPM is to restore equipment to its basic condition. But before we can get there, we need to be able to define what it actually is: what we are aiming for. Without a definitive definition, any two people will have different expectations. The argument is the same as that for applying standards. Since each company sets its own standard, some companies might choose to redefine the *basic* standard downwards as a means of controlling costs. For example, it might be decided to ignore poor or damaged paintwork. This is not unreasonable, provided it has no impact on safety or production, and it can never be seen by any potential customers. Equally, it might be decided to ignore minor defects like scratches on a gauge face, a display, or a meter cover. Once more we have the same limitations: we can ignore the scratches provided it is still possible to make accurate readings and it completely fulfills all of its functions. It might even be possible to ignore failures in features that are not currently used, even though they can become a source of future problems. A simple example might be to ignore a spare tire, since it is not being used. A more technical example would be installing new pipework, new cabling, or a new pump and leaving the old system in position. (In *Reliability Centered Maintenance* (RCM) these are known as superfluous functions.)

The specifications can also be regarded as flexible for practical reasons. Often an improvement of 50 to 70 percent will provide huge benefits. When we start improving, we would expect to make big improvements which will gradually taper off in size. There could become a point where the improvements are not cost-effective: the cost of parts and labor outstrips the returns. At this point, it might be prudent to change to another area where improvements will be more beneficial. It will still be possible in the future to return to the first task.

In many companies you will find operators stationed at a production line or a machine. They are not loading product or removing it. They are just standing there, like train spotters, waiting and watching. But what are they waiting for? They are poised and ready to leap into action, immediately—in the event of a machine breakdown. They know from past experience that the tool will break down, will stop running, or will damage product. It always does. A major goal of the PM Pillar is to eliminate this need to watch. It should be possible to reduce breakdowns and stoppages and increase reliability by enough to allow the operator to be utilised in other areas and respond quickly in the event of an alarm. The TPM target for failures is zero. If we are able to achieve *zero fails*, we will enable maximum production. By default, we also eliminate this need to watch.

According to TPM, we must repair all failures, irrespective of their importance or cost. This makes sense in a continuous flow production line, where any failure that stops the line must be avoided, but this can possibly be relaxed slightly in sequential production. In real life, though, cost control will always be an issue. There will always be a finite maintenance budget, so perhaps we will need to introduce a degree of flexibility on the "fix every issue" rule—at least initially—as we did when we chose to relax the basic condition above. A good compromise is to differentiate between what *needs* to be fixed to ensure correct operation and what we would *like* to fix if we could. These issues can be sorted out in a "fails list."

Fails can be given a reduced priority as long as they do not affect safety, operational efficiency, or product quality. In order to increase the pace of benefits and to push the positive effects of TPM on production, we should prioritize the elimination of the faults that directly affect three, deliberately chosen, key factors. These are

- Availability
- Performance
- Quality

These three factors are the component parts of *Overall Equipment Efficiency* (OEE), the key index for production. Other than safety issues, any problems that have a negative impact on OEE should be targeted first. The priority should, naturally, be those we know cause the greatest downtime or the largest loss in production rate or the largest loss of quality (or yield). It is also important not to neglect minor issues—we all have a tendency to do that. It could be that what they lack in individual impact might become seriously major if they occur frequently. So track the history and find out the recurrence rates.

Technical Standards

One key, critical, ingredient in *all* continuous improvement techniques is the use of standards. Look around your own company, are standards always used? Unless the company is a fervent follower of continuous improvement, I suspect your answer will be "No." Standards include all operating procedures and specifications. In this case, we need to find an absolute way to describe equipment condition that will help us define the basic condition. This is achieved by using proper engineering standards and tolerances. I say that as if they are something special. They are for many engineers and production people, but they should not be. Engineering standards should be part of everyday discussion. If I was to ask a production engineer in, say, a bottling area what the tolerance is on the part that picks up or supports a bottle as it races along the line, should I expect him to know? What about the same question when carrying out a product changeover? If we decide to change the diameter of the current bottle neck up to a wider neck, say 20 mm, when we exchange the parts that handle the bottle necks, what room for error do we have? If the supports move or wobble a bit when shaken, how much is permissible without it affecting the handling? Is half a millimeter too much? I suspect very few would know, not even the engineers. If I asked for the technical data, would I be able to find it? How many people do you think should be able to answer my questions?

What are the standards? We must have a way to describe the current condition of any part, including any changeover or consumable parts. If we cannot, we will never be able to predict when they are about to become unsuitable to be used for production. We must know and understand the basic condition and any tolerances that guarantee safe product handling and operation. It would not be appropriate if my tolerance was 1 mm but your tolerance was 4 or 5 mm.

Technical standards are common to TPM and RCM. In fact, they should be common to all maintenance and production personnel. They should define the target measure and any acceptable error. Just as defined in RCM, there are two types of tolerance: the absolute and the acceptable specification. Many companies have an official tolerance written in the specification, but when working and actually producing the goods, the tolerance is frequently relaxed. When restoring the basic condition as defined by TPM, we would target the absolute values on the assumption that they are correct. However, if the process allows a wider tolerance that has no detrimental effect on the product, then that value could be considered as it could be easier to maintain. If there is a compromise, it must not have a negative impact on OEE.

The following are a few examples of technical standards:

- The production rate is 1000 bottles per hour, filled to a level of 500 ± 5 ml.
- The top of the 40-mm label should be positioned 50 ± 0.1 mm from the base of the container.
- The slot width should be 1.0 ± 0.1 mm.
- The flow rate should be greater than 2.0 gal/min.
- The flow should be 2.5 ± 0.5 gal/min.
- The air pressure should be 80 to 90 psi.
- The air pressure should be 85 + 5, −0 psi.
- The vertical runout should be less than 0.025 in. over one complete revolution.
- The locating pins should limit any movement of the part to ±0.05 mm.
- The locking clamp should eliminate any movement of the assembly.
- The process cooling water should be 20 ± 0.5°C.
- The disk rotates at 500 rpm.
- The "Start Count" pulse follows the leading edge of the synchronization pulse by 120 ms.
- The track should be level to within 0.1 mm/m.

☐ When the display reads "000," the output from Pin 7 of the Dose Counter drops to logic zero. This energizes Relay K4 in the console interlock circuit.

I must have been very lucky that I have worked only with companies who produce real, professional, equipment manuals. The number of times I have asked engineers in a factory for manuals and have been told there are none defies the imagination. Just in case you have professional vendors, I recommend that you maintain a good relationship with them. The relevant standards for the equipment under review can be found in vendor manuals, customer support "handouts," and within the PM specifications. If it helps, the vendor can provide details from the factory.

If the worst happens, and you are left on your own, or, if the manuals are not detailed enough and there is no available data, it is possible to collect your own. Attach data or chart recorders and take measurements over a realistic time period. If continuous measurements are not possible, the required reference information can be sampled at regular intervals and graphed for subsequent analysis. Alternatively, there might be other tools on-site that can be used for comparison measurements. There could also be other manufacturers using the same tools that could help with information. It might be possible to set up an ongoing information exchange. The vendor will tell you who uses his equipment. Occasionally, it is possible to approach the manufacturer of the actual component—as opposed to the parent tool manufacturer. They could have useful test data. And remember, even if we do get the basic condition wrong, it is not a disaster—we can choose to learn from the mistake, make a suitable correction, and try again.

Overall Equipment Efficiency

When we buy a piece of equipment, we do so to carry out a specific function. In RCM this would be the Primary Function. The tool will produce some kind of "output unit," which can be anything from a machined part or a ceramic tile to cakes, toys, toothpaste, or bottled water. The type of tool and its design limitations will determine its "throughput," that is, how much product it is able to make per unit time. The design is often restrained by cost, which can also limit the accuracy and the quality of the output.

So, let's imagine that we have bought our new tool. We have installed it to the manufacturer's standard. Everyone has been trained. We all know how to use it, how to set it up for production, and we have started making our first product. How do we know when the tool has failed? Come to that, what is a failure? Surely it can't have failed if "bits" are still coming out of the business end?

When we bought the tool, the specification claimed it could produce 1000 biscuits every hour. We, as the users, need to consider all the different ways it can fail—or to put it another way, how many ways it can fail to fulfill the description. For example,

1. *The equipment can stop working completely.*
   That is, it makes no biscuits at all. This is the easiest failure to recognize! Especially at break times.... This is the one that no one can ignore. This is known as a *total failure*.

2. *The equipment can work slower than it is capable of.*
   The best way to get a feel for this is to substitute imaginary numbers. In the first instance we will assume it only makes 900 biscuits an hour. This one is less obvious. I have seen hundreds of machines running below capacity. I know it is possible to count the biscuits but people rarely do. Is it possible to tell—visually—if the machine is producing fewer biscuits? Probably not, but it is not impossible to have an electronic counter. Keep things in perspective: 100 less biscuits per hour amounts to a loss of 10 percent and is equivalent to the machine being unavailable for 10 percent of the time. (We will return to this point later.)
   Often the deterioration sneaks up slowly over time, but not always. There is another popular cause of speed drops, it is a deliberate, "temporary" modification carried out by engineers to compensate for another problem, normally mechanical, and it has been either forgotten about or simply never fixed. This type of speed failure is known as a *partial failure*.

3. *The equipment can lose quality.*
   There could be a multitude of ways the quality can be affected. Perhaps the biscuits become too hard or too soft or just do not taste right. Either way, quality problems result in a loss of profit.

4. *The biscuits can be the wrong size.*
   This can be defined as a quality issue. The biscuits can be too big or too small or have the wrong shape. If bad enough, the sales weight or the wrapping can be affected. If the weight is wrong, then the manufacturer will be breaking the law. Sadly, out-of-shape biscuits are not like stamps. There could never be the equivalent of the Penny Black Digestive biscuit.

Points "3" and "4" are situations that make products that are not up to an acceptable quality and must be remade, effectively doubling manufacturing costs.

Remarkably, and here I go with surprises again, but you will be flabbergasted (What would we do without a Thesaurus? And I thought they were all extinct) at how often you will find that, as long as enough *usable* biscuits are being made to supply the customers' orders, even if a pile or three must be scrapped, the tool operators and the production supervisors will be satisfied and see no need to intervene. After all, they have different responsibilities from maintenance, don't they? It is their job to get the correct amount of product to the next process and out to the customers. If it takes a longer time because the tool is running slow or some product has to be remade, it does not really matter. After all, it is not their fault if the equipment is not running at capacity.

So, it really is up to the owner of the tool to decide what he is prepared to live with: to define what degree of failure is acceptable. I'll bet you already knew that. But, do you know how many managers are aware *how much money* their underproduction is costing them? Both TPM and RCM do. They recognize the need to resolve partial failures as well as total failures.

It seems remarkably obvious that anything less than 100 percent throughput will cost the company money at the end of the day. Making only 900 biscuits in an hour is a drop of 10 percent, which has a direct effect on profits. In fact, it is the same as the machine being broken 10 percent of the time. Not only that, but extra wages must be paid to an operator to run the tool and make up the deficiency. The same goes for poor quality, which is equivalent to a throughput of zero for the duration of the run. For every unacceptable biscuit, a new one has to be made, so we are doubling the production cost. In addition to that, we have wasted power, materials, and manpower and might even have to deliver late to a customer. Even worse, what if the bad product found its way to the customer?

If the biscuit-making tool is only one step in a serial process—there might be a chocolate covering, a wrapping and a packaging stage—then, while the operator is making replacement biscuits, we are likely to be losing (wasting) money in other areas of the factory. Tools will be sitting idle down the line, while operators might be standing around, waiting for products to arrive. Again, power and facilities are being wasted. Overtime rates might have to be paid to operators or the shipping department to complete orders on time. There could be the added cost of an express delivery to get the product delivered.

If they are not already, your customers should be your number one consideration. Today's customers will change suppliers if they cannot rely on delivery times or they receive poor quality products. Both of these issues can have a knock-on impact on their customers. And remember, once a customer is lost, it can take 5 years or more to get them back—if at all.

So, the trick is to make sure that equipment is available and able to produce quality products for the maximum amount of time and that it is used properly. TPM has a measure that considers all of the above points: OEE. This measure recognizes that equipment breakdown is not the only source of production losses. Producing under capability and producing faulty goods also have a serious, negative impact.

The availability of the equipment

This is the ratio of the amount of time that the tool is capable of running quality product to the total time it could be running.

$$\text{Availability (\%)} = \frac{(\text{total time available} - \text{downtime})}{\text{total time available}} \times 100 \quad (1.1)$$

A management decision will be required to define what is accepted as downtime and to set any protocols. There is even likely to be an industry standard. I have listed a few of the options to be considered. It would be useful if they could be individually tracked, as it would help in an analysis of downtime sources. At some point you will need to know how much time is spent on tests, checks, setup, waiting for results, waiting for engineers, waiting for operators, waiting for product, running production, changeovers, and equipment downtime.

The discussions as to which losses come from which group will be entertaining. Should the performance be based on the availability for the week being analyzed or the total availability possible? If we have no operator, is that an availability issue or a productivity issue? What about utilization? How does that tie in to uptime? All these joys will soon be yours. Even if you do have an industry standard, each manager will want it to be changed to, how can I best put this ..., to make his figures look better.

I would suggest a tool is not capable of running product if the tool is down for scheduled or unscheduled maintenance. When a tool is returned from maintenance, the repair has to be confirmed. Naturally it should have completed qualification and test runs, so these losses must be downtime. Some difference of opinion arises with the use of extra test runs during production, used to confirm quality over time or to restore the tool after a setup change. TPM is interested in maximizing production and everyone in the company is responsible for that.

Other than statutory regulation, which I completely accept, the argument for routine quality test runs is a lack of machine confidence on the side of production. They either know the machine is not reliable; do not believe the machine is reliable; or, the most likely reason, they have just accepted a previous work practice. These checks have to be valid. Are they all necessary? The end goal should be to eliminate *unnecessary* testing, so we must ask, "Is the tool unreliable or does it become unstable and drift over time?" If it does, then we have a problem to resolve. Are the process limits too tight for the tool design? It is not unknown for a machine to be used for a purpose it was never designed for. Is the tool being overrun or does the tool have undiscovered performance issues? For people who run automated data collection systems, their systems could be running slow because of unnecessary reports. We might find the systems process thousands of reports, many never used as they were written by employees who have now moved on. Other reports were needed for a specific issue and not deleted. It would be a good idea to whittle them down to an acceptable number.

We also have to recognize the obvious: that different products often require setup changes and test runs to confirm quality. Lean manufacturing actively promotes frequent changeovers, where small batch production is required. This is preferred to making bulk amounts and storing the extra. What might be less obvious is the *method* for setting up and the *necessity* for test runs and the way they are carried out. Time has to be actively spent identifying ways to minimize both the setup times and the need for testing.

TPM is a cross-functional technique, not a maintenance technique. It tends to be shared by maintenance and production as they are closest to the product. But, the aim is to improve the total productivity of the tool, not just the maintenance. Losses can be due to bad scheduling or excessive setup and test losses. Any person who has a hand in getting products from the supplier to the customer will make some contribution to the productivity.

Let us consider a situation where a scheduler tells a line to immediately change from product A to product B. The run for product A is completed, the line is shut down, the changeover is carried out, and then... nothing. It transpires that one of the ingredients is missing. So we wait. The scheduler should surely have confirmed the line is capable of running product B *before* ordering it to be shut down. I would certainly think so. This is not an issue limited to schedulers; it is also a common mistake engineers make prior to, or more likely during, PMs. I would be prepared to bet there are others who make the same mistakes.

This makes it necessary for everyone to review all the issues and find ways to make improvements and develop systems—working as teams not as a maintenance group. There might be a need to pull in experience to help improve the productivity, like introducing quick changeover techniques (Chap. 7) or even input parameter monitoring, but this is for the teams and their managers to decide.

The following three points often cause disputes between maintenance and production:

- The tool is being set up for production.
- The tool is running tests.
- The tool is idling while awaiting test results.

The performance of the equipment

If the tool is running at a throughput less than its capability, then we have a loss problem. A tool running at half speed is the same as having 50 percent downtime. It is interesting that many people do not appreciate that until it is specifically pointed out. The performance of a tool is defined as the ratio of the amount of product made to the amount of product that could have been made.

For a given production uptime,

$$\text{Performance (\%)} = \frac{\text{number of units manufactured}}{\text{possible number of units}} \times 100 \quad (1.2)$$

The quality of the product

- If the quality is anything less than 100 percent usable, again we have problems.
- If poor quality or failed product gets through to the customer, we have an issue worse than just a drop in production. There is the risk of losing the customer.
- The more likely the machine is to fail or to produce substandard products, the more testing has to done to catch the fails.
- The goal should be 100 percent usable product every time.

The definition of the quality of the product is the ratio of the amount of *acceptable* product made to the total amount of product made (including any unacceptable product).

$$\text{Quality (\%)} = \frac{(\text{number of units produced} - \text{number of defects})}{\text{number of units produced}} \times 100 \quad (1.3)$$

Recognizing the importance of the above criteria, TPM utilizes a measure that is based on all three. OEE is the product of the availability, the performance, and the quality. Table 1.1 lists a couple of examples to illustrate how a loss in any one area can dramatically affect the OEE.

$$\text{OEE} = \text{availability} \times \text{performance} \times \text{quality} \quad (1.4)$$

If availability, performance, and quality are each equal to 50 percent, the calculated OEE would be

$$\begin{aligned}\text{OEE (\%)} &= \text{availability} \times \text{performance} \times \text{quality} \times 100 \\ &= 0.5 \times 0.5 \times 0.5 \times 100 \\ &= 12.5\%\end{aligned}$$

TABLE 1.1 Availability, Performance, & Quality as a Percentage and as a Probability

Availability	Performance	Quality	OEE
100% (1)	100% (1)	100% (1)	100%
50% (0.5)	100% (1)	100% (1)	50%
50% (0.5)	50% (0.5)	100% (1)	25%
50% (0.5)	50% (0.5)	50% (0.5)	12.5%
100% (1)	75%	75%	56%

From the data in Table 1.1 it can be clearly seen that even if the tool has no downtime (100 percent availability), the OEE can still be unacceptably low because of the losses in performance and quality. With no downtime and both performance and quality reduced to 75 percent, the resultant OEE falls to a value as low as 56 percent—this is almost half.

Chapter 3 will explain how to identify the current state of your own equipment and later chapters will explain how to identify *why* the tools have deteriorated and how we can improve them. There are a whole range of factors that are likely to impact the tool performance including faultfinding. In other chapters we will also cover the support systems we will need to develop a sound infrastructure: safety, risk assessment, problem solving, competency, standards, and lots more.

Natural and Forced Deterioration

Deterioration is the term used to explain the reduction in performance or reliability of equipment. As mentioned at the beginning of the chapter, it subdivides into two types: natural and forced deterioration. The names are a dead giveaway. One happens with age; the other is helped along its path. Natural deterioration is age- or time-dependent. Forced deterioration can have several causes ranging from use conditions to neglect, poor skill levels, or poor documentation.

The simplest example of natural deterioration is well known by every car owner and is no doubt regularly used as an illustration for maintenance methods. The oil in the engine will gradually become dirty and will need to be replaced. Why, because the moving engine parts constantly rub together and, despite the best efforts of the lubricant, they continue to create tiny particles that mix with the oil. Over time the oil becomes less and less effective, which causes increasing damage to the very components it is there to protect. We must use the oil, because without any, the wear would be astoundingly fast and the engine would simply grind to a stop. What else can we do?

In the best tradition of TPM, knowing the root cause and that this effect is unavoidable, designers include a filter in the lubrication system to remove the unwanted particles and prolong the useful life of the oil and the engine. They would have sampled the oil in test engines to establish the best filter size, next they would check out the filter system to confirm it worked as expected. Even though the filter does not provide a 100% solution, it is good enough for a few thousand miles. So the designers of oil products keep searching for better ways to prolong the effectiveness. The deterioration of the oil is guaranteed to happen because of the *natural* abrasion of the moving parts—hence this type of wear is natural deterioration.

Now consider what would happen if the wrong type of oil or not enough oil was used. The deterioration would be *forced* to happen faster. Luckily, there is a whole industry based on perfecting lubricants and creating specialized oils for specific functions.

Consider a tool with a maintenance schedule that is not followed, is followed incorrectly, or is only partially followed, then the deterioration that will be seen by the tool is also forced. In this case the cause is entirely attributable to maintenance—or the lack of it. Another source of forced deterioration that can have a profound impact on life expectancy is the environment in which equipment is used. Even when the tool is maintained exactly as the manufacturer's specification requires, a detrimental environment will reduce the lifetime of the tool.

A simple example of the environment causing premature failure would be two air conditioners: one in a hot environment and the other in a cool one. Which would you expect to fail first: the air conditioner in a garage in California or the one in a garage in Finland? The odds are in favor of the one in California failing first, the one in the hotter environment.

Reinforcing the first example, which one of the following would you expect to fail first: a refrigerator positioned next to an oven or one sitting in a cool area? Again, it would be the one next to the oven and the cause would be the same. The fridge that gets heated has to work harder so it would very likely fail first. The fridge in the nice, ambient temperature puts much less stress on the system. The point of this example is that, in this case, there are installation instructions for fridges that clearly advise the user not to position the unit next to ovens, radiators, washing machines, or tumble dryers. This forced deterioration is still caused by poor *use* conditions of the tool but is also a case of "operating standards not followed." The instructions were ignored. Even if the fridge does not suffer a total failure, the owner will be paying for the extra power required to maintain the temperature. I have to confess that my own fridge is badly positioned. I know it, but my kitchen layout prevents me from locating it anywhere else. So, I guess the room is also not suitable.

Use Conditions

I have personal experience of the following example of forced deterioration caused by use conditions. The issue was identified as being one of the most important failures in the group at the time. A production tool was evacuated using a vacuum pump. The pump output was fed to an abatement system that treated the toxic contents and converted them from airborne and gaseous into a liquid solution that could be disposed of safely. The problem seen was that the exhaust pipe between the pump and the abatement system would block. This either stopped the abatement unit or backstreamed into the production tool. The vacuum pump ejected a fine silicon dust into the evacuated pipeline. The dust should have been carried all the way to the disposal system. However, in this situation, the dust regularly caused the line to block as it settled in several vulnerable areas of the pipe. The users blamed the abatement system, claiming it was rubbish.

The tool and failures were systematically analyzed over a couple of weeks. To everyone's surprise and disbelief, neither of the tools (the vacuum pump or the next unit in the line) was actually to blame. It was the pipes themselves that caused the problems. Was there any way to avoid the problems? Well, yes. A look at the installation manuals showed a list of steps that should have been followed during the installation. The forced deterioration in the case of this tool was caused by "use conditions" and "operating standards not followed." Table 1.2 compares what was done with the manual's instructions.

From the pipe example it can be clearly seen that the *conditions* in which the tools are *used* can be a root cause of machine failures. Fixing the improper use will stop not only the fails, but also any consequential failures caused by the original fail.

It would seem reasonably obvious that the first place to start, when optimizing tool lifetime, is to follow the installation instructions. This often does not happen. Sometimes the error is caused by

TABLE 1.2 A Comparison of the Vacuum System's Exhaust Pipework

Recommendation	Actual situation
The pipework should be as a short as possible.	The pipeline was too long.
The pipework should have a gradient downward away from the pump.	The pipeline went vertically up to ceiling height and ran over other tools for a long distance to reach the next unit in the line.
The pipework should have no—or few—sharp bends.	The pipeline had lots of right angles.
The pipework should be heated (heat-traced) to prevent the dust in the vapor from condensing.	The pipe was not heated.

- The issue not being considered.
- Time constraints.
- Missing manuals.
  (I have a problem with this one. I hear it as an excuse all the time. I can't accept there is never a way to get a copy of manuals.)
- The tool manufacturer not being involved in the installation.
  (Often the case with secondhand tools where a competitor is used.)
- A new tool has been installed in an unsuitable location.
- The engineer knows how to do it cheaper.
- The company where the tool is being installed has a preferred supplier of support equipment (like vacuum pumps) which may not be the same vendor as the one the tool manufacturer recommends.
- The cost to move something out of the way is deemed too expensive.

A photocopier service engineer highlighted another example of forced deterioration due to improper use conditions. The photocopier was close to the coffee machine. This was not the cause of the problem, but while getting a coffee, I complained that it was a very unreliable unit. In defence of the copier, he explained that it was a low-use unit designed for a small office. This one, in an office of at least 50 people, was handling nearly 10 times the designed load. Suddenly, the tool didn't seem quite so bad!

We have shown how the environmental (use) conditions surrounding a tool will also have an effect on its life or at least its performance. A hot area can cause electronics, magnets, moving parts, and bearings to fail. If the environment is dusty, moving parts can get coated and cause increased wear. Damp conditions can cause rust and affect electrical components and connectors, particularly high voltages—as, will a dry atmosphere. Fluctuating temperatures can have a varying effect on cutting accuracy and structural stresses and can also affect the stability of electronic reference voltages. I will let you complete this paragraph: I bet you could list a few problems too.

Forced deterioration can also be caused by mis*use* of a tool. If a car is regularly run at higher revs than designed, the clutch is operated badly, or no antifreeze is put into the cooling system in winter, the time taken before the car breaks down will be drastically reduced (Mean Time To Fail or MTTF). In this case it is not a design issue but an operator issue. A tool not being used correctly could be attributed to lack of knowledge or experience—assuming that the misuse is not deliberate. In such a case, it would fall into the "inadequate skill level" category or, depending on the standard of documentation, "operating standards not followed."

The following three examples are due to "no operating standard" but only because no one would have believed them possible. The first example witnessed operators running obviously faulty equipment. Their intention was to complete their production target and pass the failure onto the next shift. At one point, the tool problem was so severe that it was likely to cause secondary damage and probably was already causing product variation. I don't need to explain the poor judgment and shortsightedness here. The next example follows the same theme but this time the problem was a sticking mechanical brake. The operator (a radiographer) was complaining that the brake had been slack for ages and was not holding. I tried to free the brake and could not budge it. The operator slipped off her shoe and gave it an almighty whack. "This is how we always do it...."

My final example of deliberate misuse happens regularly. It is one I find particularly irritating. Cryopumps are vacuum pumps that freeze particles out of the air—kind of like the way flypaper traps flies. Also like flypaper, to work properly, the surface that does the trapping has to be cleaned regularly. This involves using a different pump to suck the particles out. The control measure for a clean pump is the vacuum. A dirty pump cannot ever achieve a good enough vacuum. Herein lies the source of the problem. The vacuum must be below a certain level to ensure the pump is clean enough to allow it to be switched on. You might think this is a pretty fail-safe system, but you would be wrong!

The gauges that monitor the vacuum are often *adjusted* to trick the control system into believing the pump is below this vacuum, so the computer allows the cryopump to start. If the vacuum is not correct, oily gunge (a technical term) can build up inside the pump and it will cause the pump to fail sooner. The gunge can also reduce pumping efficiency and is a probable cause of contamination in the vacuum system and, ultimately, of the product. If the same "trick" is applied to the gauge more than once, the amount of dirt in the system is compounded.

I once pointed out this bad practice to a technician and was promptly told that to do it correctly takes too long. "This is the real world." The next day, when the system was still not working, a dirty vacuum seal was found. It took two people and a bottle of isopropyl alcohol a whole 10 min to find the leak. Yes, sadly it is the real world we are living in.

Use condition failures have also been found when tool users have switched to "second-sourced" spare parts. These are spare parts, often the same as the original vendor's parts but not always. These parts are favored because they are less expensive than the OEM parts. I have seen situations where some have caused complex process issues. It can take a long time to discover the problems they create, to analyze the system and run tests to find the cause. Often, it is necessary to track the dates of the problems against the maintenance routines. In one case, the parts were found to be made from a less pure type of graphite that produced contaminants. In another example, parts were not machined to the correct tolerances, causing positioning issues that lasted for several years. The prolonged time was due to the high investment the company made in the second-sourced parts.

I am not against the use of second-sourced parts, but I would expect them to be fully compatible in quality and standards. Follow the adage, "Let the buyer beware." Always confirm that the items are identical and ensure there are no differences that could introduce problems. If something slips through, it will be your fault, whether it is or not.

The Ideal Condition

There is a state above the basic condition, known as the ideal condition. This is the way the tool *should* have been designed! After we have used the tool for a while, its limitations begin to show up. It should have had faster throughput, used more accurate components, been easier to maintain, produced products with less variation in output parameters, should have had better process control and better remote systems analysis, and should have included self-diagnostics.

The ideal condition is the state that can be reached by redesign after the basic condition has been achieved and maintained. Once we broadly identify areas where improvement would benefit production, we set up a "focus improvement" team. The various options can be considered and the improvements evaluated. The groups responsible for implementing the improvements could also be "quality maintenance" teams. RCM can also be used to target specific problem areas and recommend improvements. Six Sigma is another useful method. All four methods are likely to come up with the same solutions.

Improvement Methodology

1. Identify the problem.
   It can help clarify the issues if it is possible to write down the problem in words.
2. Evaluate what the problem costs the company in cash and customer terms.
3. Find a root-cause solution and evaluate the cost to implement.
   It might have more than one cause.
4. Justify the expenditure.
5. Plan and carry out the fix.
6. Confirm that the fix produced the expected result.
7. Set up a system for checking, training, and maintaining the new standard.

It is important to appreciate the different causes of deterioration and the reason(s) for the failure. There is often more than one cause.

- Natural deterioration
- Forced deterioration
- Deterioration due to use conditions
- Design flaws
  (poor mechanisms, unsuitable materials, inadequate support systems)
- Lack of skill
- Poor or no procedures
- Procedures that are not followed
- Sporadic faults
  (Sporadic faults tend to be one off failures. These are not normally predictable unless there is a flaw in a new part, the design, or the way the failure is repaired.)
- Chronic faults
  (These are faults that happen regularly or we choose to live with. These faults are present all the time and would prevent the fails from ever reaching zero. Proper fault analysis will identify and eliminate chronic fails. They are targeted by TPM as an area for positive action.)

For each failure we identify, we must complete a Failure Analysis Sheet (Fig. 1.1). This is used to document the failure cause(s) in engineering terms and record the solution. It includes estimated repair costs. By documenting the failure it can be revisited at any time and the decisions can be challenged, improved, or approved.

One limitation in defining a fault is a lack of experience. I frequently mention this shortcoming. Fundamentally, the technician needs to know how a system should operate when it is working correctly. If the technician does not, how will it be possible to tell when it is not working? Normally, apart from a bit of training early on, time is rarely, if ever, taken to investigate working units to create a baseline for future diagnosis.

I remember a problem in my distant past, working on a radiation treatment machine that would not allow the operator to set the treatment dose. My colleague and I, both with several years of experience, spent a weekend faultfinding and pouring over the manuals but could not find anything wrong. While talking to one of the Vendor engineers,

Failure & PM Analysis Sheet				
Tool ID:		Tool Functional Area:		
		Module:		
Describe the fault symptoms. How should the module work? What is actually happening? *Use engineering terms and units.*		Fail Number(s):	
Reason for failure. How do you intend to fix the problem permanently.				
Repair Start Date:			Failure Date:	
Start Time:			Fail Time:	

Estimated Cost of Repair:			
Labor-hours/Tool:		Repair End Date:	
Parts/ Tool:		Repair End Time:	
Total Cost/Tool:		Total Cost for All Tools:	
Number of Tools:		Is Repair Viable?:	

Figure 1.1 Example of a failure analysis sheet.

he asked us about the Zero Achieved circuit. "The what...?" Apparently, the tool would not permit a new dose to be set until all of the displays had reset to zero and been recognized by an interlock loop. Pretty obvious function and cause in retrospect! Years later, I went back to visit some of my friends at the hospital. There was a new maintenance team. The same tool had been down for a few days. I took a sad pleasure in suggesting that the fault sounded like the Zero Achieved circuit. "The what....?"

Why include the above example? It is a good point to mention that often a complicated fault can sometimes be resolved by looking at the fault history. A vast number of faults are manmade; many have happened before. Sometimes the history can point out exactly the day a fault started and what was done immediately prior to the fault to cause it. When I worked in field service I found that a very high percentage of problems were caused by inexperience, tired night shift workers, poor shift hand-over notes, no shift handover notes, or just carelessness. This technique was key in solving the process problems caused by second-sourced parts.

While working in another customer site, I heard that an intermittent problem was breaking product. The company was not happy and was planning to complain to my bosses when they came for a visit. By looking back through the history logs, I found the first failure date. Looking back a bit further, it was recorded that a vacuum line fitting was loose because of thread damage. Since they did not have the correct fitting, a different one was used. The replacement had a right-angled bend in it, the original fitting did not.

When I investigated the tool and watched the operation over a few cycles, it could be seen that the right-angled fitting caused the vacuum tubing to stretch, stop, and then snap back to the wrong position. The movement could tolerate about 0.5 mm of error with no problems. However, on the odd occasion, it would snap back about 1 mm. Since the movement had already made the position sensor before it snapped back, the software allowed the next movement to proceed, so the arm moved forward, placing the product in the wrong place and breaking it. If the position switch had been set a bit more accurately, it would have changed state when the position changed and possibly would have inhibited the next movement. But it was never necessary to set it that precisely when the correct fitting was in place.

The cost of the fault was several wafers at a few hundred dollars each, about 48 h of lost production, the same time in labor-hours, and a lot of hassle. The cost of the wrong fitting was less than $1. What was the moral of the story? Never fit different parts without considering what problems they might introduce.

How Do We Restore the Basic Condition?

The AM and PM teams will identify problems on the equipment and issue red tags for the technical ones. Once the cause of a fault has been identified and a solution has been decided,

> Repair the failure, unless repairing first prevents proper root-cause analysis.

Use of an exchange assembly will get the tool back to production sooner and allow off-line root-cause analysis. However, it is important that the off-line analysis takes place.

- Inspect the failed parts.
- Identify *exactly* what has failed and why. *This bridge has to be crossed by many engineers. It is not so much a leap of faith as a justification as to why they need to do the extra work.*
- Identify what the mechanism or part should actually do?
- Consider all aspects of the components how they mesh together, are they strong enough, are bearings and seals functioning or are there signs of leaks or abrasion, are any seals or components hidden behind or within an assembly?

(I once worked with a part for nearly 10 years before I found out there was an oil-filled dashpot in it. It was discovered only when a technician stripped out a part for the first time and removed a screw from the base, which allowed oil to leak.)

> Ensure that the Failure Analysis Sheet has been completed. (See Fig. 1.1; This sheet is required for future reference.)

> Update the maintenance procedure to include any new knowledge learned from the fix and explain exactly how to carry out the task.

⇒ Identify any important reference points for setup procedures. Never rely on estimating positions.
⇒ Include photographs and drawings.
⇒ Be specific about setup instructions.
Use technical terms with tolerances.
⇒ Include any part numbers that will help the users.
⇒ Include references to any specific use conditions, i.e., air pressures, cooling temperatures, water flow rates, etc.
⇒ If the setup can be improved by the use of a jig make one and ensure that everyone knows what it is and how to use it.
⇒ Purchase any tools that will improve the setup (torque spanners, levels, set squares, feeler gauges, lighting, micrometers).

> Monitor for the failure returning to confirm that the solution was the correct one. It should be flagged as a red tag if it comes back.

Monitoring enables the teams to identify the issue sooner. With the old system, a recurrence could be missed if it did not happen on the same shift.

- Identify a PM frequency. The intention is to eliminate the unscheduled failure.
- Look for ways of simplifying checks. Rather than take the time to read and record a value from a gauge, is it possible to use visual markers like maximum and minimum levels? Visual markers allow checks to become a yes/no choice. Once set up, an operator need only tick if between the marks. Levels can be applied to gauges, meters, flow meters, oil inspection glasses, and dials; expected readings can be identified on a sign next to the check position.
Consider the check being transferred to AM.
- Train all of the technicians on the new procedure.

Chapter 2

TPM Jishu-Hozen— Autonomous Maintenance

How often have you washed your car and been surprised to find dents, chips, scratches, rust, or even that someone has stolen the bonnet badge? Fairly often, I would bet. As it happens, most types of equipment would benefit from the same level of inspection. There is only one problem—who would we get to do it? Is it reasonable to have highly qualified and highly paid technicians and engineers actually cleaning equipment, when they often can't get the time to carry out permanent, root-cause repairs? Yes, I think so and so does *Total Productive Maintenance* (TPM).

Equipment cleans in TPM are officially known as clean and inspect routines. We clean the equipment at close quarters to help us clear away any debris that acts as camouflage and prevents us from seeing and detecting any lingering or looming problems. As I will elaborate later, it is not only our eyes we use to find the problems, it is all of our five senses. The specific act of cleaning enables us to locate the source of any problems, which will be recorded for data analysis using F-tags and will later be prioritized for action. The exercise is very similar to the application of 5S, except where 5S tends to be external to equipment, *Autonomous Maintenance* (AM) and *Preventive Maintenance* (PM) tend not to be limited by boundaries, other than those that might be necessary to avoid safety issues.

If we graded the teams in increasing skill level, we would have 5S, AM, and then, the most technical, PM. In all cases, the work is carried out by teams. However, there is a difference in the makeup of the teams: AM, the subject of this chapter, is predominantly operator-driven, whereas PM, which tends to be a more technical team, has clean and inspect duties similar to AM, but also has the responsibility for the resolution of problems. This necessitates an engineering element in addition to operators. How, then, does this cleaning help in addition to fault identification?

When I was a customer support engineer I would maintain equipment at the customer's site anywhere in Europe. The first task I would perform on taking up a maintenance contract was to assess the condition of the tools under my care. This normally involved a quick review of the fault history, as that paints a picture of the standard of maintenance the machine has undergone. I would also visually inspect and operate the equipment to see how it performed relative to its specification and watch how the machine was handled by its operators. Next, I would create a fault list, prioritize the issues, and then systematically restore the tool to the best condition I could. This hopefully would have been to a level that would exceed the customer's expectations and, of course, satisfy my own boss. If the customer was kept happy, the contract might be renewed and that was good for business and job security.

Imagine then a situation where a potential customer arrives at a company. His purpose is to assess the organization and its manufacturing capability and to decide whether or not he can entrust his business to them. His host will be busy trying to impress him. He is taken on a guided tour of the factory. For the sake of argument, let's assume that the factory can only be in one of two conditions. We will ignore the gray areas in between. In scenario one, the company has equipment that looks scruffy and gives the *perception* that it might not work too reliably. Scenario two is the opposite: it has a clean, well-maintained, professional-looking kit with obvious external signs of maintenance and attention to its appearance. Which company do you think would get the new business?

I think the question would be ideal for a premium rate, multiple choice, mobile phone quiz. You know, the ones where the correct answer is obvious and the other two are ridiculous:

"What do fish swim in?"

Text: "a" if you think the answer is beer.
        "b" if you think the answer is water.
        "c" if you think the answer is chicken soup.

Way back in 1986, I knew some guys who were being made redundant from a manufacturing company that was closing down. They offered to buy as much of the old equipment as they could afford and set up their own new factory. Without having any experience in TPM, the first thing they did was to respray all the equipment panels to make the tools look as good as they could. They then set up the production area to look as professional as possible. Why?

Text: "a" if you think the answer is . . . .

The answer above is too obvious—everyone knows it. So, why then do so many factories allow their equipment and production areas to deteriorate to such a degree that it is likely to put customers off? Even if the equipment works well but just looks bad, it is the perception that the equipment promotes—the real condition is unimportant. These days business is way too competitive, with the trend likely to get even worse. To attract or retain customers, companies need all the edge and positive influence they can get. It simply does not make any sense to take such an unnecessary risk, particularly when the situation can be so easily prevented. Bear in mind, it is not only the customers who can be influenced by equipment that is in a good state of repair: when a company's own engineers work on a well-maintained tool, their mind subconsciously sets an equivalent standard for their own work. A bad job on a well-maintained tool would stick out like a box of chocolates at weight-watchers club. Furthermore, working on an oily, sticky, dirty tool is not a pleasant task—to say nothing of safety.

The TPM Initial Clean and Inspect and F-Tagging

The initial clean is the first practical equipment task for every TPM team, whether it is an AM or a Zero Fails (PM) team. The idea of cleaning often puts people off—particularly the operators. For some reason, they fear that their entire working life is about to change for the worse and that they are about to be turned into cleaners. They are concerned that they are about to be dumped with all the tasks that the technicians don't want to do. This could not be further from the truth. The ultimate goal of TPM is, or should be, to upskill everyone as far as they want to or are able to go and offer them the possibility of an alternative career path.

There is a basis of truth in the operators' ideas about the types of tasks, at least initially. The tasks they will be given do tend to be very simple ones—cleaning equipment, checking gauges and levels, etc. Often they are tasks that the technicians are responsible for and they should be doing, but with all of their other, more technical and important duties, the technicians often neglect them because they are seen as a low priority. Experience has shown that no one would bother if such tasks were missed out. If the tasks did not impact equipment operation, they could be given a low priority. Besides, with no formal system to pick up their shortcomings, who else would know?

Many of the cleaning tasks the AM teams will initially have to carry out will be caused by the very faults that the technicians did not feel had to be fixed yet. These would be minor issues—leaking oil from motors, water leaks, broken connectors, and cable supports. They would be "unchecked deterioration" that might eventually lead to complete failure of the parts in question, but so far have not. Every member of the team will be involved in the initial clean. Even the technicians have to take part. There are no exceptions.

Although called the initial clean, and the teams literally clean the equipment, the exercise has a bit more complexity. The teams are also faultfinding the tool, although, not in the same way that an engineer would normally look for a fault. Having said that, he does use the same methods sometimes, but in a more general way. The test equipment that the teams are using is their five senses: sight, touch, smell, hearing, and taste. These faculties are often overlooked, but by using only their five senses, the AM teams will be able to

1. *See* obvious areas of damage and contamination.
   For example, the dripping water leaks that have never been repaired; the stains from the oil or grease that has been slowly seeping from motors and valves—which point to potential failures; the odd screws, bolts, and washers from incomplete jobs, and so on. Do not undervalue the faultfinding advantage of sight. Even qualified electronics technicians, when looking for problems on a circuit board, will visually inspect it. They look for broken components or tracks, damaged connectors, heat damage to components, dry joints, and melted and discolored tracks often below components or heat sinks. It is also possible that they may see mechanical movements not following their designated paths or conveyer belts that have unusual rippling patterns or broken links.
2. *Touch* surfaces as they are being cleaned.
   Touch is a very powerful tool. Feel rough surfaces, flaking paint, vibration, and loose linkages. Feel the knocking of faulty electric contactors or solenoids. Perhaps an area of the tool might be warmer than expected, because of friction from a failing, rotating component. Air lines or connectors might be loose and become disconnected while cleaning. It might even be possible to feel air or nitrogen blowing through a pinhole in a tube. Lock nuts on sensor assemblies might have been left loose, causing the operation to be unreliable as a result of excessive wobble or vibration. Cooling fans might be clogged and have no or a reduced air flow. Surfaces might have jagged edges that can cause moving products to stop.
3. *Smell* leaks and spills.
   This is not possible in situations where respirators need to be worn, but otherwise there can be the smell of nontoxic oils, vinegars, overheated motors or relays, and so on. It is also possible to detect unusual smells from process chambers that might suggest something

has changed. All of which will initiate a more technical investigation to help identify issues.

4. *Hear* noises.
The section on P-F Curves in Chap. 10 explains how noise is often a precursor to failure as is heat. Failing motors can knock or squeal, worn moving parts might make unusual clicks or beats while moving, and bearings can also knock or squeal. Leaking air lines can hiss as the air escapes, as can their manifolds and pipe connections. Mechanical assemblies can slam into end-stops if their damping systems have failed. Noise (and vibration) can also be generated by out-of-balance rotating parts.

5. *Taste* ...
Well OK ... that is a hard one ... but it does happen, particularly with airborne vapors. Not a lot tastes as bad as a leaking nitric acid bottle.

Upskilling all of the AM team members on their own equipment will give them an increased feeling of ownership. It also reduces the fear that the operators are simply becoming cleaners. After all, how many cleaners

- Know how some really complex tools actually work.
  They can tell you all of the major modules within the tool and what they do. They also know the processes and how the modules interact. They will not necessarily know the recipes they follow in detail but will understand roughly what the recipe means and what it does to the product. They will eventually learn details like how to recognize the symptoms of failures caused by incorrect setup.
- Know all of the potential hazards of a tool and how to neutralize them.
  It is essential that the operators are confident about their equipment. They must be in control of their own safety.
- Know, understand, and can carry out risk assessments.
  They might be limited in their technical knowledge, but they will understand the method of evaluating risk and the kind of issues to look for. They will be able to recognize things that *might* be a risk and will know how to pursue further information. I guess they will be treated like trainee technicians in terms of training. Besides, why create two courses when one will cover both operators and trainees.
- Know which areas they and their teammates have been approved to work in and have had the required training to enable them to actually do the work.

✴ Know that they can keep on learning.

AM is an area of continuous improvement. Once a company starts down the path, they get used to seeing the benefits. The same goes for (most) team members. Some team members will not want to embrace the change—usually about 20 percent of them. Their careers will probably stagnate unless they find a way to develop themselves in other areas. But it will no longer be enough to be an operator only. The title "Manufacturing Technologist" echoes around the halls. Either way, the operator of the future will be more technical.

The flow diagram in Fig. 2.1 is a visual summary of the steps taken by both the AM and Zero Fails teams when carrying out an initial clean routine. The basic sequences are identical, which is why they are being handled together. The diagram shows the training sequence and all of the aspects of the performance of the equipment that the teams need to be taught: basic process, maintenance, and operation. The next stage moves into the safety training. Operators are effectively trained in the same way and to the same degree that a trainee technician would be. This will include how to understand and create cleaning maps, hazard maps, risk assessments, and the allocation of the areas of responsibility for the initial clean and subsequent inspections. It also shows the planning, safety checks, F-tags, data recording, and the procedures to carry out the clean—including the actions to take in the event of problems being discovered.

Notice that safety appears in two places: in an initial set of training and procedures, maps, and risk assessments, followed later by a final confirmation to make sure nothing has been missed and the teams are fully prepared. Remember (*and I am beginning to get worried about the emphasis I am putting on this*), operators need special consideration to make doubly sure that they are capable of working safely, particularly when they are on their own (autonomous).

The preparation and paperwork required before the machine is touched is pretty vast. Both the AM and the PM teams have to be totally conversant with them all—or, if you have different procedures, they must be fully trained in the safety procedure setup used by the organization they will be working under.

The procedures contained in the book include

- *Area Maps*
  A drawing that gives functional areas a defined border.
- *Hazard Maps*
  A list of which hazards are in each area.

Autonomous Maintenance 49

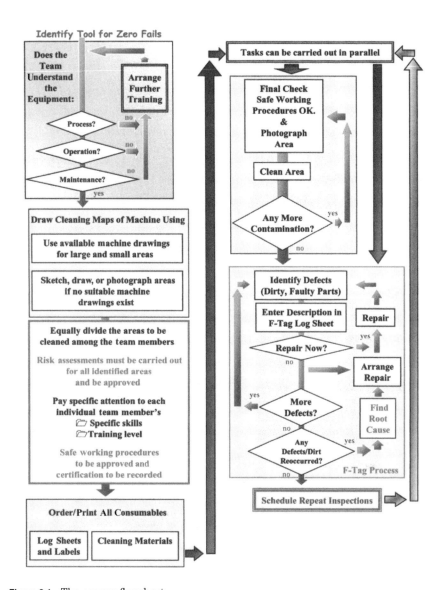

Figure 2.1   The process flowchart.

- *Cleaning Maps*
  A drawing or photograph showing details and special comments about which parts are to be cleaned.
- *Risk Assessments(RA) and Safe Working Procedures (SWP).*
- *F-tag Log Sheets*
  The data describing the F-tag.

50   Chapter Two

📖 *Task Certification Sheets*
The proof that the team is competent and the associated risk assessments and safe working procedure references.

The Cleaning Map: What and Where to Clean

Figure 2.2 is a simple cleaning map that focuses on one area of a tool. It is the area where the operator normally loads the product. (Interestingly, it is the same type of tool that is used to illustrate the *Reliability Centered Maintenance* (RCM) process.) Notice that it also has a close-up in a side window. Maps do not have to be photographs; they can also be scale drawings of the tool, similar to the plan diagrams of the implanter we can see throughout the book. Just as in Fig. 2.2, they should include close-up details of areas that are not visible on the large drawing. The positions of the close-up sections should be identified on the large map to maintain continuity. Blue dots (which appear as black in the figure) are added to show contamination areas.

Bearing in mind the kind of tool a furnace is, the entrance to the furnace should be reasonably clean and, depending on the length of time from the last clean, would likely be due an AM clean in the near

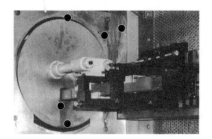

Tool must be switched off to gain access for cleaning the buildup of the process residue.
Risk Assessment & Procedure:
RA - Furn-R12
SP - Furn - R12

● Area Cleaned

Figure 2.2 Illustration of a cleaning map of a part of a tool.

future. The red area (which appears as light gray in the figure) that can be seen to the left of the blue dots is the element, which is probably sitting somewhere around 800°C. This high temperature is the reason that the area cannot be cleaned while the furnace is powered on.

The numbered "blue dots" are found at contaminated areas that are planned to be cleaned. More than one dot positioned in the same area highlights that the tool has a recurring issue and that a root-cause, cost-effective solution needs to be found. The source of the contamination, if known, should also be shown on the diagram. In our case it would be the accumulation of process materials.

Never forget safety issues. If the area requires any special handling, any preparation (like switching off the furnace), or there is any protective clothing (PPE) that needs to be worn, this should also be shown in the map. In addition, the appropriate risk assessments and safe working procedures for the task should be referenced. The operators will be expected to clean everything in their designated area, including all the places that have been contaminated.

The initial clean is like giving the tool a coat of "clean" paint. Dirt can and will hide the fact that a real problem exists. It is the black cat in a coal cellar scenario: how could you spot an oil leak if the floor or tool base plate is already covered in oil and grime?

Before the cleaning starts, the areas of contamination need to be identified. They should also be documented with a "before" photograph that can be displayed on the activity board to promote the success of the team. Risk assessments must be carried out to cover the cleaning task—with consideration given to any surrounding areas the operators might be in contact with. The same assessment can be used to cover similar tasks in identical areas that have been made safe by the team, provided the steps for confirming that the areas have been made safe are clearly documented. (See Chap. 5 for information on modular risk assessments.) If an area of contamination is found and the contaminant is unknown, it should always be assumed to be hazardous until proven otherwise. Once the contaminant has been identified, the risk assessment and safe working procedure must be revised to include the identification of the material and any relevant safety data from the *Material Safety Data Sheet* (MSDS). It should also clearly state the cleaning method and materials to use and the source of the contamination—for example, it might be oil from a leaking pump sited nearby or dripping from a feed pipe.

It should be remembered that the aim is not simply to clean equipment but to eliminate the sources of contamination, thereby making each subsequent clean easier. For example, in the case of the furnace contamination seen in Fig. 2.2, would it be possible that reducing the process gas flow or increasing the waste gas extraction might reduce the amount of cleaning necessary? It would make sense that if there is "leftover" gas at the end, there might be too much going in. This would have to be checked by experiment. At all times, however, it is important to consider the effect a potential solution might have on the existing process and the quality of the product. If we looked at the furnace problem from another perspective, as if we were spraying the product with paint (as opposed to using gas), then is it possible that using less paint might change the color of the product or make the coating too thin?

This is why it is very important to calculate the total dollar cost of the time taken to plan and carry out the experiments. Take decisions based on data, not on gut feelings. Don't ignore gut feelings, but prove that they have validity, or experiment to prove them. Remember that for each experiment we carry out, we are

- losing several hours of production.
- utilizing several hours of manpower.
- paying for the cost of the test product and the materials.
- utilizing postexperiment time to check and to process the results.

This is not to say that time should not be allocated to find a solution. An expensive fix does not need to be dismissed; it can be managed into suitable time slots. What you could do is carry out a quick RCM-type evaluation to compare the dollar cost of a solution to the cost of the maintenance, remembering to take into consideration any possible product loss or other consequence of the contamination. Just remember, if all experimentation was judged on cost only, we would never have research, product development, or, indeed, any new products. It only needs proper budgeting and planning to make the best of resources.

F-Tags: How to Record Fuguai

The Japanese call contamination or abnormalities *fuguai*, which explains the name F-tag. F-tags are labels that are physically attached to the contaminated area and are used to identify it. It does not have to be a real label. Sticky dots can be used to identify points within small areas where labels are not suitable. Sometimes red or white electrical tape is used. The tape can be numbered using a permanent marker and be attached to areas where a tag cannot physically be tied to a unit. The dots and labels must be numbered in sequence with no duplication.

The F-tag is color-coded to identify the complexity of the task. The recognized color coding is

- *Red tag:*
  This requires a degree of technical knowledge and a technician would carry out the fix. The knowledge requirement is not simply for carrying out the task but the technician must be capable of handling any unexpected events that might occur while the task is being carried out. There are any number of things that can go wrong during maintenance. A simple example might be a valve leaking or a gas pipe being ruptured. Even a small nitrogen leak in an enclosed area has to be properly handled.

  It would not be possible for an operator to be allocated a task that has a red tag. In time, as the operator's skill increases, selected tasks can be considered for transfer to the operator, provided the appropriate conditions can be met (see Chap. 4, AM Task Transfer)

- *White tag:*
  This requires much less knowledge than a red tag and can be carried out by an operator after instruction and testing to confirm understanding. White tags require very little special training. Initially the tasks are based on cleaning, from simple dusting to cleaning contamination. There will always be debris, a *maintenance residue*, to be picked up and screws to be replaced. The target is to restore a tool to its basic condition and keep it there. Although the complexity of white tags will start low, AM and Zero Fails teams are regularly trained to support any new skills they might need, and hence the opportunity to increase their experience will always exist. There is even the likelihood that the same task could be a red tag for one team and a white tag for another, if the team members are suitably competent.

Each tag will have a unique number and can contain all the same data that is recorded on the log sheet. The advantage is that anyone looking at a tagged item will be able to tell immediately when the tag was issued, why the tag was issued, who issued it, and when it is expected to be repaired. If the tag only has a number—and that has its merits too—then the F-tag log sheet has to be found before it can be used to find out the same information.

Every F-tag must be recorded and identified as red or white. The information is entered on the F-tag log sheet. Figure 2.3 is a sample sheet, but you can develop your own as long as it contains the correct data. The forms can be used to record the data in real time as the teams work or, as some prefer, notes are taken as the work is carried out and the final data is transferred to the sheet following the clean.

The numbering system shown in Fig. 2.3 looks more complicated than it really is but it provides all the information needed to describe the file.

Tool Area	F-Tag ID	Date	Task Description	Risk Assessment ID	Risk Assessment Approved	Safe Working Procedure ID	Repeat Problem?
1	R1	10-Jan-04	Clean oil pool below RP1	RA-Tool1-R1	1-Feb-04	SP-Tool1-R1	Y
1	W2	10-Jan-04	Clean dirt on source insulator	RA-Tool1-W2	24-Jan-04	SP-Tool1-W2	Y
1	W3	10-Jan-04	Replace screws on panel above source	RA-Tool1-W3	24-Jan-04	SP-Tool1-W3	Y
2	R4	10-Jan-04	Repair electrical connection to Ion Gauge IG2	RA-Tool1-R4	6-Feb-04	SP-Tool1-R4	N
5	W5	10-Jan-04	Replace drive belt on Input Track	RA-Tool1-W5	28-Jan-04	SP-Tool1-W5	Y
	R6						
	W7						

Figure 2.3 An example of an F-tag log sheet.

Some companies might prefer a more simple numbering system. The log sheet should be completed as follows:

➢ *Tool area:*
This is the area of the tool in which the task was discovered. See Chap. 5 and its *Area Map* section to get an appreciation of the concept of areas and modular risk assessments.

➢ *The F-tag ID:*
This is also known as the Task ID. The tag number can incorporate R or W to signify a red or a white tag or it can simply be left as a number. If the tag is labeled as white or red, it is easier to get an immediate appreciation of the complexity of the task. It also avoids any misunderstandings as to which group the task can be allocated to. If the tag is chosen to be a number only, then a new column must be added to the spreadsheet to identify the tag type.

➢ *Date:*
The tag creation date must be recorded. If the date is not recorded, it will be impossible to assess if the tags are being resolved and whether or not they are being resolved quickly.

➢ *Task description:*
The task description should be summarized, but it must be ensured that it has enough details to make sure everyone knows what it means. This does not only mean on the day it was written, but also

later when we might look back—in the event the task was not tackled immediately.
- *Risk Assessment ID:*
  Each risk assessment will have a unique identification number. Consider a format like RA-Drill3-W6.
  ⇒ (a) The RA stands for risk assessment.
  ⇒ (b) Drill3 is the tool ID number.
  ⇒ (c) W6 refers to tag number 6, which is a white F-tag.
  No work can be carried out until a risk assessment has been completed. It will also form the basis for the safe working procedure.
- *Date:*
  Record the date the risk assessment has been approved. The safety committee (if there is one) or some other designated group must approve the risk assessment.
- *Safe working procedure ID:*
  A safe working procedure, with a unique ID, has to be written for every task. Consider a format like SP-Drill3-W6.
  ⇒ (a) The SP in the ID stands for safe working procedure.
  ⇒ (b) Drill3 is the tool ID number.
  ⇒ (c) W6 refers to tag number 6, which is a white F-tag.
- *Repeat problem:*
  Record if the task is recurring. Naturally, this will not be known initially. One of the advantages of creating the F-tag log sheet as an Excel spreadsheet is that it can be sorted into areas and, if the task has been tagged more than once, it will show up in the sort.

The team members for the areas that include the task should be trained on the basis of the risk assessment and the safe working procedure. They must also be tested on the details of the procedure and the cleaning prior to beginning the task. All training and test details must be recorded. If the team member passes the test, the details must be recorded on the task certification sheet. Figure 2.4 shows a suggested format. This sheet identifies which team members are competent (certified) to carry out any specific task and the date of certification. Much of the information on this sheet is the same as that on the F-tag log sheet, so the two sheets could be merged together or the data can be cut and pasted onto a spreadsheet or, even more cunningly, added to a database.

An Excel spreadsheet would be better to use than a table in Word. The Word table does not enable sorting. The spreadsheet flexibility will also highlight recurring tasks and enable sorting.

Chapter 5 has details on the training and testing methods.

Tool Area	F-Tag ID	Task Description	Risk Assessment ID	Safe Working Procedure ID	Team Members Certified	Certification Date
1	R1	Clean oil pool below RP1	RA-Tool1-R1	SP-Tool1-R1	Technician 1	2-Feb-04
1	R1	Clean oil pool below RP1	RA-Tool1-R1	SP-Tool1-R1	Technician 2	2-Feb-04
1	R1	Clean oil pool below RP1	RA-Tool1-R1	SP-Tool1-R1	Operator 2	3-Mar-04
1	W2	Clean dirt on source insulator	RA-Tool1-W2	SP-Tool1-W2	Technician 1	26-Jan-04
1	W2	Clean dirt on source insulator	RA-Tool1-W2	SP-Tool1-W2	Technician 2	26-Jan-04
1	W3	Replace screws on panel above source	RA-Tool1-W3	SP-Tool1-W3	Technician 1	2-Feb-04
1	W3	Replace screws on panel above source	RA-Tool1-W3	SP-Tool1-W3	Technician 2	2-Feb-04
1	W3	Replace screws on panel above source	RA-Tool1-W3	SP-Tool1-W3	Operator 2	3-Mar-04
2	R4	Repair electrical connection to Ion Gauge IG2	RA-Tool1-R4	SP-Tool1-R4	Technician 1	6-Feb-04
2	R4	Repair electrical connection to Ion Gauge IG2	RA-Tool1-R4	SP-Tool1-R4	Technician 2	6-Feb-04
5	W5	Replace drive belt on Input Track	RA-Tool1-W5	SP-Tool1-W5	Technician 1	29-Jan-04
5	W5	Replace drive belt on Input Track	RA-Tool1-W5	SP-Tool1-W5	Operator 2	29-Jan-04
	R6					
	W7					

Figure 2.4 Task certification sheet example.

The following list is a summary of the knowledge required before the team members can begin the initial clean.
   They must

✓ Understand how the tool works.

✓ Understand all of the different areas within the tool, the area maps, the cleaning maps, and which areas they have been approved for access.

✓ Know all of the potential hazards and how to neutralize them.

✓ Know and understand all of the risk assessments and safe working procedures for their tasks.

✓ Know how to use the various log sheets so far.

All that has to be done now is the actual clean itself, which is covered in Chap. 4. Cleaning will not be started until all of the safety and training data has been confirmed. This stressing of safety might get boring and repetitive, but we are using operators in a way that was probably never considered when they started working. Any accidents could have serious repercussions and, not to undermine any injury to the operators, an accident might affect the implementation of TPM.

The entire machine will be cleaned by areas and every fuguai (abnormality) will be recorded. The team leader will schedule the areas to be tackled and will decide how the complete machine will be tackled. There are not really a huge number of alternatives. The choice is really only between completing one area before starting the next or rotating the team between areas. The operators will be supervised by the allocated technical support person until they are deemed competent to work unsupervised. The engineering support will still be required to complete F-tags, to give extra training, and to ensure that the correct practices are still being adhered to. Refer to the section on assessing competence in Chap. 5.

	Actions
2-1	☺ Create the training documents. The training documents will be created by the technician or the equipment engineer.
2-2	☺ Train the team members on the tool.
2-3	☺ Create initial safe working procedures and risk assessments.
	☺ Create an initial cleaning map of the tool. (To be carried out by the technical support plus operators.)
2-5	☺ Create an F-tag log sheet. (See Fig. 2.3)
2-6	☺ Arrange to have a pile of "blue dots" and F-tags (white and red).
2-7	☺ Number the dots and the tags as ONE sequence of numbers.
2-8	☺ Identify and record the obvious areas of contamination. Do not clean at this point.
2-9	☺ Modify the risk assessments on the basis of the contamination seen and the team members who will be working in those areas.
	☺ Ensure that they are suitably skilled for the clean.
2-10	☺ Create or modify the safe working procedures to include the specific cleaning tasks.
2-11	☺ Train the team members on the tasks.
2-12	☺ Complete the task certification sheet. (Fig. 2.4)
2-13	☺ Supervise the operators until the team is deemed competent to work unsupervised.

Discovery of a Serious Fault During the Cleaning

There is one issue that has no boundaries and will affect both AM and PM (or Zero Fails) teams: What do we do if we find a serious fault while carrying out the cleans? It is important that this question should be resolved before the teams begin. If you wait till it happens, Production will likely want the issue to be scheduled for later so that they can have the tool back. On the other hand, maintenance will want an immediate repair—we would hope! As is always the case, the solution to the problem will lie somewhere in between. The defining clue is in the description "a serious fault." We need to ask, "Will the fault have an immediate effect on the tool or product quality?" If the answer is "Yes," then there is no choice—it must be repaired immediately by the technician just as he would any other fault.

The fault would still be treated as a red F-tag (given a label and logged) but the repair would be carried out under the standard risk assessments and procedures that the technician would use normally to faultfind the tool. Since the team will have only been allocated a limited time for its initial clean, and the clean will still have to be completed, extra time will have to be provided to the teams on a flexible or an as-required basis to allow for any unexpected repairs.

If the fault does not have imminent consequences, which means it will not immediately damage the tool and the tool is likely to be able to run for a while without affecting the product, the problem should be assessed and the repair coordinated with Production. The fix should be given a higher priority than normal; it is weighted higher because it was found by an AM team, if it was ignored it has the potential to demotivate the team.

Tracking the Progress of the Initial Cleans

The initial clean is really just the start of a series of cleans, but it is the one where most errors might happen and team confidence builds. Once completed, the initial clean becomes a routine clean and inspect cycle. The cleans will then be scheduled as if they are routine maintenance. Like the cleans in 5S, each time the tool is cleaned, the time taken to complete the tasks should decrease because of increasing experience and the elimination of recurring F-tags. If the teams graph the times taken to carry out the clean and inspect routines, they will find that it is a good measure of their progress and demonstrates the degree of success of the team.

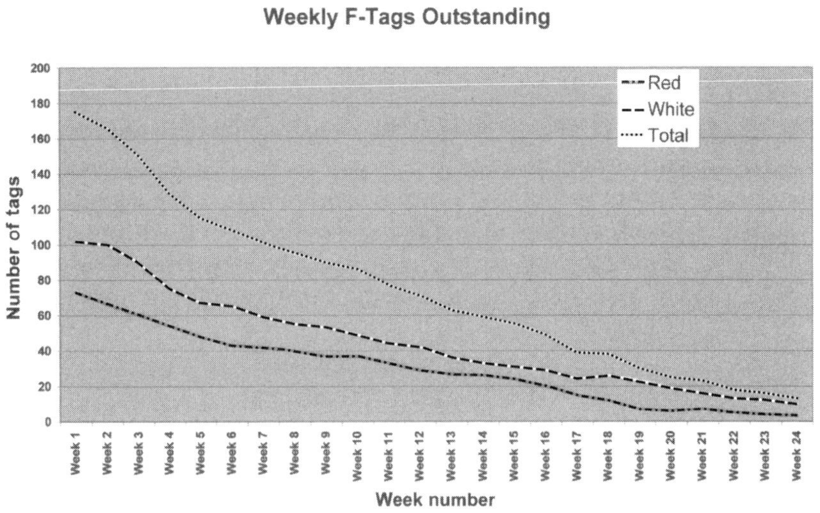

Weekly F-Tags Outstanding

Figure 2.5 Example of the drop-off of F-tags.

There are a few graph formats that can be used; the best one is a simple line graph with the cleaning times plotted on the y-axis and the dates of the cleans plotted on the x-axis, with a continuous time series. Figure 2.5 shows the most suitable type of graph to use.

Another useful graph to demonstrate team progress is the F-tag distribution. This is a triple plot of the total number of F-tags, the number of red F-tags, and the number of white F-tags. The graph (Fig. 2.5) will highlight the improvements being made. Success results in a decreasing number of new tags.

The rate of decrease of F-tags will fluctuate. Initially, we should see a rapid fall as the bad working practices stop. These include the same, very simple F-tags that we have highlighted previously—dropped screws and washers, missing screws and panels, doors that will not close properly, drips, and oil spills. The recurring tasks will take a little longer as a root-cause fix will be required. As time progresses and the easy tags are removed, the remaining, more difficult tags will take longer to resolve because of their complexity. There might also be delays caused by scheduling on the basis of available funding, or where new parts are needed, there could be budgeting constraints.

Remember, F-tags are issued every time a problem is found until a root-cause fix has been carried out.

		Actions
2-14	☺	Create a skill log/task transfer sheet. (Chaps. 4 and 5)
2-15	☺	Assess the skill level of the team members based on the individual tasks.
	☺	Enter the skill data into the skill log/task transfer sheet.
2-16	☺	Create a chart to monitor the progress of the F-tags. (Fig. 2.5)
2-17	☺	Create a defect map showing the areas of contamination. (Chap. 3)
2-18	☺	Learn the categories of deterioration.
2-19	☺	Create an F-tag category spreadsheet. (Chap. 3, Fig. 3.2)
2-20	☺	Learn root-cause analysis and the "5 Why's" technique.
2-21	☺	Create a "5 Why's" analysis sheet. (Chap. 11)
2-22	☺	Analyze the F-tags, categorize them, and enter the data in the spreadsheet.
2-23	☺	Create a defect chart of the F-tag data. (Chap. 3, Fig. 3.3)
2-24	☺	Update the activity board with the new charts and data.

Chapter

3

TPM—Analyzing and Categorizing the Failure Data

Imagine what it would be like if money was no object. I don't mean that we wouldn't have to work to make a living, but that when we were out there earning our living, we would not have the usual fiscal limitations that keep us fighting to stay on top. Sadly this is never going to be the case. Mind you, I am sure there will be some companies that are very profitable and their maintenance department will have a proportionally bigger budget than most, but I suspect if you looked back, into their organization in their precontinuous improvement days, they would probably have an even higher percentage of $waste than many of the poorer companies that survive with lower budgets.

Being a profitable company has other benefits too. They tend to spend more on training, and having the extra money enables them to have the freedom to try out latest ideas in business techniques. It is usually around this point, when they start sampling the available offerings, that the light comes on and they suddenly become aware that their profits could be even better than they are now. All they have to do is invest in improving their maintenance organization and eliminate productivity waste. Then, when they decide which direction to proceed, new money is made available to enable the improvements. The cash is not shaved from other areas. This spending is not a benevolent gesture, they realize fully well that there will be a real return on their investment. For the bulk of us, however, we have a harder time. Competition is our motivator. We must compete with prices and quality to survive.

Many companies are still not aware how inefficient their own maintenance is, or if they are aware, they might not want anyone else to know. I have heard of situations where maintenance managers leave meetings when people start discussing topics like *Total Productive Maintenance* (TPM). They have never given a second thought to the possibility that they might have a need for it. In fact, they are often proud of the way things are, but then they have nothing to compare to. It is a bit like the way people talk about how the poor were brought up with all the sacrifices, but often the poor were not aware they were poor. They had nothing to compare *their poverty* to usually because their friends and relations were also poor. For the less affluent companies who start using continuous improvement, their situation gets worse before it gets better and it starts with the managers who have to juggle budgets, manpower, equipment, and production time.

When we start the analysis of the equipment, it will be a revelation how many things we find that are wrong. For each tool, we will have to find the money and the manpower to put the problems to bed. This cost will be greatly increased if the factory has multiple tools, as each fix has to be repeated on every tool of the same type. TPM targets *zero fails* and so does not prioritize on the basis of cost. When you think about it the reason is obvious: if we did prioritize, then the less important jobs would never be tackled, particularly if their cost was high. At least initially, I would say that it is important that the cost of each fix be estimated and equated to the cost of the consequences (over a time period). This will help to control the funding and give a return on the money spent.

So, costs have to be recorded and the best value for money found. The savings will act as a motivator for the teams and the other managers. Involve the whole company in the introduction of improvements. If a purchasing department is available then they can help with the high cost and multiple purchase items. Even if it is decided to fix all of the issues, it is always a benefit to monitor costs.

This is a good point to remind you of one of the reasons for *Autonomous Maintenance* (AM). It is finding suitable tasks that are currently carried out or ignored (or overlooked) by the technicians/engineers but are suitable for allocating to operators. This will free up some of the technicians' time and make them available for the technical fixes and development that will restore the equipment to its basic condition.

		Actions
3-1		☺ Create a spreadsheet to monitor the costs of bringing the tool back to its basic condition.
		☺ Include a column for the number of tools that would need to be restored.
		☺ Include a column for the estimated cost of repairs.

F-tags, The Machine History Log, and Minor Stops or Unrecorded Losses Categorizing

To carry out a proper analysis of equipment performance, we need to review how the machine has behaved over a reasonable period. Three months is an acceptable analysis time. As we proceed through the stages of TPM, we will collect our failure information data from three main sources.

1. *F-tags*
   These are faults that have been discovered by the AM and the Zero Fails teams during their initial clean and the subsequent routine inspect and clean cycles. It is also possible that a few tags will come from issues discovered while analyzing the minor stops. These faults are categorized into groups of faults based on root-cause reasons for the failures. The topic of F-tags is covered in more detail in Chap. 2.
   The record sheets used for F-tags are
   ⇒ *Cleaning Map*
   A diagram used to visually highlight the areas of contamination on the tool.
   ⇒ *Defect Map*
   A diagram used to show the distribution pattern of all the F-tags on the tool. It is limited in accuracy but can be increased by the addition of "close-up" photographs or drawings of the areas where the detail is needed.
   ⇒ *F-tag Category Spreadsheet*
   A log sheet that is used to collate the five different types of JIPM (Japan Institute of Plant Maintenance) fault category.
   ⇒ *Defect Chart*
   This is a graph that shows the F-tags broken down into the JIPM categories and plotted as a histogram.
   ⇒ *F-tag Log Sheet*
   A spreadsheet used to identify and record the F-tags and link the tasks with their risk assessments and safe working procedures.
   ⇒ *Task Certification Sheet*
   A spreadsheet used to identify who is allowed to work on each task.
   ⇒ *Failure Analysis Sheet*
   This is used to document the repair and provide the repair data (times) for analysis.
2. *The Machine History Log*
   Every fault must be recorded to monitor equipment performance. TPM treats these issues as F-tags. They are given a number, recorded on the log sheets, and categorized to identify the failure reasons. For

the analysis, they are recorded over the same time interval as the minor stops data (3 months). The record sheets used for the machine history are

⇒ *Costs Spreadsheet*
Used to track the expenditure and estimated cost of repairs.
⇒ *Categorization Log Sheet*
A spreadsheet used to identify the faults and record the analysis data.
⇒ *Malfunction Map*
This is a visual summary of the location of all of the three types of fails.

The machine fault history data must cover the same period as the minor stops data. If a different time is used, the machine performance will be skewed. Initially carry out the data analysis as a complete team. This will set the standards and teach all the other team members how to use the data. To avoid the need for all the work to be carried out by the technicians, confirm that all of the team members are comfortable with the task. When they are ready and with a view to minimizing the disruption to production, the task can be shared and rotated between pairs of the members (one technical and one production). Take care when numbering the F-tags for the groups; keep to the same number sequence used to cover all three sets of data. Try to ensure there is no duplication as you proceed, but if you do make a mistake it is easy to change a duplicate when it is discovered.

There is certain repair data we need to record to ensure we can get the best results from the analysis. For each fault, we should collect

- The F-tag number.
- The date and time of the failure.
- The initial failure symptoms.
- The time the repair began.
  For response times and a review of the manning levels.
- The time the repair was completed.
- The time the tool returned to production.
  This will give information on the actual downtime and the repair time and requalification time.

Some fault-tracking systems provide performance data automatically. For all faults, we should be able to retrieve and graph the following, although the reader will select the parameters he prefers.

- The number of failures.
- The tool uptime.
  There are several performance graphs that would be useful: uptime, total downtime, scheduled downtime, unscheduled downtime, availability, and utilization.

📖 The Mean Time Between Failures (MTBF) and the Mean Time To Repair (MTTR).

The MTBF is a measure of the reliability of the equipment and the standard of any maintenance and repair work done. The MTTR is a measure of the support systems: the skill of the technical staff, the complexity of the failures, and the availability of spare parts.

How the data is accessed will be company-specific. It can be manually recorded in a logbook and transferred to a spreadsheet or logged directly to a PC. Although all faults should be recorded, it is not unusual to discover that they are not, but that is often due to someone "saving time" by not logging them. This is not a good practice and must be discouraged. We can only analyze the data we have available.

☞ It is essential that a clear set of instructions explaining how to access the data is created.

☞ The instructions should be written to enable *any* team member to collect the data. This way the task can be shared and rotated between all the members of the team.

Actions
3-2 ☺ Write the instructions for accessing the machine fault history.
3-3 ☺ Train all of the team members on how to access it.

The data fault will be displayed on the activity board. Details can be found in Chap. 12.

3. *Minor Stops or Unrecorded Losses*

If I asked you to tell me about the equipment in your factory, I would bet that for most machines you have responsibility for, you could list a number of faults that happen daily. On the plus side, you will quickly add that, fortunately, they are quick and easy to resolve. They would have to be because they happen so often that no one even bothers to try and fix them any more. These are the faults companies have chosen to live with: the minor stops. TPM recognizes these faults as a major source of cumulative downtime and a significant drain on the engineers' time. They are singled out and targeted for elimination. Categorize them if possible, but it might be difficult because of the types of problems they are.

The record sheets used for the minor stops are

⇒ *Area Photographs or Drawings*

Used for placing yellow and blue dots to identify the locations and the duration of the fails.

⇒ *Minor Stops Log Sheets*
Used in parallel with the photographs to record the dot number and to give a brief summary of the initial date discovered, the fault symptoms, how often it happens, and when it was completely resolved. It also enables monitoring and counting of the number of fails.

The team must analyze each of the three F-tag sources individually to see if there is a pattern to the failures. Their distribution might highlight unexpected shortcomings in the equipment or possibly a lack of specific knowledge within the maintenance group. All three sets of data should also be reviewed as a group, to get an understanding of how the faults might have developed and get a better idea of the spread of issues.

The impact of the minor stops is often seriously underestimated. In many companies they are not considered at all. This is a mistake. They are virtually never recorded because they only take a few minutes to fix. In fact, many technicians claim that it can take more time to formally log the failure than it could take to fix the tool. Minor stops tend to be made up of sticking parts, minor drifts in calibration, adjustments to compensate for variation in product, poor product positioning linked to a lack of precision or sensors in the equipment that can cause an automatic system to stop, a drive belt falling off, worn movements, a poor vacuum seal, a hole in a pneumatic air line, a wobbly component, etc. We have all seen them. Yet, even though they are not recorded, they happen often enough to really irritate the operators and the technicians. Because no one actually knows how many times they fail—they don't know how much they cost production.

Why not just ignore the minor stops? No one would ignore minor stops if they appreciated how much impact they have on production. I was told about a case on a continuous production line where the same fault was repaired regularly, around five times a week, 52 weeks a year. The five times amounted to about an hour's downtime, with a production loss of *more* than £5000 per hour. This is a minimum of a quarter of a million pounds each year. In addition to the financial loss, if we have a machine that is so unreliable that an operator has to stand and watch it all day waiting for production to stop, we are actually preventing that same operator from being productive elsewhere. He/she might, for example, be able to run product through more than one tool. I am not suggesting that they should be worked like beasts. In fact, from my perspective, I cannot think of many things more frustrating than just standing and waiting. The ideal situation for a production machine would be that the operator could load the tool with a batch of product and then confidently leave the tool, only coming back to remove the completed product or load a fresh batch.

I once spent a couple of weeks in Germany watching a tool and waiting for it to fail. Not because it was a minor stop and acceptable, but because the company expected to load the product, press "Start," and the operator would then be free to leave the area for up to 2 days, at the end of which she would return and set up a new batch of product and disappear off again. I would imagine the operator might have looked in on the progress if she were passing, but it was not mandatory. I was there because on two occasions a product positioning arm moved too slowly to be seen by the electronics and the system stopped. On each occasion, which was several weeks apart, they lost about 24 h of production time. The machine specification claimed that the system should load 1000 product units with no operator "assists." The company (quite rightly) took this to include not even having to press a "Restart" button.

What was the cause of the failure? The loading system was sticking because of lack of use. It would operate 13 times and then do nothing for about 6 h, at which time it would move another 13 times and so on. The machine had followed this pattern for around 6 months. The Japanese software, at that time, was a basic Go-No Go logic. They would never have considered the software waiting for even as little as 1 s and then checking the position sensor again. (I asked!) Besides, the Japanese believed the mechanical assembly should be reliable. I simply had to identify where the slow movement was, adjust the position of one of the sensors, and speed it up a tad. This was my first experience with a company that expected a system to work to this standard of reliability. To be honest, I have not worked with many others since then either.

What is the impact of minor stops on technicians and engineers? Let's assume we have a single unit of product out of position and stopping the entire line.

- ⊗ The first step is to find the engineer, usually by paging.
- ⊗ The engineer has to stop whatever he is doing, irrespective of the complexity of the procedure he is working on.
- ⊗ Answer the pager to find out where to go.
- ⊗ Walk to the unit.
- ⊗ Manually reposition or move the blockage.
- ⊗ Walk back to the job he/she was working on.
- ⊗ Start working again and not make any errors despite a break in continuity.

Then he has to repeat the exercise the next time the same fault happens. The problem is not the time it takes to clear the fault ... Well it is one of the problems. Did you notice that I said clear and not fix? However, I will return to that issue later. The real problem is the cumulative number of times the system fails and their impact on the rest of the production system.

Let's consider the financial cost of the minor stops. If a tool has the same 5-min failure, say, six times in one day, then we lose

- ☹ Half an hour per day
- ☹ 3.5 hours a week
- ☹ 15 hours a month
- ☹ 180 hours in a year *180 hours is*
    - ➤ 22.5 full 8-h shifts
    - ➤ 15 full 12-h shifts
    - ➤ 7.5 days of lost production
    - ➤ 4.5 working weeks for someone on a 40-h shift

Most tools have their own "routine" faults. Ones that are regarded as not "worth" fixing and tend to be accepted as normal. Sometimes this is because the faults have been considered to be too costly or of too much effort to fix. However, Zero Fails targets the complete elimination of all faults, so these minor fails have to be recorded and analyzed for causes. At the very least, any decisions to live with the faults should be based on real cost information and not just gut feelings. TPM has a nice, simple way of recording these minor losses.

TPM, like me, loves visual displays. They provide an immediate grasp of the size of a problem without having to crunch numbers. The areas of the tool that exhibit minor stops issues should be photographed or drawn in the same way as a cleaning map. The images should be displayed on a wall or panel as close to the problem area as is practical. These areas are most likely to be where operators work, not because the operators cause the faults, but because this is where the product is found.

Now comes the high-tech part. Sticky dots, with numbers written on them, are placed on the pictures as close as possible to the point where the failure occurred (see Fig. 3.1). The dots are color-coded to reflect the amount of time lost. Although it is possible to choose any dot colors and time intervals, it is best to select one set and make it a standard. In my case, 10 min was chosen as the time interval.

- Yellow dots (which appear as white in the figure): For times less than 10 min.
- Blue dots (which appear as gray in the figure): For times greater than 10 min.

Figure 3.1 A product handling system showing the initial failure area.

If it is not possible to place a dot at a fail position because it is not visible in the photograph, then take a new photo or make a new drawing and mount it next to the main photograph. To assist in understanding the cause of the fault when it is being analyzed later, a quick written record of any new fault should be made the first time it appears (see Table 3.1).

When Figs. 3.1 and 3.2 are compared it is clear that a pattern has built up over the 3 months. The dot clusters clearly show a need for active intervention. After analyzing the symptoms and making investigations at the tool, a reasonably good assessment of the problems in the area can be made.

Many of the issues are due to "inadequate skill level" and are resolvable by training and/or by the creation of specific documentation covering the areas with the problems. The company that manufactures this tool may provide some of the best documentation and many of the essential reference and setup data points, but, for some adjustments, it could be improved by the addition of a bit more detail. There are also a large number of issues caused by "unchecked deterioration," i.e., issues with parts for which there were no

Figure 3.2 A product handling system showing the 6-month failure areas.

scheduled PMs covering the failure areas. The solution here is much the same as inadequate skill level except a PM has to be designed from scratch, checked to confirm it works, have a maintenance or inspection frequency established, and then be put in place.

In order to solve many of the positioning problems, new mechanical assemblies would be required as the existing ones are at least 12 years old and have served their owners well. If the manufacturer

TABLE 3.1 Note of Initial Fault Symptoms Illustrated in Fig. 3.1

Fault	Yellow dots (less than 10 min to fix)	Blue dots (more than 10 min to fix)
Wafer not leaving cassette	III	
Flat Finder vacuum not releasing	I	
Flat Finder not at correct position	I	
Wafer not picked up by input arm	III	
Wafer not picked up on unload arm	II	
Output track belt fallen off	I	
Wafer unload pad out of position		II
Exchange misaligned		I

was to develop the same equipment today, the system would be replaced by a robot that would eliminate the need for many of the complicated movements. The approach the PM teams should take is to evaluate the $cost of each of the failures in turn, including the cost of lost production and manpower. Once that is known, the cost of repair or upgrading can be directly compared with the downtime.

To get a fair comparison we must cost the failures over a number of fails and a realistic time, say 6 to 12 months. It should even be possible for faults that return frequently, to plot cost against time and get a visual representation of the increasing losses. An Excel template could be put together and the relevant values entered for each failure. It goes without saying that if the cost of failure rapidly surpasses the cost of a solution, then the decision should be made to proceed with the solution.

It would be unwise to choose to repair only a few of the issues as this would act as a major demotivator for the teams. Before the introduction of TPM it is likely that they already have a perception that their suggestions will only be ignored. It is reasonable to expect some outlay to be spent to improve the situation. This decision should be shared by the teams who work with the problems and would be the ones who have to live with the decisions.

	Actions
3-4	☺ Get lots of yellow and blue dots.
3-5	☺ Identify the team for the tool where data are being collected.
3-6	☺ Teach the team the concept of minor stops.
3-7	☺ Let the team decide the areas to be analyzed.
3-8	☺ Take the photographs or make the drawings of the areas.
	☺ Identify the location(s) for displaying the pictures.
3-9	☺ Create the log sheets.
	☺ Note the start date and begin recording the data.

Finding Out the TPM Causes for the F-tags to Help Find the Cure

1. *Basic Condition Neglect*
   This occurs when a tool is not maintained to the proper standard (basic condition). The lack of maintenance is often not accidental; the users are aware that a task has been left incomplete or they have simply ignored it. Many of the early F-tags tend to be made up of missing screws and panels, broken covers, leaking water lines, and missing cable clamps.

These issues do not sound serious, but can be if the screws affect high-voltage areas or the water leak drips on electronic components or cables become damaged, and so on. . . .

2. *Operating Standards Not Followed*
   This is where there is a procedure or specification for carrying out a task but it has not been followed or has been followed incorrectly.
   ⇒ Perhaps the wrong type of o-ring has been fitted and it leaks when the system is fitted.
   ⇒ Perhaps the old o-ring was reused, even though it had obvious signs of damage, despite an instruction to use a new one every time.
   ⇒ Possibly the wrong screw has been used. It could have the wrong thread, be too long, and could stick out on the inside; perhaps it has a worn head and might be difficult to remove at the next PM and needs to be drilled out and retapped.
   ⇒ Possibly a bearing or a joint that should have been lubricated has been missed.
   ⇒ A lock nut might have been left slack allowing excessive movement.
   This has been found to be the cause of many sensor problems.
   ⇒ The problems can often be more complicated. For example, a slot has to be positioned "precisely 23 mm from the right-hand side panel," but 25 mm is the distance to the center of the assembly, so the technician just "eyeballs" the position. In this example, the error limits the tool operation to the extent that it has to be shut down for correction. It takes around 5 h to shut down the tool, correct the setting, and qualify the tool again.

3. *Unchecked Deterioration*
   This is usually confused with basic condition neglect because both involve tasks not being carried out. However, in this case there is no recommended scheduled PM to prevent the failures or any checks to detect problems. The issues tend to be modules that have never been considered as sources of failure. They are
   ⇒ Moving assemblies that become worn over the years perhaps exhibit backlash or wobble and they cannot be relied on to stop at critical positions.
   ⇒ Pumps that have ran so for so long they have worn internal parts that now rattle or vibrate.
   ⇒ Motors that have worn brushes or failing torque.
   ⇒ Drive belts that are frayed or have damaged teeth and are ready to fail.
   TPM recognizes that all of these types of issues exist and introduces a new PM task for routine checks as soon as the issue is discovered.

4. *Inadequate Skill Level*
   This is exactly what it says... The fault is often not entirely due to the technician or the operator, because they are only doing the job as they were trained to do it. Often, though, it can be due to lack of skill.
   For complex issues, training has to be more thorough and be based on correct procedures and good notes. Skill level should be confirmed for all required tasks (see Chap. 5).

5. *Design weakness*
   Again, self-explanatory.
   ⇒ Compromises are often made in design to save cost, just as they are in everything else in life. This can mean that the best materials or mechanical arrangements are not always used, which can often lead to problems developing in a moving assembly.
   I have experience of two parts that did the same job: one was designed by an American company and one by a Japanese partner.
   The American part cost around $20,000. It was fine for the money and enabled the tool to work to its specification. The way it was designed meant that movements followed curved paths as opposed to straight lines and the business end tended to vibrate a little, which limited the positioning of the part in use and affected the operation slightly, but it was not an issue.
   The Japanese used a similar part, but wanted increased machine performance and this part was critical in achieving that aim. It was redesigned to incorporate a movement like a microscope table, smooth and perfect in three axes. Beautiful stainless steel bellows assemblies were used to maintain the internal vacuum and still allow the three-axis movement. It had incredibly precise positioning and had virtually no vibration, which naturally would have affected the position. It was very impressive. I think the cost was around $200,000 (give or take a bit).
   ⇒ Other issues can be due to mechanisms that are too complex for the task (overdesigned). We rely on manufacturers to evaluate the reliability of their equipment and they will carry out performance checks, but how long should the parts be tested for when the failures might not appear for at least a couple of years?
   Root-cause analysis will identify any weakness in recurring failures and improvements should be considered. If parts are simply restored to their original design, they will fail again. Evaluating the consequences of the failure and their frequency against the cost of redesign or replacing the model with a better unit is essential here.

6. *Unknown*
   There were too many failures that did not fit into any of the previous five TPM categories. It was difficult to find specific causes for the

failures, but not for the reasons they were unknown: poor records. The main causes were

⇒ Poor fault logging.
⇒ The lack of root-cause faultfinding.
Root-cause faultfinding is not the norm in many workplaces. The primary aim of many sites is simply to return to production as quickly as possible. This can mean that, if a complete unit has been replaced, the real failure category cannot be known unless the old unit is stripped down and analyzed—and that rarely happens. The normal procedure would be just to service the unit or return it to the manufacturer. Failed units that have been returned should be analyzed by the manufacturer and a written failure reason supplied. Again, many companies fail to see the value in this—including many of the manufacturers.

Now, that TPM is being used, there should no longer be a need for this category.

The F-tags must be categorized as explained above to identify what the actual causes of the problems were. A spreadsheet format is shown in Fig. 3.3.

Using the spreadsheet format will allow the sorting of data which will make it easier to analyze the results:

➤ We need to be able to count the number of F-tags in each category.
Large numbers of tags of any one type could imply there is a company problem in that area. If we discovered we had an excessive number of *operating standards not followed* F-tags, it could suggest that the standards are not well written, they are too confusing, they are out of date, or there is no system for controlling compliance. It is sometimes necessary to audit compliance and confirm that the standards are being followed at all times and are not being ignored.

➤ We need to know when the tags are resolved.
This date allows the weekly Tag Count graphs to be compiled.

➤ It is important we review the time taken to resolve the issues.
We have the date the tag was initiated and a resolution date. There will be a priority system for selecting the tags to fix but we have to be certain that the motivation is not dwindling. If we recorded the date the team started working on each tag, we might be able to establish their technical skill level through the MTTR. A poor resolution rate could be flagging the need for extra training or support.

➤ The ratio of white to red tags.
Some tools will be F-tag magnets where as others might have relatively few. The ratio gives a means of comparing different tools.

TPM—Analyzing and Categorizing the Failure Data   75

Tool Name: Location:

F-Tag Number	Description of Fault	Possible Root Cause	Further Investigation Planned	Basic Condition Neglect	Operating Standards Not Followed	Unchecked Deterioration	Inadequate Skill Level	Design Weakness	Unknown	Countermeasures	F-tag Type (Red or White)	Date of Detection	Date Closed Out	Closure Approved By:
1	Clean oil pool below RP1	Leaking pump seal		Y						Exchange pump and arrange repair	R	10-Jan-04	17-Jan-04	
2	Clean dirt on source insulator	Using dirty gloves when fitting			Y					Change spec to check that: Parts are clean before assembly Clean gloves are used when assembling Final clean is made before task completion	W	10-Jan-04		
3	Replace screws on panel above source	Unfinished job		Y							W	10-Jan-04		
4	Repair electrical connection to Ion Gauge IG2	Cable support not suitable	Y					Y		Replace support for improved style	R	10-Jan-04		
5	Replace drive belt on Input Track	Incorrect belt size			Y					Check stores stock and compare with correct size	W	10-Jan-04	10-Jan-04	
6	Flag Faraday mis-aligned	Incorrect position set			Y					Check spec to confirm correct positioning stated; Talk to tech who did job and discuss	R	17-Jan-04		
7														
8														
9														
10														

Figure 3.3 Example of a possible F-tag category spreadsheet.

Over a period of time, particularly on a large tool, a plethora of maintenance shortcuts can be taken. (Yes, I saw Blazing Saddles too!) Fortunately most of them can be fixed easily and cheaply. Also, as soon as TPM is introduced on the equipment, the tool users become aware of these bad practices or at least become aware that they will now be visible to everyone. If they continue they will be identified as new F-tags in every analysis, so it will not take long before they disappear. At this point in our analysis, we are only considering the initial clean F-tags and so basic condition neglect is high. When we add the tool history data and the minor stops to the pot, the distribution of the categories changes. When the initial clean is complete, the clean and inspect is now, effectively, a new PM task and is embedded into the routine. Once the data has been recorded, it can be analyzed into a defect chart (see Fig. 3.4).

If the categorization log sheet is located on a spreadsheet application like Excel, it is easy to sort the data and make it easier to count or

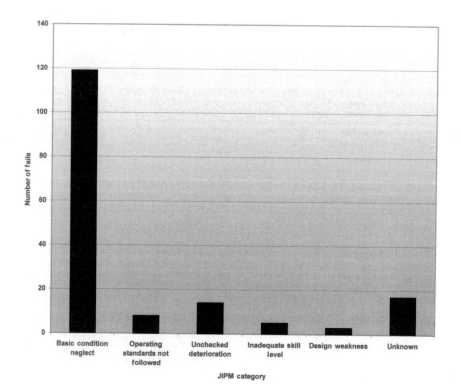

Figure 3.4 A standard defect chart—to visually display the JIPM categories.

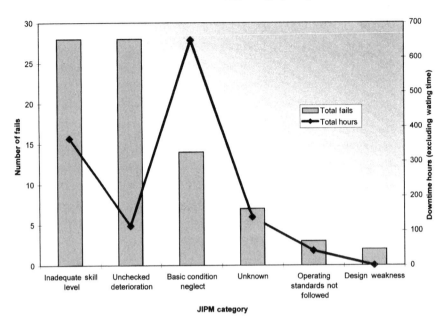

Figure 3.5 Modified defect chart showing the historical fails—sorted as number of fails. A second axis has been added to show the fault downtime. Table 3.2 is the data source.

it can do the counting for you. The way the data is charted helps to visualize the issues. Figure 3.5 has been modified by the inclusion of the Unknown category and the double axis. Notice the size of the Unknown category. It is a significant percentage of the *fails* data and shows how unreliable the logging systems can be if unsupervised.

When we only count the failures as illustrated by Fig. 3.5, inadequate skill level and unchecked deterioration are the two categories with the largest impact. Next in line is basic condition neglect. The Unknown category is 7 percent of the total fails but, since we don't know what they are, they cannot be fixed other than by proper faultfinding and reviewing the logging system. Before deciding which fails to work on first, it is worth considering the complexity of the fixes. There is no rule that says we must repair the categories one at a time. If there was, unchecked deterioration or inadequate skill level would be the first categories to target, followed by basic condition neglect. The best idea is to target the failures that will give the biggest improvement.

Everyone has their own favorite way of making charts. I like to get as much data on one graph as possible. When plotting TPM data, the two criteria I like to keep track of are the number of failures and the

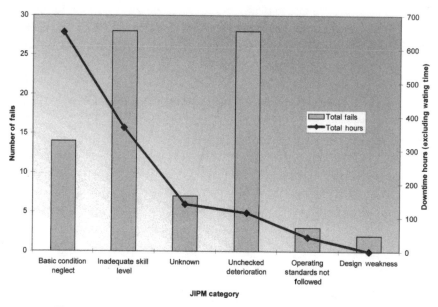

Figure 3.6 Historical fails data sorted as repair time. Table sorted as hours.

downtime hours. The downtime hours give a better idea of how the failures are impacting production.

Notice that the impact order changes in Fig. 3.6 when hours of downtime is used to sort the categories rather than number of fails. This time, although inadequate skill level had the largest impact in terms of number of fails, it has now become second when the time taken to carry out the repairs is the main consideration. This is followed by Unknown. Basic condition neglect would be the first one to target when sorted this way. To solve these issues we need to look at the PM steps and find out the areas that are letting us down. Only when we understand the failure mechanism the appropriate solutions can be initiated.

Solving the inadequate skill level issues will be achieved by rewriting procedures, and then using them to train *and* test the technicians. They should not involve any complex technical solutions. So, to get immediate benefits and put a big dent in the numbers, these are the first fails that I would target. At one site, the benefits of rewriting and training only the four major failure issues on one toolset increased the availability on the toolset by 10%. When creating the new procedures there is an opportunity to make them *best practice* standards by incorporating the best experiences of the whole maintenance group.

TABLE 3.2 Sample Data as Number of Fails and Hours of Repair Time

JIPM category	Pareto total fails	Pareto total hours
Unchecked deterioration	28	366
Unknown	28	114
Inadequate skill level	14	650
Basic conditon neglect	7	139
Operating standards not followed	3	42
Design weakness	2	0

The unchecked deterioration tags will be resolved by putting PMs into place, creating the procedures, and then training. The distribution of the failures and the modules that have been failing might give a clue as to other parts that have no maintenance procedures too. If the tags all belong to a particular area, perhaps a review using RCM methods would help evaluate the parts that are most likely to fail.

	Actions
3-10	☺ Create a categorization log sheet (Fig. 3.3).
3-11	☺ Review the minor stops and machine history and identify the categories. (Initially analyze as individual source groups and combine later.)
3-12	☺ Populate the sheets with the data.
3-13	☺ Sort the data into categories and count the numbers of fails in each.
3-14	☺ Chart the data.
3-15	☺ Update the activity board with the new charts.

TABLE 3.3 Data as Percentage of Fails and Percentage Hours of Repair Time

JIPM category	% Fails	% Hours
Unchecked deterioration	34	28
Unknown	34	9
Inadequate skill level	17	50
Basic condition neglect	9	11
Operating standards not followed	4	3
Design Weakness	2	0

TABLE 3.4 Data as a Pareto Chart—Sorted against Fails

JIPM category	% Fails	% Hours	% Fails Pareto	% Hours Pareto
Unchecked deterioration	34	28	34	28
Unknown	34	9	68	37
Inadequate skill level	17	50	85	86
Basic condition neglect	9	11	94	97
Operating standards not followed	4	3	98	100
Design weakness	2	0	100	100

Pareto Charts

Rather than list the actual data, a Pareto chart uses percentages and cumulative sums for each category. Using percentages is a way of standardizing the data.

$$\text{Percentage Fails} = \frac{\text{category fails}}{\text{total fails}} \times 100\%$$

Table 3.2 uses the same data values as used for the graph in Fig. 3.4, but the categories have been changed. In Table 3.3 the data has been converted from numbers to percentages. If this was graphed, the relative heights of the peaks for the "fails" would be the same, but the scales would be changed to 0 to 100 percent.

The data in the first two columns of Table 3.4 is the same as you would find in Table 3.3. To create a Pareto chart we need to manipulate the data to make it the same as the data in columns 3 and 4 (see also Table 3.5). Each cell in a Pareto chart is the cumulative sum of the percentages up to 100 percent. So, the values entered in each row are as follows:

1. Unchecked deterioration: 34% and 28%
   The same values as in row 1.
2. Unknown: 68% (34 + 34) and 37% (28 + 9)
   The values entered in Pareto row 2 is the sum of % rows 1 and 2.
3. Inadequate skill level: 85% (68+17) and 86% (37+50)
   The values entered in Pareto row 3 is the sum of % rows 2 and 3.
4. Basic condition neglect: 94% (85+9) and 97% (86+11)
   The values entered in Pareto row 4 is the sum of % rows 3 and 4.
5. Operating standards not followed: 98% (94+4) and 100% (97+3)
   The values entered in Pareto row 5 is the sum of % rows 4 and 5.
6. Design weakness: 100% (98+2) and 100% (100+0)
   The values entered in Pareto row 6 is the sum of % rows 5 and 6.

TPM—Analyzing and Categorizing the Failure Data 81

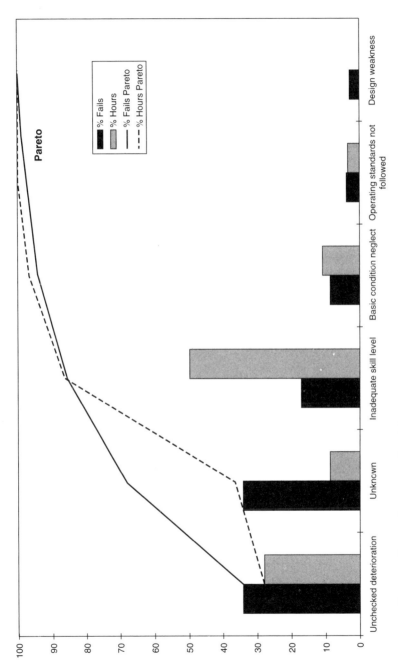

Figure 3.7 Data as a Pareto chart—sorted as fails.

82   Chapter Three

TABLE 3.5 A Simpler Method of Explaining the Pareto Data Calculation

Row number	% Data	Pareto value, %
1	A	A
2	B	A + B
3	C	A + B + C
4	D	A + B + C + D
5	E	A + B + C + D + E

In Fig. 3.7, the data has been sorted against percentage fails. Pareto charts normally only have one variable, but I kind of like keeping track of the other one too. I do not really use Pareto charts; my preference is the style of the other charts used previously. Having said that, many people do like the Pareto chart, which is why I have included it in this chapter.

This is an example of what a full size defect map would be like.

Notice that it shows only clusters of dots with no precise positioning.

These areas should be photographed to give more detail on where the tags are and the dots positioned with more accuracy.

Before and after maps can be very impressive.

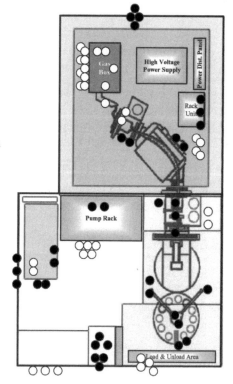

Figure 3.8 Defect map example.

The Defect Map

The defect map (Fig. 3.8) is a visual representation of the location of all the F-tags found on a tool. Notice that the base diagram is the same as many of the other diagrams, only the information on it has changed.

Figure 3.8 is a simplified representation of a diagram: a real map could have hundreds of tags. TPM defines five failure modes to cover the reasons for failures. The dot clusters are fairly accurate for this tool. The red tags (which appear as black in the figure) on the perimeter identify faulty door interlocks. They can be set up properly, but it might be a relatively cheap improvement to change them for a different style; perhaps one that has a lamp to say when it is *open* or *closed*. Finding a better interlock will not be too easy, so someone would have to make the effort and commit the time.

The defect map shown here should be read in conjunction with an area map or a layout diagram. This would enable discussion of the clusters as it would be possible to identify them. If we used an area map we could also sort the F-tag category sheet by area and look for common causes.

	Actions
3-16	☺ Create a defect map. (Fig. 3.8)
3-17	☺ Analyze the results as a team. Try to identify clusters that can be resolved by common solutions.

Chapter 4

TPM—Creating Standards and Preparation for Autonomous Maintenance

Once each of the three sources of data have been collected for a minimum of 3 months, all of the minor stops, the machine history, and the F-tags will have been reviewed and documented. It is at this point we collate the data. It is then divided into white or red tags, which are compiled into two lists, one for the *Autonomous Maintenance* (AM) teams to tackle and one for the *Preventive Maintenance* (PM) teams, with the PM teams getting the red tags. In Fig. 4.1 the AM teams and the AM component (operators) of the PM teams will follow the left side of the chart and the Zero Fails (PM) teams will follow the right side.

The AM teams should have undergone all their training and preparation and, after the appropriate period of supervision, they will be almost ready to start to take full autonomy (responsibility) for their own tasks. This does not mean that they are about to be abandoned. Far from it, they will still require ongoing support from the technicians and managers. For example, if it is necessary that *lock out tag out* (LOTO) measures be applied to make an area completely safe for the AM team to work, it will be the technicians or engineers who carry out the LOTO tasks. The AM teams will still have the responsibility for organizing the equipment time with production and coordinating with the technicians to ensure the work gets carried out. These appropriate steps will be included in the AM task's safe working procedures and risk assessments.

Although really an imaginary split, the PM tasks are subdivided into two further groups: a master failures list (to prioritize all of the repairs) and a PM tasks list. The PM task list is intended for the issues that can either be

86   Chapter Four

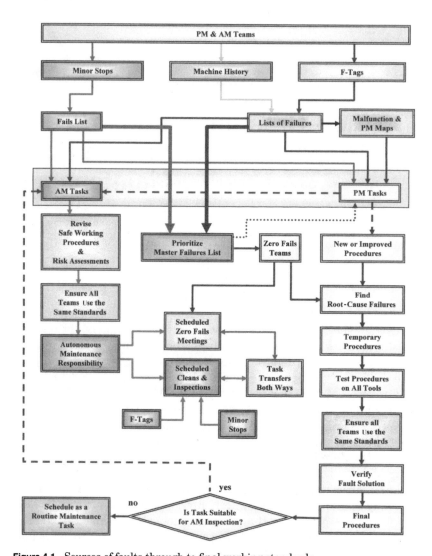

Figure 4.1 Sources of faults through to final working standards.

- Resolved quickly.
- Require immediate intervention to avoid recurrence.
- Likely to affect the equipment performance or product quality.
- Having either no procedures or need new procedures.

Creating procedures provides a fast return for effort. All that is required are

- A first draft.
- Initial training for the users.
  During the training, each class should be encouraged to recommend improvements to the method. This is the time to tap into the experience of the operators and technicians.
- Making any approved corrections to the procedure.
  Notice I said approved corrections. Sometimes there is a tendency to make a change based on the input of any one person, irrespective of whether or not it might result in a negative change. Even though the idea has some merit, check it out and make sure it is a real improvement. It could be the basis for a change. Involve the other technicians; see if the idea can be adapted or improved further. Remember, the goal is a best practice method for everyone to follow and the best way to ensure compliance is to involve "everyone" in its creation.

Chapter 5 has a section illustrating a highly effective method for creating safe working procedures. It is based on two columns, the left side having the instruction steps and the right side having diagrams and advice. Use the column on the right side to include some of the experience gathered during training: "Ensure the torque spanner is used. If the drive assembly is overtightened at the Schwarzenegger joint, it will slow down the movement and will probably damage the bearing."

- Retrain the technicians on the changes.
- Formalize the procedure.
  It is now the official standard. Take away any old procedures (particularly handwritten ones) from the technicians. There is only one standard.

The master fails list will now contain the more complex problems. The order of the tasks will be prioritized by the Zero Fails teams. It will be their responsibility to find the solutions to any problems; they will also carry them out and then create and test any documentation. The Zero Fails teams must also stay in regular contact with the AM teams, helping them with any issues and deciding on task transfers. Often, the technicians and operators will work in pairs, one AM and one PM; this arrangement has a few advantages including improving AM training, breaking down barriers, and removing the need for two technicians to work on a single issue—to name but a few.

Immediately below the AM tasks list, there is a short safety and training revision step for the AM teams. These pre-AM split checklists are intended to confirm the status of safety and training. It should be supervised by both the technical support group and the safety department. The contents to be reviewed can be found in List 4.1. *Total Productive Maintenance* (TPM) targets zero accidents and the AM teams, who might have nontechnical backgrounds, rely on management to ensure their safety. This check is fully documented and proves that all necessary training and safety precautions have been complied with. To this end, we must ensure that all the log sheets are up-to-date and correct and that all the competencies are met.

List 4.1: Pre-AM Safety Checks.

- ✓ The team member's tool-specific training log.

    This covers how the tool works, what it does, and what all the bits are called.

- ✓ The team member's training sheet.

    This covers the general training about TPM, Zero Fails, and AM. It also includes the general safety data, meeting skills, and so on.

- ✓ Confirm that everyone in the team understands hazard and area maps.

- ✓ Confirm that everyone in the team understands and can use risk assessments.

- ✓ Confirm that everyone in the team understands and can use safe working procedures.

- ✓ Does everyone in the team know what to do if they don't understand something?

- ✓ The Area Responsibility and Certification Table.

    This is the sheet that says who is competent to work in each area of the tool and the makeup of any teams.

- ✓ The cleaning maps, the F-tag log sheets, and the task certification sheets.

    The F-tag log sheet identifies which risk assessments and safe working procedures must be followed when carrying out specific tasks.

    The task certification log sheet identifies who can do the task and when they became authorized.

- ✓ The skill log/task transfer sheet

    This assesses the degree of competence.

    Remember that the Safety Committee must approve transferred tasks.

- ✓ Check that the standards of the competency tests are valid for the tasks.

Task Transfer: Red to White F-Tags or PM to AM Tasks

Task transfer refers to the redefinition of a tag. It can be in either direction but the most frequent is from red to white. White tags are not complex, certainly in the beginning. Normally, they require little or no technical expertise. The minimum criteria for red tags are that the task is deemed to be too complex for the AM team to carry out. This is not a static definition. As the AM team gains experience their skill level will improve, eventually reaching a level that could allow them to carry out some of the less complex red F-tags. The complexity of the tasks being passed over is limited only by the skill and capability of the AM team.

Transferring a task is not simply passing it over to the operators. It can be permitted only if the standard safety criteria can be met. Safety is absolutely critical and, consequently, too much care cannot be taken to ensure this.

➢ Each task that an operator will carry out must have a theory and a practical test before the operator can be certified to carry out the task.
➢ Tests and results must be retained for future reference.
➢ For more complex tasks, the "fuel gauge" system of the skill log/task transfer sheet (Chap. 5) must be observed. The five levels are
  1. No knowledge
  2. Knows theory
  3. Can do with supervision
     (there will be a minimum number of times the task must be carried out before the next step is allowable)
  4. Can do unsupervised
  5. Can teach
➢ The safe working procedures and risk assessments for the tasks must include any safety preparation that is required prior to the AM teams starting work. They will also specify who is responsible for carrying out the actions.

Figure 4.2 shows the embedding and responsibility record. It includes the task information, the proposed "maintenance" frequency, and the person responsible. The inclusion of an "Area" column enables sorting the fails and making modular risk assessments easier, as the same area and AM or PM tasks can be grouped. Sorting also highlights recurring faults and the team members certified to work in these areas.

This is a good time to review any "hard-to-access areas" and look for improvements that would give time-saving benefits. Hard-to-access areas are, as the name suggests, difficult to get to. If we need to take a reading from a gauge that is obscured from sight, depending on its

Chapter Four

Tool Name: Location:

Fail Number	Tool Functional Area	Description of Fault	Cleaning Task Responsibility	Inspection Task Responsibility (Movements Positioning Max/Min Limits)	Operating Standard	Embedding Conditions	Current Frequency	Task Time in Minutes	Possible Frequency Change	Responsible for Embedding Task	Date Completed
1	1	Clean oil pool below RP1	AM		No oil contamination	Add to Area 1 Clean & Inspect	2-weekly	5		Operator #1	
2	1	Clean dirt on source insulator	AM		No dirt on insulator	Add to Area 1 Clean & Inspect Routine	2-weekly	5		Operator #2	
3	1	Replace screws on panel above source		AM	All screws in position	Add to Area 1 Clean & Inspect Routine	2-weekly	5	Monthly	Operator #2	
4	1	Repair electrical connection to Ion Gauge IG2		PM	Cable support in good condition. Plug and connections made	Add new task to PM#7	2-weekly	5	Monthly	Technician 2	
5	5	Replace drive belt on Input Track		AM	Belt in place and not slipping	Add to Area 5 Clean & Inspect Routine Add Part Number of belt to Safe Working Procedure	2-weekly	3	Monthly	Operator 3	
6	2	Flag Faraday misaligned	PM	PM	Should rotate to a position where the entry slot is perpendicular to the beam direction. A set square must be used for alignment and rotation confirmed	Add step to PM#3, Flag Faraday procedure	Monthly	10		Technician 1	
7	8	Input air pressure low		AM	Pressure gauge within the max. and min. levels	Confirm that max. and min. points have been marked on the pressure gauge Add to Area 8 Clean & Inspect Routine	Weekly	2	2-weekly	Operator 2 support from Technician 1	
8	2	Flag water cooling low		AM	Flow rate within the max. and min. levels	Confirm that max. and min. points have been marked on the pressure gauge Add to Area 2 Clean & Inspect Routine	Weekly	2	2-weekly	Operator 1 support from Technician 1	
9											
10											

Figure 4.2 F-tag embedding and responsibility spreadsheet.

Figure 4.3 Improving hard-to-access areas by modifying a panel.

location, there are a few simple modifications we can make. Provided all safety implications are satisfied, we can cut inspection holes or access doors into panels (Fig. 4.3). If it would be beneficial to see more of the area, we could replace complete metal panels with Perspex ones. It is also possible to reposition the gauges, indicator lamps, switches, etc. To check or change the oil on a hard-to-access pump, we could consider moving or rotating the pump, even though the pipework would need modification. Moving and rotating tools can solve a whole range of issues.

Do not limit thinking, encourage it. Consider a task like an oil change in a pump; we know that it can be modified to use quick connect fittings and be filled using special automated trolleys complete with fresh oil, waste oil reservoirs, and pumps. Would it be possible to extend the pump's fill and drain "ports" to improve access and also use an oil change system? Think of cost-effective improvements that can be tried out to make access easier.

In common with many machines, to take a reading from this gauge, the panel would have to be removed. In this instance, the AM team had the idea of cutting a hole in the panel to make the dial visible without the necessity of removing it. The Perspex cover fully restores the functionality of the panel and speeds up the check. Note also the maximum and minimum levels: they do away with the need to take an absolute reading except in the cases where trending levels is required.

Each team will have its own areas of responsibility, the boundaries being set by the TPM management team. One of the purposes of TPM is empowerment of employees: moving decision making down to the lowest, most practical level. The power need not be absolute; all that is necessary is a simple mechanism for approving decisions in the shortest possible time. Upskilling the operators is intended to enable shifts in responsibility. List 4.2 makes a suggestion as to what responsibilities might be applicable. It is not comprehensive.

List 4.2: Suggested AM Team Responsibilities

1. Scheduling routine meetings and clean and inspect routines with the technical support and production personnel.
2. Identifying new faults, fitting, and logging the F-tags.
3. Initial classification of the F-tags.
   Care has to be taken here. Lack of knowledge could lead to mistakes: red classified as white. The classification must be a team task in which everyone participates, including the manager. Each task will be considered and allocated a classification.
4. Informing the technicians/engineers of any new red F-tags.
   The AM team has a dedicated engineer for support, but if the task requires more engineering support, a system for requesting help can be developed.
5. Maintaining the minor stops system.
   This includes adding and removing the dots when either a new problem has been found or an existing problem has been resolved.
6. Analyzing the data.
7. Creating the graphs, diagrams, and updating the activity board.
8. Ensuring the technical resolution of red F-tags and chronic and recurring faults via the technicians and engineers.
   The operators will work through the maintenance department and the system to ensure that tags are not dismissed. It would be nice to think that everyone will willingly take on all of the tasks found, but it will not happen. The mechanism must be such that the operators can forward their prioritized issues to engineering to be resolved. In the event of a road block, the team manager has to become involved. It might be necessary to move a level up the management tree to find a solution.
9. Ensuring that the technicians/engineers are in a position to allocate time to resolving tasks.
   This is channeled to management via team meetings.

The technicians are not free of the F-tagging, but their attendance will be reduced. It has to be, they have their own responsibilities and also need to continue to provide the technical support and guidance for the AM teams as specified in points 1, 6, 7, 8, and 9 in List 4.2. This will include group meetings, fixing red F-tags so that they are permanently resolved, making modifications to tools, training, assisting with data analysis, and developing new ways to overcome hard-to-access areas.

There will be red F-tags that repeat or will likely repeat if there is no preventive intervention. This makes it necessary to develop routine PMs for these tasks. Once identified, the tasks will be logged on the embedding spreadsheet (Fig. 4.2) to ensure the documentation is closed out and not left incomplete.

It might be worth mentioning here that although the teams are working on the basic condition of a tool, I would not recommend restricting them to only equipment repairs that will improve *Overall Equipment Efficiency* (OEE). There will be issues that are not equipment-related. Consider the benefits of all issues uncovered by the teams. While analyzing the machine performance or during their brainstorming sessions, they might come up with obvious, loss-making issues that indirectly affect their OEE.

The "7 Wastes" recognize waiting as a source of losses. Consider how much waiting can be caused by inefficient production schedulers: the people who set the daily routines? If I am running product A and I have been told to switch to product B, have the schedulers thought of the consequences of their change and how many departments it affects? They should confirm that the line has completed the planned product A runs and have used all of the materials that cannot be reused. This is simply to avoid unnecessary scrap, which might have a disposal cost in addition to its own material value. It would also be reasonable to assume the line should not have to wait for the raw materials to arrive. One minor administration task they might want to consider is that, before ordering a change, they might ensure that the raw materials are on-site and ready to run. It might not be so obvious to the planners though, who often seem to forget they too are a production activity for the whole company. I know: who would order a change, have all the changeover work done, and then have a line standing by, doing nothing except waiting for product?

One other source of possible OEE and equipment losses is purchasing. Every department in a company has their own spending—or, to be more precise, not spending—targets. They have a responsibility to cut costs, but they too must have standards to operate to. If they buy the cheapest materials, it is reasonable to expect that less than 100 percent of it will be usable. In some cases, their lack of knowledge of the equipment, coupled with less than fully experienced engineers, leads them to purchase second-sourced parts that have been manufactured using lower quality materials. The quality reduction does not have to be huge, just enough to mess up the functionality of the product. This purchase should always be a joint decision with engineering, but where the driving goal is cost reduction, there can be a bit of snow blindness. Different parts that are superficially similar (an o-ring is an o-ring right?), thinner labels, and so on must be checked for all tolerances, strength of mechanisms, and quality of raw materials. Purchasing should insist on supplier quality standards and checks and, of course, they should get their own Quality Control people to police the incoming goods. Never accept untested goods with the intention of returning them if they are discovered to be out of specification. Why? Because the production line will be ready to go and will have to stop until they get a usable batch of goods. So, not only has company money been wasted buying inferior parts, but they act like Semtex to high OEE figures, by stopping production and requiring baby-sitting by engineers to overcome the issues they cause with the equipment. If it is at all possible, get the supplier to agree to compensate for loss of production if the loss is due to their product being out of specification.

It is important that the equipment tolerance is always considered when buying parts. When purchasing an item with a looser specification be certain that even though the tool might be able to accommodate the difference, it can do it without the need for constant adjustment. If the variation in size is unit-to-unit and not batch-to-batch, it might require constant resetting to keep the tool running. Batch errors could require less adjustment, but that depends on the size of the batch; but if the batch changes during a run, there might be some disruption for a time while the initial setup incompatibility surfaces, causing jams and misalignments before it has been recognized that the dimensions have changed.

All departments in a factory should be trained in continuous improvement and the plans the company is making to improve efficiency. It is not a good idea to have some departments in the factory targeting improvement and others (unwittingly) planning the opposite. Saving £50 or even £500 on buying cheaper parts might look good to the purchasing department when they analyze their spending figures. But, does it look as good when the thousands of pounds that are being lost every hour in reduced production are brought into the equation? Of course it is not a fair exchange. Inadequate skill level and lack of standards are not limited to production groups. The walls between different departments must be broken down—or at least have windows fitted—if there is to be better understanding of the needs of each section. All decisions that could have an impact on production equipment and productivity must be reviewed by multiple departments.

Explanation of the Embedding and Responsibility Spreadsheet

This sheet is designed to track and ensure the resolution and follow-up actions of an F-tag. It also assigns responsibilities to different team members. What information is it looking for? The following points are not column headings, but could be.

- The fail or F-tag number.
- The area of the tool where the fault appeared.
  This helps with sorting and looking for tasks that can be grouped into cleaning or maintenance schedules.
- Team responsible: AM or PM.
- F-tag description.
  Because of limited space on a spreadsheet, this tends to be a brief summary of the problem.
- The standard to which the task is to be completed.
  As explained above, we need standards to ensure that tasks are carried out properly by everyone.
- Current frequency.
  How often do we carry out the clean and inspect routine? It is likely to be every 2 or 3 weeks initially. It will be a management decision based on the availability of the workforce.
- Possible new frequency.
  There is no point in carrying out a task when it is not needed. We need to optimize the frequency just as with condition-based maintenance.

  The column "Possible Frequency Change" is intended as a reminder to review the checks. If we assume that a task is scheduled to be carried out every second week and find on inspection that there is never any dirt or deterioration, the time interval should be revised. Consider extending the scheduled time to 3 or 4 weeks. Over the next few inspections, confirm the condition and if all is still acceptable, extend it further. Just remember to consider how the machine is being used (use conditions) over the assessment time. Be certain the part is not clean just because a new or different process is running or the tool is running at reduced capacity.

  It would take only one failure to revise the time downwards.
- Person responsible for embedding the task.
  If no one has overall responsibility for a task, it will never be completed. The tags should be allocated to team members in small batches, possibly sorted into a grouping decided by the team.

It is not a good idea to allocate all the tasks at once; it might prove to be disconcerting.

- Completion date.
  The date when the task will be completely formalized. At this point it no longer will be seen as an add-on, but an essential routine step.

Actions

4-1 ☺ Review the list of tasks that the operators are certified to carry out unsupervised.
Are they capable of carrying them out safely and properly?

4-2 ☺ Review all of the safety points listed at the beginning of this chapter.
This is linked to point 4-1, but is intended to confirm the safety and that the records are complete.

4-3 ☺ Set up the official AM schedule for the inspect and clean routines.
Remember that it will still require some setup time and some supervision by the technicians for those tasks that require it.

4-4 ☺ Define the task as a check, inspect, or clean in the embedding and responsibility sheet (Fig. 4.2).
A task can be a combination of all three.

4-5 ☺ When setting the AM check times, consider linking with a standard tool PM, or scheduled production break.
It will still be the responsibility of the operators to choose when it can be carried out. Linking is an option that might reduce the overall equipment downtime and maximize production.

4-6 ☺ Check with the safety department to confirm there is one master directory on the PC for storing soft copies of all procedures, training documents, spreadsheets, and risk assessments.
If there is not a master directory, there should be.
There can obviously be subdirectories for each toolset and so on.
Initially, there will be copies of the files scattered across vast numbers of PCs. This is the time to get the filing up-to-date.

PM Teams (Kobetsu Kaizen)

The goal of the PM or Zero Fails team is to improve the production efficiency of the tool. This is why it is a cross-functional team. The measure is Overall Equipment Efficiency (OEE).

It is the responsibility of the PM teams to resolve all of the red F-tags (equipment failures), throughput, and quality issues. This is likely to be a fairly significant change in their working pattern. I am uncertain as to how many groups/companies follow minimum maintenance procedures and rely on reactive maintenance, but it is enough to continually surprise me. Even then, it is often not really reactive maintenance they follow but quick fixes that get the tool back on line as soon as possible. This is seen, and rewarded, as highly skilled technical behavior—even though I believe it is not!

How can I make that claim? Let's consider a hypothetical example: an *employee* has fractured a bone in a leg. Naturally, we visit the doctor to fix it. But what would we say if the doctor only put a bandage on it to take the strain and prescribed a *huge* course of pain killers. Would we be surprised if it never got any better? I suspect the initial time off work would be very short, because the employee could limp along until it becomes impossible to walk any more or the pain is too great to bear. Then it is a case of, "Help, page the doctor." The doctor manipulates the leg a bit more, applies a few more layers of bandages, maybe even adds a splint, and prescribes stronger pain killers. Then the employee will go back to work and into the same cycle of events. The problem will certainly return again and again until real corrective action is taken.

How does the doctor's repair differ from the maintenance repair? Both *repairs* overcome the immediate problem and get things *running* (or at least limping) again. Both repairs will return again and again until the fixer provides a more lasting solution. It seems to everyone that both examples are exactly the same. So why is it that the solution is acceptable for one group (maintenance) and not the other (doctors)? Standards: or to be more precise, the lack of them.

It will be the PM team's responsibility to develop and operate the technical standards. They will be the ones who have to define the repair standards for the F-tags and make the improvements. They will discover a few new F-tags that might have been ignored previously. Most of these will be productivity-related: the speed of loaders, throughput problems, and *muda* (waste). These will be partial failures, low-performance rates, which would rarely have stopped production in the past. Referencing Fig. 4.1, we can see that the teams will follow a different path to the AM teams. Technicians will follow the steps on the right-hand column, starting with the list of fails that they have to resolve to return the tool to its basic condition. For each task they will follow virtually the same steps. The one difference, as we have already mentioned, is that some F-tags will be resolved simply by the creation of procedures and training.

Their first step for the more complex ones will be to establish root-cause solutions for the failures. This is frequently a change to normal working practice. For many people root-cause solutions create a barrier that has to be overcome, a leap of faith, before their advantages can be appreciated. The strange thing is that these same people would never accept a television or a car repair that continues to break down after it has been fixed. It has to be appreciated that each completely resolved fault will free up time, manpower, and money that can be used for other problems. Only when the solution has been found will the team create a first draft of a new PM procedure. The procedure must include all the technical standards and definitions of the part's basic condition necessary to ensure the task can be carried out correctly by anyone and must have enough detail to enable it to be completed to the same standard with no variation. In the event that the procedure can be applied to more than one tool, it must be confirmed that it is 100 percent compatible and does not require any modifications. Remember, not all tools of the same type or model are completely identical. The procedure will have to be amended to compensate for variations in equipment revisions, software operation, etc.

The solution must be verified. Apparently we engineers don't do this even though we know we should. Perhaps it is overconfidence in our abilities or perhaps we really do check and no one notices. In any event, if we have developed a new procedure and identified the correct PM frequency, then the fault should never return. We must follow Deming's "Plan, Do, Check, Act" cycle. When we are satisfied that the solution works, we can create the final procedure that will be used for training.

Once the procedure is complete, it can be reviewed by the team to see if it can be transferred to the AM teams. If not, a permanent PM slot will be chosen and embedded. The PM teams will need to establish a PM interval or, better still, use RCM (*Reliability Centered Maintenance*) methodology to find some kind of on-condition monitoring that will warn when the PM is due.

List 4.3: Suggested PM Team Responsibilities

➢ Root-cause solutions.

➢ Redesigns to overcome failure prone components.

   Redesigns have to be fully evaluated in terms of practicality, effectiveness, and cost. The cost must include all the production

variables that would be used to evaluate an RCM task. $Cost vs. $Consequences.
- Establishing the standards for the toolset.
    Common to all continuous improvement techniques. Get the manuals, talk to vendors. How else can we ensure that everyone carries out the same task to the same quality and accuracy other than by applying technical standards.
- Creating procedures and risk assessments.
- Establishing and optimizing a PM interval or method of self-diagnosis.
    Follow the techniques for establishing time-based maintenance intervals (the bell curve) or, if suitable, consider on-condition maintenance (the P-F curve).
    Do not forget the need for failure finding.
- Clean and inspect routines—although their participation will be less frequent.
    The PM team should regularly check on the progress of their AM team counterparts and support them with any issues.
- Technical support for the AM side of the teams and attendance at the Zero Fails meetings. In addition to the meetings that must be attended, the technicians' time should be concentrated on problem resolution but never to the extent that they appear unapproachable for assistance by the AM team members.
- Maintaining a PM section on the activity board.
- Monitoring the progress of the PM team.
    Charts comparing the weekly data on Mean Time To Repair (MTTR), Mean Time Between Failures (MTBF), and number of fails per tool and as a toolset. OEE will eventually be displayed on the board. These charts will show the improvements and pick up any new trends that are appearing on the tool. This was once the responsibility of the equipment engineer.

The master failure list, as shown in Fig. 4.4, is another way to view the different failures on a weekly basis so as to track the number of times they have repeated. Using a stackable bar chart also enables a weekly total to be seen. Initially there will be a high number of fails that should taper off with time, as the F-tags are resolved. New faults will appear as the tool continues to be used. The cycle will have to be followed until it stabilizes by which time TPM methodology will be embedded in the maintenance system—at which point the graph will have two axes with no data ... I wish.

Chapter Four

Tool Name: Centrifuge 1

	Total Failures	HGIG Connection	Flag Faraday Alignment	Fault 3	Fault 4	Fault 5	Fault 6	Fault 7	Fault 8	Fault 9	Fault 10	Fault 11	Fault 12
1	32	2		4	9	1	8	5	3	1			
2	39	1		3	7	3	6	7	6	4	2		
3	36		2	2	2	3	9	8		8	2		
4	18		1		6		5	4		2			
5	21			3			9	7		2			
6	13						9	4					
7	15							7	8				
8	23						8		7		7		
9	12						10				3		
10	6						3				3		
11	25						6				8	5	6
12	24						6				3	6	9
13	0												
14	0												
15	0												
16	0												
17	0												
18	0												
19	0												
20	0												
21	0												
22	0												
23	0												
24	0												
25	0												
26	0												
27	0												
28	0												
29	0												
30	0												

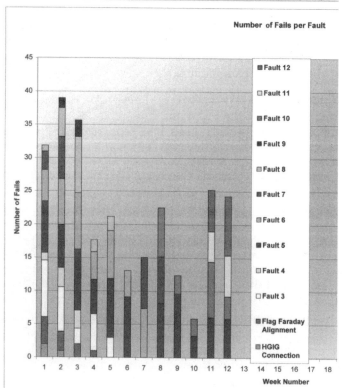

Figure 4.4 Example of a master fails list and weekly chart.

	Actions
4-7	☺ Create a master fails list. A line graph might be preferable to a stackable bar chart. I find the stackable bar charts difficult to compare week by week.
4-8	☺ Populate the list from the categorization log sheets. If the categorization log sheets are sorted by tool area and then by symptom, the number of fails can be counted and and the directly entered into the master fails list.
4-9	☺ Create the weekly fails graph.
4-10	☺ Prioritize the fails on the master fails list. The goal, initially, is to reduce the number of fails as fast as possible. There are two initial options: to select the faults with the highest number of fails to fix first or to consider the possibility that some of the less frequent faults might be easier and faster to resolve. If it is possible to fix more of the less frequent faults in the same time as it would take to fix a complex issue, the total will decrease faster. Unless there is a compelling reason, design problems should be resolved last as they take much longer to resolve.

Chapter

5

TPM: The Education & Training and Safety Pillars

There is one area that contributes significant benefits to maintenance and productivity but is often severely underappreciated by companies, if not virtually ignored. It is one of the subjects of this chapter: Education & Training. I have seen training on just four tasks improve a machine's uptime by 10 percent. I think it is remarkable that throughout my working life (so far) I still reckon that at least 70 percent of problems are caused by the TPM (*Total Productive Maintenance*) category "inadequate skill level"—It is important to realize that this is not the same as bad engineers. I have found, as have many of my colleagues and all of the continuous improvement practitioners that I have met, that even engineers with 30 or more years of experience cannot be relied on to do jobs correctly. Why should this be the case?

I am a bit of a believer in precourse testing. I like to get an idea of the standard of the people attending a course. It is the only way to find out for certain if I have actually taught them anything. Many of the courses I have run have been attended by managers, engineers, technicians, or operators, some for the first time, others as refresher courses. In other cases, the training has been specially modified by the customers to suit their specific needs and fill gaps in training, or arranged entirely to improve performance in a specific, single problem area of a tool. Sometimes, I have had to instruct people who have had the same *length of time* experience as I have had and often they feel they do not need to be trained by me, but rather by someone with more "time" experience. Some people still confuse years of experience with job content. It is possible to have 10 years experience of doing the same things year after year or 10 years of different, challenging experience.

In any event, very few people in the pretests score more than 50 percent. Most test scores are much lower. The tests are always fair: they rarely contain trick questions. If I think something should be known by the technician to enable them to do their job better, I will ask a question about it. On the other hand, if it is a purely academic question I will not ask it. For example, I would ask the gap size for setting spark plugs in a car, but I would not ask the thread size the plug screws into. A mechanic would need to set the gap regularly, but would rarely have to drill and tap the engine block. If he found that he needed to carry out this task, then there should be an in-house technical library that stores all the procedures he is likely to need and one of the procedures will include all the information required to carry out the task. In this case, all he would need are the basic mechanical skills. The main thing to remember about tests (and interviews, for that matter) is they should be designed to see what people know and not to show (others) how clever the guy is who set the questions.

Right then, so why is the pretest knowledge so low? As I get older I find I am not as fast at answering as I used to be, and I have often wondered how well I would do in my own tests. I believe that in most cases the low scores are caused simply by the lack of *proper* training. Besides, to use a vendor for training costs a lot of money. This is especially true when it is possible to learn from Bob; especially when he attended the course that Fred ran 5 years ago. Besides, Bob's course is free. And Fred, you must remember Fred: he came here as an apprentice 20 years ago and he worked with Big Jimmy for a few years. Big Jimmy, he was brilliant, he did the vendor training course in his last job, just before he came here... Would you be happy if Fred did your heart transplant? Big Jimmy's cousin was a doctor and he let him watch a few operations during the summer he spent in London. They used to chat about the procedure in the pub when the operation was over...

On-the-job training can definitely be effective but it has some serious limitations. It should have the following characteristics:

☺ It should not be 100 percent casual.

☺ The person giving the training should actually know what he is talking about (less than 50 percent is not a good starting point).

☺ The trainer should have access to the appropriate training documentation and standards for the topics being covered.

☺ If the trainee takes handwritten notes, they should be checked by the trainer and then the student should copy the correct information into a proper workbook.

What I am saying is the training should always be planned and considered. Even when people attend a vendor course, they are only being primed for when they get back to the factory. On the course, they *might* work on each practical module once, and most of the preparation steps will be missing. It is also possible that any complex and expensive parts that could be damaged by their inexperience might be left out. What did they learn then? On their return home, the technicians will be at least familiar with the parts and how they should be maintained and they will know where to find the instructions. They will not be 100 percent capable. It takes a person about 17 times, on average, to learn how to carry out a job perfectly every time. It could take up to a year of reasonably continuous exposure for an engineer to learn a tool and get used to its idiosyncrasies.

I attended a meeting somewhere... I can't remember where, but I do remember an HR training person from one major company believing that his guys were experts, because they had attended the vendor course. This type of misconception could be part of the reason for "inadequate skill levels." There is one aspect of training that I find to be a problem, even in vendor training; it is that they always show you the correct procedures for doing things but rarely how to get out of the holes that open up when things don't go as planned. Surely most of the early-day traps are well known, catch all's, and could be passed on to beginners? Another thing that gets me is the trainers who ask: "Are you going to be able to do the PMs when you get back to work?" The response is usually various forms of yes, absolutely, and that they will have no problems. Maybe I am a bit slow, but I never feel like that, I just feel even more aware of all the things I still don't know.

I worked with one company as a field engineer for more than 10 years and I felt I was still learning when I left. I had visited companies, with highly skilled engineers who serviced the same equipment. One part of the tool was quite complex to set up. It had two critical setup references: one, a single point in space, determined the setup of the rest of the tool; the other, the product handling and processing system, where everything should be set up referenced to the horizontal. Not one engineer knew where the reference point was even though it was mentioned in the manual. In addition, not one of the almost 30 engineers had a spirit level in his toolbox for setting the handling up. Not one! Their technique was the *bending of the part with the spanner method*, to visually set the alignment.

More than 30 years ago, TPM recognized the shortcomings in, and the advantages of, training to such an extent that they gave Education & Training its own pillar. It is one of the key elements in setting up a TPM system.

The TPM Education & Training Pillar

TPM teams will cross functional borders and team members will work with many people, all with different experience and backgrounds. This means the members will need to be trained in a range of subjects. The teams are required to maintain training records and be assessed for competence before being allowed to work unsupervised. The goal of the Safety pillar is zero accidents, and one of the most effective ways to avoid accidents is to ensure that people are well trained on the tasks they need to do.

The range of work experience of the Zero Fails team members will give each of them a different perspective of the equipment to be worked on. The team can consist of a mix of operators, technicians, equipment engineers, production engineers, and managers. More adventurous teams (possibly the best ones) might involve personnel from the administration departments like planning, purchasing, quality, or stores, even if only on a temporary basis. All team members must be well-enough trained to enable them to work as an efficient group. Initially, they will only need to understand each other at a basic level, but as time goes on, their skill level will increase dramatically.

Team training is split into three parts:

- *General*
  This will be the same for all teams.
- *Equipment training*
  This gives details about the tool they will be working on. This will have to include operation, maintenance, safety, and production.
- *Tool-specific training*
  This covers more in-depth details on individual areas that have problems.

In short, the team should be trained in everything they need to know to enable them to work safely in their designated area and participate within the team. Every team member should feel able to discuss any issues involving the tool and to understand them. They should be able to identify areas in their own knowledge or skill base that need improvement and seek out that knowledge. List 5.1 summarizes the kind of general training all teams must have.

I would just like to point out that when I talk about training, I do not mean "heads up" lectures. Too many people think that training has occurred when someone stands up and rattles off a few slides, gets an attendance sheet signed, and leaves. Trainees love that kind of lecture too. It is effort-free. They don't even need to listen, especially if the training room has an outside window. Proper training should encourage interaction and generate a bit of debate to reinforce understanding. Ask people questions, especially the ones that are yawning. If the training is important and the team is expected to know and remember it, I would always recommend *before and after* testing. Testing is the only way I know that can confirm if the class has learned anything. If it is unrealistic to expect them to have memorized all the information, it can be an open book test. An open book test, as the name suggests, is one where the class can look up the manuals for the answers they don't know.

List 5.1: Prerequisite Training for TPM Teams

The main points are listed below. Many of them will be revisited later.

1. An introduction to TPM.
   This includes planned maintenance, zero fails, and autonomous maintenance.
2. Leadership training for the team leader and team manager.
3. Team-building skills for the whole team.
4. How to carry out training: train the trainer.
5. How to run and participate in meetings.
   Use the standard meeting format of outcome agenda rolls and rules (OARRs).
6. Root-cause analysis for faults.
   Initially using "Why Why Analysis" (see Chap. 11).
7. The "office" software that is used in the organization (e.g., Microsoft, Lotus, etc.)....
   This enables report writing, simple spreadsheets, graphing, and some presentation skills for reporting progress.
8. General equipment safety and risk assessment procedures.
   This knowledge will be the foundation for team learning.

The team leader will be responsible for creating and maintaining the training records (Fig. 5.1). He will work in conjunction with the training department. He should have the authority to delegate any tasks within the team once the systems are established, but if he delegates training, he must always be aware of the training status of his team, as poor training could compromise safety.

More advanced training cannot be planned until it is decided where or on which tools each team will work. It is often simple to identify areas for 5S and SMED improvements as these areas will likely have obvious issues. For the more complex Zero Fails teams, it is tempting to choose an easy tool as a starting point. I can see the merits of the argument, but I think I would recommend a choice in the opposite direction and tackle

Team Member Names	Team Leader (Lead Hand)	Lead Hand (Different Section)	Tech #1	Tech #2	Tech #3	Source Builder	Source Builder	Team Manager	Operator #1	Operator #2
TPM Introduction										
PM Introduction										
Zero Fails										
AM Introduction										
Zero Fails										
Team Leadership										
Team Building										
Running Meetings/OARRS										
"Office" Software										
Root-Cause Analysis (Why/Why)										
Accessing Equipment Histories										
General Equipment Safety										
Risk Assessment Training										
Understanding Costs										

TPM Team Members Initial Training

Figure 5.1 Sample training record as would be used on the "activity board."

the worst equipment first and get some benefits. Most users already "know" which tools are problems but rarely know precisely why. As a rule, technicians and engineers might simply claim the tool is "crap"—a technical term frequently used to explain poor performance. Often the cause is the skill of the other shift. This has to be true, everyone tells me so. A real, unbiased review of the reliability of all the equipment will generate a priority list of issues to solve. Teams should be allocated a tool from the top of this list. So it goes without saying that the ideal team membership will contain mostly operators and technicians/engineers with experience of the tool they will be allocated, but there should also be room for people with other experience who can add to the total skills of the group. For example, there could be someone who is expert in faultfinding or mechanical engineering.

All training must be documented. It might be necessary to prove the quality of the training in the event of an accident. When carrying out safety training, I recommend all safety information should be verified by the safety department as being correct. I know this one will sound a bit obvious, but many managers want to be minimalists: the training should be as long as it needs to be for the level of understanding required. An example of a tool-specific training record is shown in Fig. 5.2. List 5.2 contains the sort of topics that need to be covered. Notice it also includes functions and features that are specific to the process. Total Productive Maintenance is not only about maintenance, although some see it as such: it should have been called *Total Productive Manufacturing*.

List 5.2: Equipment-Specific Training Required

1. What the tool does, i.e., the process and its production capability.
2. How it actually does it.

Team Members Tool Specific Training

Team Member Names	Team Leader (Lead Hand)	Tech #1	Tech #2	Tech #3	Team Manager	Operator #1	Operator #2
Tool Function and Process							
Tool Operation							
Tool Hazards & Safety Information							
One-Point Lesson							
One-Point Lesson							
Understanding Vacuum							

Figure 5.2 Example of tool-specific training records.

110   Chapter Five

3. How the product is tested for quality.
4. The operation of the tool, including any relevant software or hardware interlocks.
5. Safety that is specific to the tool.

Notice Fig. 5.2 contains a simple technique known as the "one-point lesson." It might be used for specific features that need to be understood and remembered. The one-point lesson, illustrated in Fig. 5.3, is similar to an advertising poster, except that it is used to convey essential information.

TPM teams and committees always have an overlap of members to avoid discontinuity of management. Figure 5.4 illustrates the structure of TPM teams that is favored by the Japanese. Using Fig. 5.5 as a reference, a Zero Fails team will have four members from level one and a supervising manager in the group. This manager will also be a part of a team of four managers at his own level, with one manager from the level above. This pattern is repeated all the way to the top of the tree. The beauty of using this structure is that it allows the manager in the team to disseminate his experiences to his peers, his managers, and his subordinates.

Figure 5.3 Example of a one-point lesson. This one is about safety, but it can be about anything the team needs to know.

Team Manager	Section Manager/Shift Supervisor/Equipment Engineering level. Should be able to remove roadblocks.
Team Leader	Lead Hand/Foreman/Senior Technician
Member	Technician
Member	Technician
Member	Operator
Member	Operator
Facilitator/Mentor	If required to assist on TPM/Team process problems

Figure 5.4 Zero Fails team composition—membership.

The team mix of job experience and perspective helps develop an understanding of each other's priorities and how they (or their group) are affected by costs, production problems, safety, or technical issues. Always remember that the team has its own limited autonomy, defined by the management team, and can call on help from any section within

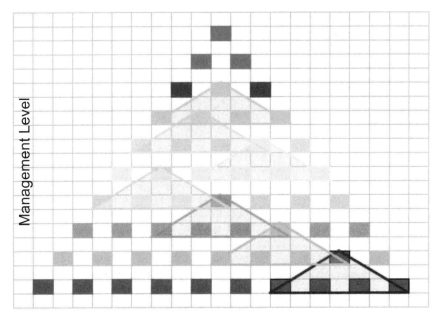

Figure 5.5 Zero Fails team composition—overlapping management.

the organization as needed, but it should always be considerate of the needs of the other departments. For example, the team could ask for specific information in advance to enable the temporary member to prepare or they should not expect an immediate response when a help request has been made.

I sometimes feel we should all be trained in diplomacy. I have had situations where I have gone to ask for some information either to validate that a problem exists or to quantify how often the problem occurs. I try to explain what I am looking for and ask if they are able to help. I try to explain why the problem matters and how we should all be trying to find a root-cause solution. I now believe that as soon as I recognize a defensive attitude and posture, I should get out and come back later, preferably with a familiar face from the company concerned. The clues are pretty much always the same:

- I don't have time to do that...
- The operators/engineers/purchasing department/suppliers are untrained...
- ...they always come here with problems but the vendor sets it up fine...
- It never used to be this way...
- The management in here is always changing...
- It doesn't matter what you do, no one appreciates it...
- I am looking for another job...

Sometimes they give you the appearance that they are settling down and everything is normal, and then when you leave someone calls up to complain about the person who was accusing their department of creating all the problems. Next thing is everyone thinks you are out to get them.

Another advantage of the supervisor/manager being in the team is that because he has a supervisory responsibility, he can assist in allocating support when needed. TPM uses manpower and takes up valuable production time, which some managers might prefer to see being used only for running product. The need to drive production often conflicts with the early stages of the introduction of TPM, at least until the benefits start being seen. This brings us back to the advantages of training, because it is the drive for production that creates the atmosphere that drives quick fixes as opposed to proper, root-cause solutions. Often, the drive for quick fixes is imagined, so to speak. No one has actually asked for a patched-up job: the perception has taken control. The solution might involve a quick fix the first time, to keep production running, but a plan to introduce a more permanent solution should be in the pipeline waiting to be applied at the first opportunity. When the manager is involved, and has been trained, he is less likely to promote the quick fix.

Naturally, there will be situations where production must be given priority but these must be monitored and controlled. If the management is serious about TPM, the teams must be given the time they need to do the job. Consideration should be given to including the progress of the teams into the manager's performance appraisal. This is not often an initial decision, but can arise later if a lack of progress is regularly reported. What has to be done is to develop a process that will be followed when a conflict arises. The process would have to include an increasing weighting based on the number of times a project has to be canceled.

	Management Actions
5-1	☺ Select the tools to be improved.
5-2	☺ Identify the training needs for the teams.
5-3	☺ Do you want a pilot team of managers?
5-4	☺ How many teams do you want to begin with?
5-5	☺ Who you want on the teams for each tool selected?
5-6	☺ Begin the general training program.
5-7	☺ Create the general training record sheets (Fig. 5.1).
5-8	☺ Create the tool-specific training record sheets (Fig. 5.2).

Equipment training

The introduction of TPM is a long-term plan. Five years is not an unreasonable time to get it going. Some companies are more ambitious and try to make it faster. This is not necessarily a bad idea, provided the standards are maintained. The old adage is still true when it comes to this: "You only get what you pay for." To enable a Zero Fails or an AM team to function, they need to know how the equipment actually works and what it does to the product. This is a lot of knowledge, so training must be staged. The initial explanation should be basic and limited to about 1 or 2 h. It is intended to be a general description that will give the whole team a common starting point. After all, it is easier to give directions to someone who knows the area. As in all training, the explanation should be targeted toward those with the least knowledge. Later, as the team develops and progresses into new areas, the training detail will be increased.

The team's trainers will need to have certain knowledge and skill:

- The equipment training should be carried out by a technician or an equipment engineer.

  Normally one of the qualifications for promotion to senior grades is the ability to train. Besides, "train the trainer" is recommended as a prerequisite to be included in the general TPM training. Vendor trainers and service engineers can be used where the in-house knowledge is insufficient, but beware of escalating costs if external training is to be used.

  Another point to remember is that the training will have to be highly diluted. It has to be based on the minimal technical experience of some of the team members. It will be extremely important to watch for problems in understanding and to be able to explain concepts in a simple way when the need arises. It will also be essential that the teacher/student barriers are broken down as quickly as possible to encourage interaction and the members to feel able to shout "Stop" when they don't understand or when they want to know more.

  Remember, all nontechnical personnel and those who have little or no experience of working with the equipment must have technical supervision until approved. Approval will depend on experience and competency. Responsibility to work unsupervised will be limited to approved tasks in safe areas.

- The technician or engineer should test his training courses and technique prior to the real class.

  The test class should include technicians and engineers with no knowledge of the tool. Their limited knowledge prevents them from making allowances for poor explanations and will enable them to point out where extra detail or better explanations are needed. Those who do understand the tool should look for mistakes and omissions. Everyone should help with advice on the presentation and the documentation.

  Follow any constructive advice that comes from the audience. Think about why the advice has been made and whether it is correct. If it is correct, go for it.

- Initial training should be in a classroom, using photographs, drawings (including tool layouts), and one-point lessons.

  I often think I am alone in my belief that training should not always be carried out at the tool and be 100 percent practical. I am a great believer that before the teams are taken to any part of the tool (except maybe the operating console), they should have a good idea of what to expect. They should know if there are any potential dangers near to or beside the tool (like an acid bath or a fork lift route); they should also know what might be dangerous on the tool and what not to touch.

I try to cover all the equipment basics and layout using diagrams and photos—a bit like a tourist guide. The classroom lets everybody see what is under discussion and when we do visit the machine, they get to hear it all again. This time it is familiar and is reinforcing previous learning. I also believe that the more relaxed atmosphere of a classroom makes for better learning. This is all enhanced by the fact that no protective clothing (except maybe earplugs) needs to be worn, there is no background noise to shout over, and there is the freedom to have coffee when the class feel they need a break.

If I have to explain a technical task, I will go through the instructions on-screen first using photographs and drawings as illustrations. If possible, I would have a setup in the classroom where the actual parts can be worked on and be clearly viewed by every person in the class. If that was not possible, I would go to the tool after the class training and carry out the practical exercise there. Where a part is known to be the cause of a number of problems, I have used a classroom environment to get every person to carry out the task, in turn—and when one person is carrying out the task, the rest all watch it—repeatedly. The class will hate it, but they will never get it wrong again. It is a bit like the 100% Proficiency technique of repeating and memorizing a task till you can't fail to get it right. I usually explain the task to the first candidate, who then explains the task to the second candidate, who explains it to the next, and so on.

The classroom training is only the introduction. It is not the whole training.

- Copies of the training notes should be issued to each team member.

  Distributing the notes before the training gives the class a chance to prepare and should increase the training experience. It also gives the members a chance to request subjects for inclusion in the training.

- The team should visit the tool after the classroom introduction and be taken through all of the key points one more time.

  As I mentioned above, some trainers prefer all the training to be carried out at the tool. They believe classroom training is inferior to "practical" tool training. This is not my experience. I find it is much safer for everyone to know what they are expecting to see. I have seen trainees open panels "just to have a look" or stick fingers where they shouldn't (for the want of a better phrase...). I have even seen them press emergency off switches (EMOs).

- If any points are particularly important and *must* be remembered and understood, then either a written or an oral test must be carried out.

    The complexity of the task will dictate which kind of test should be used. If unsure, discuss it with the teams. Document the results.

- Update the tool-specific training matrix.

    Highlight any failures to help remember that extra attention might be needed when considering their competency.

Initially, when writing this book, I considered using a range of different tools as examples to emphasize that TPM can be applied to any tool. Eventually I decided that by continually referring to the same few tools, the reader would develop a familiarity of them and will be able to follow the message without constantly having to absorb new examples. Readers will also appreciate that the training method works. When applying TPM in their own site, the reader should consider exchanging these drawings with tools their employees will be familiar with. The physical act of creating the diagrams or adding extra ones will cause the reader to improve his/her understanding of the process. Familiar diagrams will also help the teams with their own understanding.

A sequence for training equipment

Begin the training by explaining the main functions of the tool. Just as a domestic gas cooker is for heating food, it also has a number of functional areas. One area is for heating pots; grilling or toasting bread and doubles as a small oven; another is a main oven. It also has a control area that switches the gas on and off, a timer, a temperature control system, and a series of safety interlocks. This is the same for a process tool.

Break the tool down into functional areas. Use drawings and photographs to explain what the process actually does to the raw materials. This is not a test of the reader's or the team member's drawing skills: the vendor manuals or training course materials can usually be adapted. If you need additional information, the manufacturer or service engineer will probably be able to supply it. Manufacturers are well aware that the better you understand their equipment, the better the tool will perform, the happier you will be as a customer, and the more likely it will be that you will do more business with the company in the future.

Move on to the explanation of the layout of the equipment. The tool shown in Fig. 5.6 is as complex and dangerous a tool as you will be likely to meet. It has more ways to kill than the *Terminator* and it will take more than one diagram to give a trainee a good level of understanding.

TPM: The Education & Training and Safety Pillars 117

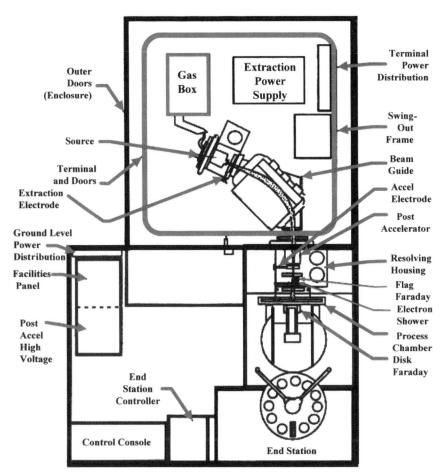

Figure 5.6 A layout drawing of an ion implanter (NV10-160 high current implanter) mentioning the main modules.

The tool is an ion implanter. When training on equipment, one fundamental rule to remember is to always use the correct names for parts, functional areas, and systems. This ensures that everyone is talking a common language. Using the same terminology is particularly useful when communicating to the vendor by telephone or e-mail. Include common names as well as formal ones. For example, in Fig. 5.6, the End Station is more commonly known as the AT4.

To help trainees remember and identify parts, I have been known to attach sticky labels to them. (To the parts not the people.) This is particularly useful for anonymous-looking parts like power supplies, pumps, and pneumatic valves.

118   Chapter Five

Eventually you will need to cover safety. Only the main hazards should be highlighted at this point, since we are only learning about the equipment at a basic level. Unless the "class" prove to be fast learners, are showing a high level of interest, or are asking questions, the more specific hazards should be covered only as required. Always be honest and straightforward when making explanations. I have found that some trainees and operators can be very suspicious that dangers are being concealed from them. One company wrote an excellent safety book on chemicals using very simple English and medical data. The operators thought the "bad stuff" was being missed out because it was too easy to read!

Explain the function of each of the modules on the layout diagram. I would mention any hazards they present at this point, although I would not expect them to be remembered. For example, the gas box in Fig. 5.6 can be at a voltage as high as 160,000 V and use several toxic gases. For basic training, I would mention the names of gases, that there is a special toxic gas extraction system for increased safety, and how the gases can be detected. Use easy reference tables similar to Fig. 5.7 to summarize the data. They can also be used for future

Gas	Detection	Characteristics	Hazards
Boron Triflouride BF_3	Has a strong pungent smell which is suffocating when inhaled.	Invisible in a vacuum but produces white clouds (HF) when it is exposed to air. It is heavier than air so it can accumulate in enclosed areas.	Toxic when inhaled. Extremely corrosive to mucous membranes (mouth and nose linings) and also skin and eyes. Skin contact with the vapor can cause serious burns.
Phosphine PH_3	Smells like decaying fish.	Pyrophoric—it is likely to combust spontaneously. It has no color. It is heavier than air so it can accumulate in enclosed areas.	Extremely toxic. Highly flammable. Will affect kidneys, heart, and brain.
Arsine AsH_3	It has a garlic-like smell, but is difficult to detect.	It has no color. It is heavier than air so it can accumulate in enclosed areas.	Extremely toxic. Highly flammable. A nerve and blood poison.

All the gases used in Fabs are diluted with 85% hydrogen to limit the toxicity. However, this means that they will rise (as the mixture is lighter than air), but can drop as the mixture disassociates.
Leave the area on detection of any dangerous gases.
Never work with these gases without breathing apparatus and training on how to use it.

Figure 5.7   Information on the process gases used.

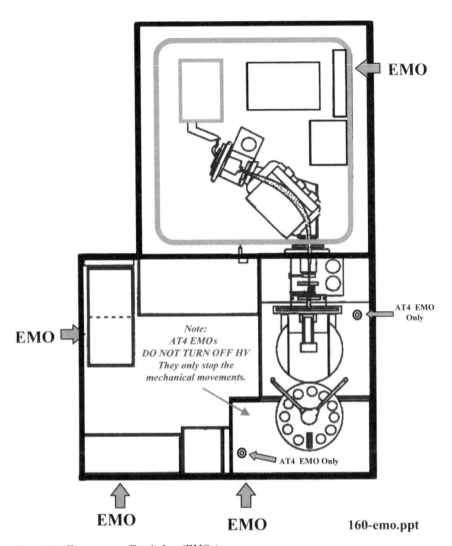

Figure 5.8 Emergency off switches (EMOs).

reference. If unsure whether to mention a detail at this point, always err on the safe side and include it. Essential safety features must be highlighted. Notice that the layout in Fig. 5.6 makes no mention of any EMOs, which are a particularly important detail that every team member must know, so a second diagram will be essential (see Fig. 5.8).

EMOs are located in the positions shown in Fig. 5.8.

There are two *pneumatic* EMOs located in the AT4. Unlike the other EMOs, *they do not isolate the power to the tool*, but only stop movement of any parts in the End Station (AT4).

They have a different physical appearance from the other EMOs.

Explain high-level safety functions only. The location of the following safety features can be identified by referencing Fig. 5.6. The team must know how they operate. The following is an idea of the level to which you might want to train: "The outer doors are lead-lined to reduce any x-rays to virtually zero. Both the outer and inner doors are interlocked. If any door is opened while the tool is running:

✓ The high voltage and the Terminal Power will switch off.
✓ "Drop bars" will ground the Terminal and Gas Box.
✓ The process gases are switched off when a door is opened...."

To ensure safety, which is fundamental to the success of the TPM process, it will be necessary to test the team for understanding before allowing any work on areas of the tool. For some of the simpler sections, oral questions can be used. More complex sections require a written test to be used and/or blank diagrams can be labeled in conjunction with a written test to establish knowledge of the location of modules and safety features. Multiple choice questions can also be used, but they have never been a favorite of mine. In all cases, keep records of the test questions, the correct answers, the date of the test, and the names of those tested. The summary information must be retained on the appropriate training record sheet.

One piece of advice I have found helped trainees (and myself) stay safe is, "Always assume everything is dangerous" and proceed from there. For example, never assume a valve is closed or that something has been switched off.

Competency: How does TPM assess the skill level of the team members?

When evaluating the competency of a person on the basis of their ability to perform tasks, the skill level of that person falls neatly into five different levels that suit TPM.

Each task can be assessed to one of the five levels. The levels can be graphically represented as shown in Fig. 5.9. The "fuel gauge" level is vertical and the shaded area corresponds to the knowledge level. From Fig. 5.9 we can see that the subject has

📖 No knowledge of Task #2.

📖 Understands the theory of Task #4.

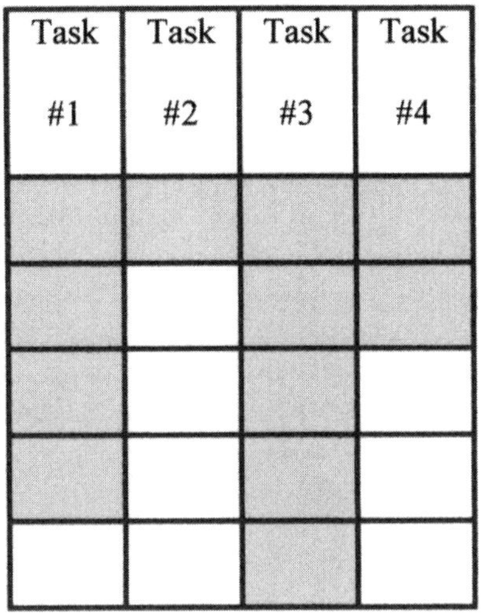

Figure 5.9 Competency levels for four tasks.

- Has no knowledge of the Task.
- Understands the theory of the Task.
- Can carryout the task if supervised.
- Can carryout the task unsupervised.
- Knows the task well enough to teach it.
- Can carry out Task #1 with no supervision.
- Knows Task #3 well enough to teach it.

The Autonomous Maintenance (AM) and Zero Fails teams need to record more detail than simply the level. They also require the date the theory test was passed, how many times the task has been carried out under supervision, and the date the practical test was passed.

Do you remember the tool we used for the layout diagram in Fig. 5.6, the ion implanter? It has a component called the *Source* that is regularly serviced because of deterioration. The Source is replaced every 2 days and takes about 3 h each time. Although it is complex, it does not really require a highly skilled technician to do the job. In fact, it could be a suitable task to transfer to operators with the required basic skills *if and when* they become more skilled and competent. Initially, it would be designated as far too complex for operators, but the operator would be taught the theory from the safe working procedure and permitted to watch the task being carried out. Each time the operator watches, his understanding of the task and its surroundings increases. Eventually the operator would be permitted to carry out the task under supervision, the number of times being recorded. When the supervision is no longer felt to be necessary, the operator would be tested on the task.

When agreed by the team that the operator's skill level is suitable, and is approved by the safety committee, the task can be "transferred" to that operator only. The technician will still have the responsibility for making the working area safe. The other team members have to be approved in the same way (as would trainee technicians). The task will not become a fully *autonomous* task until all of the AM team members responsible for that area have approved.

I am particularly pleased with the type of spreadsheet format shown in Fig. 5.10. It uses some nice Excel features and groups columns to give the "fuel gauge" or "graphic equalizer" effect and expands to give extra information. In this spreadsheet the gauge fills from the bottom up. This is the opposite to what is shown in Fig. 5.9.

Referencing Fig. 5.10 the information to be recorded would include

- The task description.
- The task number.
- The skill level.
  The skill level is the fuel gauge and must be shaded using the "fill color" of your choice.
- The date that the theory test was passed.
- The number of times the task has been carried out under supervision.
- The date the practical test was passed.

Notice that experienced technicians are automatically approved. For the simple examples shown it is understandable. Naturally, there will be tasks that not all technicians are able to carry out and so they will be expected to follow the procedure for approval. It is worth remembering that wee thing we discussed at the beginning of the chapter: a great many issues are caused by technicians with insufficient skill or inadequate training. Even though they might believe they are doing a task correctly, experience has shown that they might not be and can create a large amount of downtime. TPM will eliminate them as a source of downtime because the root cause for the failures will be correctly identified and a standard procedure will be developed that will be followed every time.

When testing, be sure you test only on the things they need to know. For example, we know the tool generates x-rays, but there is no gain in asking what the x-ray energy is. For the user, what is important to know is where the x-rays are generated and the potential danger of bypassing interlocks and running the machine with the doors open—because if they do, they will be exposed to x-rays. In addition, it

TPM: The Education & Training and Safety Pillars   123

Skill Level "Fuel Gauge" - Key:	
Can Teach	
Can Do Without Supervision	
Can Do With Supervision	
Theory	

Task Description	Task Number	Name				Team Leader				Tech 1				Tech 2				Operator 1			
		Skill Level	Date Theory Test Passed	Number of Times Supervised	Date Practical Test Passed	Skill Level	Date Theory Test Passed	Number of Times Supervised	Date Practical Test Passed	Skill Level	Date Theory Test Passed	Number of Times Supervised	Date Practical Test Passed	Skill Level	Date Theory Test Passed	Number of Times Supervised	Date Practical Test Passed				
Clean oil pool below RP1	R1				02-Feb-04				02-Feb-04				02-Feb-04								
Clean source insulator	W2				26-Jan-04				26-Jan-04				26-Jan-04								
Replace screws on panel above source	W3				02-Feb-04				02-Feb-04				02-Feb-04								
Repair electrical connection to Ion Gauge IG2	R4				06-Feb-04				06-Feb-04				06-Feb-04								
Replace drive belt on input track	W5				29-Jan-04				29-Jan-04				29-Jan-04								

Figure 5.10 Example of a skill log that can be used for transferring tasks to operators.

would not do any harm if the team was taught to use a radiation detector. Tests should ask questions like, "Which sensor is calibrated first?" or "What happens if the lock nut on the positioning arm is not tightened?" Keep the questions simple. Just because the tests should not include academic questions does not mean that the more theoretical information should not be included in the training. If that happened, I would imagine the training would be really depressing.

The most common equipment failures should also be covered. This will make the equipment more real and introduce the operators into the idea of discussing faults and the way they affect the tool. The most common faults are usually due to moving parts, so use a step-by-step drawing of the failure areas to illustrate what is happening. Highlight how the machine tracks where the product is. Explain what signal the control computer is looking for before it carries out the next step. Watch videos of the tool during operation. Some of these failures might eventually become suitable for AM teams.

Figure 5.11 is part of a drawing of a resist spin track process. I put it together in less than an hour using PowerPoint. Physically, the drawing is nowhere near as complex as it looks. There are only a few different shapes and lots of use of the "Copy, Paste, and Group" tools. PowerPoint is an excellent drawing tool, with a good range of shapes, lines and arrows, color fills, and simple text boxes. If you saw a real spin system, it will not look like the drawing, but the component parts will still be recognizable and the detail is easily good enough for the purpose. As part of the training, the real system would be demonstrated and the purpose of all of the sensors and drive mechanisms would be explained. A video of the handling system could also be used for training. It would be nice if it could show some of the problems actually happening, but that could take hours of taping until a failure occurs, and then some editing time. It would, however, be worth it.

Many of the problems experienced on the tool are due to sensor setup. Since every step is controlled by sensors, poor positioning and detection can cause the movement to stop if the computer cannot "see" the wafer. Often, these "minor stops" are not fixed properly, since it is very simple to clear the cause (or symptom) of the fault. If a wafer has jammed, it can be assisted by giving it a push. Even the sensor can be manually tricked to think the wafer passed the checkpoint.

Although the following might be of no particular interest to the reader, they are included to illustrate the type of faults the team would be told about. It is because the following problems happen often enough to become recurring F-tags that the team should be aware of: process materials getting under the wafer causing vacuum seals to fail, drive belts falling off or becoming too worn to move the wafer, drips from the spray nozzle affecting the surface by making it lumpy, the spray nozzle

TPM: The Education & Training and Safety Pillars 125

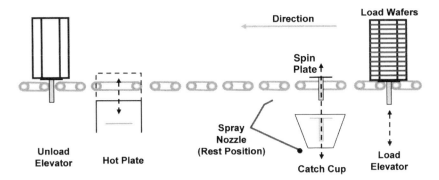

Step 1: Reset using a set of microswitches and fixed pneumatic pistons; the system is driven to the starting (Initialize) position. Sensors confirm system OK.

Step 2: Place a cassette of wafers on the load position and an empty cassette on the unload. Sensors confirm cassettes in place. Press START.

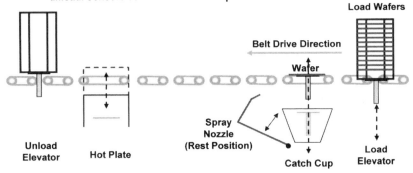

Step 3: Load Elevator lowers until sensor detects wafer on belt; switches belt ON and drives until wafer detected at Catch Cup position.
The Spin Plate rises between the belts and centers the wafers.
A vacuum holds the wafer in position and a vacuum sensor confirms the hold.

Step 4: The Spin Belt mechanism opens to allow the Spin Plate to drive down to the "Spray Position."
A sensor confirms the position has been achieved.

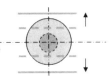

Step 5: The Spray Nozzle is moved to the Catch Cup and the Spin Plate motor starts. When 500 rpm is achieved, the Resist Spray is turned on for 3 s.

Figure 5.11 A spin track system, drawn using PowerPoint.

blocking, the reservoir running out of resist, incorrect resist thickness occurring because of the spin speed being incorrect, and incorrect hot plate temperatures. The degree of detail to be explained should depend on the audience and whether or not the system has issues and needs more detail.

	Actions
5-9	☺ Identify the tool for training.
5-10	☺ Select the person to create the course and carry out the training.
5-11	☺ The trainer should talk to the operators before creating the course and find out what they already know about the tool and what they would like to know. If possible, include it in this course.
5-12	☺ Create the training.
5-13	☺ Confirm that the safety department is familiar with the concepts of TPM. ☺ Check the safety content with the safety department.
5-14	☺ Plan how the tool can be made absolutely safe for the training that is carried out at the tool.
5-15	☺ Create the tests for understanding.
5-16	☺ Confirm that the training record sheets are available, if not they must be created.
5-17	☺ Carry out the training, record and file the results of the tests, and store the originals.

The TPM Safety Pillar

Today, safety must always be the number one consideration when carrying out tasks in any industry. Even without the legislation, regulations, and law suits, tasks must be designed so that they can be carried out with the minimum of risk or, better still, with zero risk. TPM sets the ambitious target of zero accidents.

In this chapter we will consider five key actions that will help keep the teams ahead in safety methodology:

1. How to identify the risks that are present on a specific piece of equipment.
    - How to use an area map.
    - How to use a hazard map.
2. How to identify the risks associated with each task to be carried out. There is a significant added complication to be considered for AM and Zero Fails teams: operators will eventually be performing minor technical tasks. This was rarely a requirement when they were initially employed, so they will probably need to develop new skills. Consequently, extra special care must be taken to compensate for any initial lack of technical experience they will have.

The operators will begin their autonomous maintenance with very simple tasks. To maximize safety, they will initially be working with technicians on cleaning and inspections, in controlled areas that have been made extra safe (lock out tag out (LOTO) must be used).

Risk assessments will be carried out for every task and must be written as if the least experienced team member was carrying out the task.

3. How to carry out a risk assessment.
   All of the team members must be trained in evaluating risk assessments and reading hazard maps. They do not have to be experts, but should be good enough to identify a potential problem. It is important that everyone understands how the risks are evaluated and agree on the results. Being able to evaluate a risk causes team members to think in a different way when they are working.
4. How to apply control measures to minimize risk.
   Control measures are actions that are taken to eliminate or minimize risk. They will be discussed more fully later.
5. How to create safe working procedures.
   Safe working procedures should become the document of choice. Specifications normally give instructions and refer to relevant other documentation. A good procedure should contain all the information required.

The area map

The area map is the first step toward identifying the hazards in a tool. It is a drawing of the tool, derived from the layout diagram, which has been divided into smaller, practical or functional areas. Figure 5.12 has been split into nine areas. In this instance, the split is the same as that used by the vendor to define the separate functional modules, but it does not have to be. The drawing can be two- or three-dimensional, as long as it clearly shows the areas. As always, additional drawings must be used if needed to show areas not visible on the main diagram.

One advantage of the book being published in black and white is that the reader can visualize how much of an advantage it would be if he invested in a color printer for use by the teams. A word of warning, however, is to beware of the cost of the inks. The printers might appear cheap to buy, but the replacement ink cartridges probably will not be. Running costs can escalate rapidly. At the very least, ensure that a printer with individual ink colors is used and not one cartridge containing three or more colors. There is also the option of using compatible inks. In my experience these can be of a lower quality than original

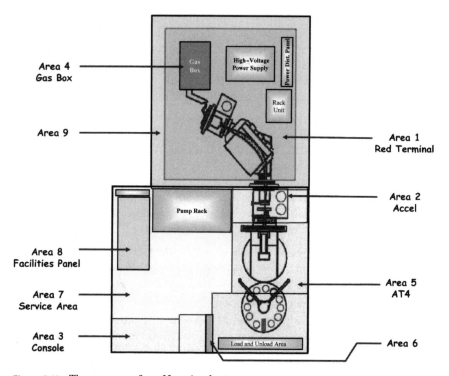

Figure 5.12 The area map for a Nova implanter.

cartridges but are good enough for general work. They can also void the printer's warranty.

TPM promotes the involvement of operators in the maintenance of their own equipment (Jishu Hozen). Not surprisingly, most operators have different skills than maintenance staff, so steps must be taken to avoid their exposure to hazards.

In the tool shown in Fig. 5.12, before starting TPM, operators are limited to Areas 3 and 5. These are the control console, where the process is run, and the AT4, where the product is loaded and unloaded. It is relatively simple to make these areas safe for minor maintenance by the operator. However, if the operator was to carry out tasks in Area 4, the gas box, where there are several potential hazard sources, significant precautions, training, procedures, and supervised experience would be essential.

This book recommends a modular system for generating safe working procedures and risk assessments, not just because I was involved in their development, but I like the way it simplifies everything. It is not essential to use this system for TPM to work; naturally you can still use your own systems, provided they are able to make the tools safe enough.

I just like to simplify things as much as I can, being a tad lazy.... Going back to Fig. 5.12, if we want the operator to work in Area 4, which is virtually in the center of the tool, it is necessary to pass through two other areas to get to it: Area 9 (which is the space between the door enclosure and the red terminal) and then Area 1 (the red terminal).

Using the modular system is a bit like giving directions in chunks. Rather than list every step in traveling from Glasgow to Munich, it could be roughly summarized as follows:

1. Go to Glasgow Airport.
2. Catch the flight to Heathrow Airport.
3. Transfer to the Munich flight.
4. Get a taxi to the hotel from outside the Munich terminal building.

For each of the four steps above, there will be a detailed list of instructions. The beauty of the modular system is that if the final flight destination was suddenly changed to Australia, the first two instruction sets would stay the same. Only Steps 3 and 4 would change: the transfer step in Heathrow and where to catch the taxi at the destination airport.

This method simplifies writing a safe working procedure and risk assessment by minimizing repetition, because the author can refer to the prewritten procedures for making Area 9 and Area 1 safe before proceeding with the detailed steps for Area 4.

The hazard map

Ah, there's nothing better than a trick name, because it is not a map at all. It is a table that lists all the hazards present on a tool and where they are located. If any person has to access the tool, these are the hazards that he would expect to find behind any given door or panel. Hazard maps are like the signs you get in large department stores that tell you what you can buy on each floor (only more accurate).

When creating a map, if unsure whether or not to include a hazard, always err to the safe side and add it. The hazard map is also useful for outside services, like firemen, who can use them to assess potential dangers in an emergency. From a user point of view, in modern equipment, most hazards are protected or interlocked. For example, the vacuum ion gauges listed in Areas 1 and 2 of Fig. 5.13 present three potential hazards: heat, electric shock, and they can implode. For safety, they are fitted with covers that also have warning notices, making it very difficult to touch the gauges unless, of course, the cover has not been fitted. This is not as unusual an event as it might seem.

Hazard	Area 1	Area 2	Area 3	Area 4	Area 5	Area 6	Area 7	Area 8	Area 9
Electrical	High-Voltage Power Supplies Ion Gauge 200A Magnet and Rack Mounted Power Supply Vacuum Roughing Pumps Power Distribution Panel Extraction Electrode	High-Voltage Power Supplies Ion Gauges	High-Voltage on PCBs	High-Voltage 200A Filament Supply Arc Voltage	Servo Motor and Power Supply High-Voltage Supplies	High-Voltage	High-Voltage Power Supplies Electron Shower Accel Electrode Roughing Pumps Power Distribution	High-Voltage Power Distribution	High-Voltage
Mechanical	Vacuum Roughing Pumps	Flag Faraday	N/A	Moving Parts	Chamber Door Spinning Disk (Potentially) Disk Exchange Arms Wafer Handling System	N/A	Terminal Grounding Bar Roughing Pumps	N/A	Terminal Grounding Bar
Gas	Nitrogen Argon Boron Trifloride Arsine Phosphine Compressed Air	Helium in Cryopumps and Compressors Compressed Air	N/A	Nitrogen Argon Boron Trifloride Arsine Phosphine Compressed Air	Compressed Air	Compressed Air	Compressed Air	Compressed Air	N/A
Toxic Materials	Phosphorus Boron Arsenic Graphite	Phosphorus Boron Arsenic Graphite	N/A	N/A	Phosphorus Boron Arsenic Graphite	N/A	Phosphorus Boron Arsenic Graphite	N/A	N/A
Pressure	Pneumatic Valves	Pneumatic Valves	N/A	Pneumatic Valves	Pneumatic Valves Hydraulic Pistons	N/A	Pneumatic Valves Hydraulic Valves	Compressed Air Inlet to Facilities Panel Process Cooling Water Inlet to Facilities Panel	N/A
Temperature	Diffusion Pump Roughing Pump Ion Gauge Source Housing	Ion Gauges	N/A	N/A	N/A	N/A	Roughing Pump	Transformers	N/A
Magnetism	Analyser Magnet Source Magnet	N/A	N/A	N/A	Disk Faraday	N/A	N/A	N/A	N/A
Radiation	X-rays	X-rays	N/A	N/A	N/A	N/A	X-rays	N/A	X-rays
Biological	N/A	N/A	N/A	N/A	N/A	N/A	N/A	N/A	N/A
Slips and Trips	N/A	N/A	N/A	N/A	N/A	N/A	Oil/Water on Floor (Potentially)	N/A	N/A

Note: This is not a comprehensive list of hazards found on this tool, but is an example of how a Hazard Map is made.

Figure 5.13 Sample hazard map.

	Actions
5-18	☺ Arrange risk assessment training for the team members.
5-19	☺ Arrange safe working procedure training for the team members.
5-20	☺ Create an area map for the tool.
5-21	☺ Create a hazard map for the tool.

Risk assessment

Every maintenance task probably has some sort of risk if it is not carried out correctly. Some have risks even when they are. It is the responsibility of managers to ensure risk to employees is minimized. To evaluate a task, consider it in detail to see what is actually being done, assess the working environment, and try to anticipate what could go wrong. Then we need to take suitable precautions to avoid any problems. It is important to include any potential secondary consequences that could happen while working on or removing parts. What if a valve is removed in a water-cooling loop? Will any other part of the tool be damaged when the water is turned off? Is it possible that someone else could turn the water supply back on while the task is being carried out? If so, what damage could the water cause as it sprays out of the open system, could it lead to electrocution?

Lock out tag out (LOTO) is a simple way of preventing problems. In the valve example, a mechanism with a lock is placed over the main control valve to actively prevent any possibility that it can be reopened. A personal ID is often used to identify who has locked out the system. A warning sign is also placed on the valve of the tool being maintained. There are mechanisms available for locking out electrical isolators, switches and breakers, connectors, gas bottles, and other devices. When setting up a system, make sure there is a backup system for removing padlocks in case the technician goes home and forgets to pass on the keys. And, don't forget the obvious, easy safety devices like a "Machine Under Maintenance" sign.

A nice risk assessment example is the installation of a standard light switch in the bathroom. What would happen if it gets wet? Easy, the user gets electrocuted. To avoid this outcome we need to use a switch that distances the user from the electricity. One method is a switch that operates by pulling a cord, because the cord insulates the operator from the electricity. There are other possibilities, including locating the switch outside the bathroom, which would ensure there would not be any problem at all. The health and safety regulations for buildings and electrical installations would cover this type of task. It is important that

TABLE 5.1 Three Natural Levels of a Risk Assessment

Level 1	Operator level
	This covers the areas the operator would normally be exposed to: the control area, loading area, and nearby floor.
Level 2	More complex for tool cleaning
	If an operator or team has to work in an area, it has to be made safe to work in.
Level 3	Maintenance tasks covering main modules and requiring high technical skill content.

anyone writing risk assessments for a task is aware of all the relevant safety regulations that apply to the equipment and the type of task. If they do not know what the regulations are, the Health and Safety Executive can be asked for advice.

Risk assessments can all follow the same format or they can be split into levels of complexity. There are three natural levels that a risk assessment can be divided into, as are shown in Table 5.1.

If we had to carry out a complete risk assessment for every task, not only would it take a very long time, but it would be an unnecessary and inefficient waste of manpower. What we need is a way to simplify them: by eliminating repetition. The method used is kind of like the same technique used by SMED solutions where we put external parts on a trolley and wheel them in just before we need them. The only difference is we are wheeling in prepackaged risk assessments, like the chorus of a song, that make areas safe to pass through or work within. Making risk assessments modular is an elegant solution (the method is illustrated in Fig. 5.14) and will be described in detail later in this chapter. In the same way, risk assessments can be divided into levels that are related to their complexity. This too has the effect of making their creation more efficient and yet not diminishing their effectiveness in identifying potential hazards.

The first level of a risk assessment, Level 1, an example of which can be seen in Fig. 5.15, ensures the tool is safe to be used by the operator for production and to load the product. I dislike making global statements, because there will always be an exception hiding around the corner or lurking in the back of a storage area. But here goes any way.... Equipment manufactured in recent times has had safety designed into it. Not everywhere, but certainly focused in the area of operator protection. This effectively means that any (unsafe) areas of a tool that are not necessary to be accessed by an operator will be enclosed, so making the risk assessment that much simpler. There will also be a safe working procedure for the operator to follow when performing his/her duties, but to avoid the possibility of creating any opportunity for errors through duplication, it could be incorporated in the process or the operating specification.

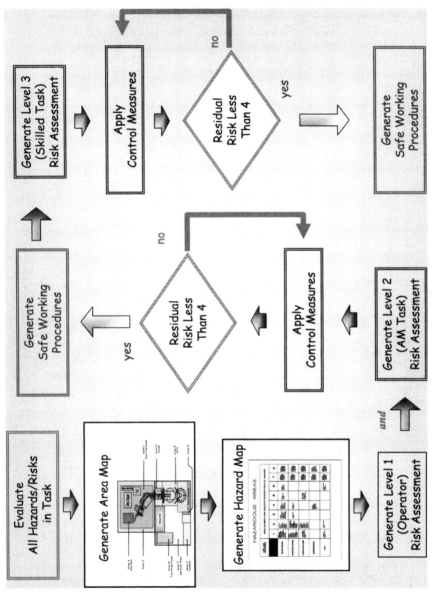

Figure 5.14 The sequence of steps for developing modular risk assessments and safe working procedures (see also the "Quantifying Risk" section).

134 Chapter Five

Tool Identity:	Implanter 3	Level #1:	Yes		Risk Assessment Date:	02/03/00
Tool Location:	Bay 19	Level #2:			Risk Assessment Team:	James Dean
Area Within Tool:	Terminal	Level #3:				Dean Martin
						Martin Short

Task	Generic Hazard Type	Initial Risk Evaluation — No Countermeasures			Final Risk Evaluation — With Countermeasures	
		Worst Case Injury	Risk Calculation		Risk Calculation	
Machine Operation	Electrical	Death	Severity	3	Severity	0
			Likelihood	3	Likelihood	0
			Initial Risk Evaluation	9	Final Risk Evaluation	0
		Countermeasures to Reduce Risk			Responsibility for Countermeasures	
		No exposed sources of electricity. Safe Working Procedures. All operators trained in operation and safety of system.			Name:	James Dean
					Complete By Date:	10/03/00
					Approved Date:	12/03/00
					Approved By:	Names of approvers
Task	Generic Hazard Type	Initial Risk Evaluation — No Countermeasures			Final Risk Evaluation — With Countermeasures	
		Worst Case Injury	Risk Calculation		Risk Calculation	
Machine Operation	Mechanical	Crushed hands	Severity	2	Severity	0
			Likelihood	3	Likelihood	0
			Initial Risk Evaluation	6	Final Risk Evaluation	0
		Countermeasures to Reduce Risk			Responsibility for Countermeasures	
		Loading area has interlocked screen to stop movement when open. Guard plate to prevent access to loading when moving. Chamber Door & Disk exchange protected by doors and inner panels. Safe Working Procedures. All operators trained in operation and safety of system.			Name:	James Dean (Equipment Engineer)
					Complete By Date:	10/03/00
					Approved Date:	12/03/00
					Approved By:	Names of approvers
Task	Generic Hazard Type	Initial Risk Evaluation — No Countermeasures			Final Risk Evaluation — With Countermeasures	
		Worst Case Injury	Risk Calculation		Risk Calculation	
Machine Operation	Chemical	Death	Severity	3	Severity	0
			Likelihood	3	Likelihood	0
			Initial Risk Evaluation	9	Final Risk Evaluation	0
		Countermeasures to Reduce Risk			Responsibility for Countermeasures	
		Operator does not come into contact with any areas of the tool that are likely to be contaminated. Wafers handle automatically, or manually with vacuum wands and operator wears gloves. Safe Working Procedures. All operators trained in operation and safety of system.			Name:	James Dean (Equipment Engineer)
					Complete By Date:	10/03/00
					Approved Date:	12/03/00
					Approved By:	Names of approvers
Task	Generic Hazard Type	Initial Risk Evaluation — No Countermeasures			Final Risk Evaluation — With Countermeasures	
		Worst Case Injury	Risk Calculation		Risk Calculation	
Machine Operation	Biological	None	Severity		Severity	0
			Likelihood	3	Likelihood	0
			Initial Risk Evaluation	0	Final Risk Evaluation	0
		Countermeasures to Reduce Risk			Responsibility for Countermeasures	
		Not Applicable.			Name:	James Dean (Equipment Engineer)
					Complete By Date:	10/03/00
					Approved Date:	12/03/00
					Approved By:	Names of approvers

Figure 5.15 Example showing part of a Level 1 risk assessment.

The Level 2 risk assessment (Fig. 5.16) defines how to make an individual area of the tool safe for work and defines which tasks can be carried out in that area. For example, if we had a block of cages housing tigers, the Level 2 task might be the way to clean the cage. By moving the tiger into another cage, the entire cage becomes safe to clean.

TPM: The Education & Training and Safety Pillars 135

Tool Identity:	Implanter 3	Level #1:		Risk Assessment Date:	02/03/00
Tool Location:	Bay 19	Level #2:	Yes	Risk Assessment Team:	James Dean
Area within Tool:	Terminal	Level #3:			Dean Martin
					Martin Short
		Initial Risk Evaluation – No Countermeasures		Final Risk Evaluation – With Countermeasures	
Task	Generic Hazard Type	Worst Case Injury	Risk Calculation		Risk Calculation	
Clean outside of gas box to remove dust.	Electrical	Death	Severity	3	Severity	0
			Likelihood	3	Likelihood	0
			Initial Risk Evaluation	9	Final Risk Evaluation	0
		Countermeasures to Reduce Risk		Responsibility for Countermeasures	
		Shut down machine, Control Power Off and remove Control Power key.		Name:	James Dean (Equipment Engineer)
		Open Outer Doors at Gas Box and Ground Red Terminal using Earthing Rod. Ensure grounding bar works OK.		Complete By Date:	10/03/00
		Open Outer Doors at Power Distribution Panel and ensure grounding bar works OK.		Approved Date:	12/03/00
		Open inner Red Terminal doors at Gas Box and ensure grounding bar OK. Ground Gas Box & Source using Earthing Rod and leave rod grounding source. Open inner Red Terminal doors at Power Distribution Panel and ensure grounding bar OK. Turn Off Breaker "Gas Box" and "Extraction Electrode" and Lock switches Off. Technician must work with operator. Follow the Safe Working Procedure. Train all users in procedures and MSDS for IPA. Wear PPE for cleaning.		Approved By:	Names of approvers
		Initial Risk Evaluation – No Countermeasures		Final Risk Evaluation – With Countermeasures	
Task	Generic Hazard Type	Worst Case Injury	Risk Calculation		Risk Calculation	
Clean outside of gas box to remove dust.	Chemical 30% IPA solution		Severity	2	Severity	0
			Likelihood	3	Likelihood	0
			Initial Risk Evaluation	6	Final Risk Evaluation	0
		Countermeasures to Reduce Risk		Responsibility for Countermeasures	
		As above		Name:	James Dean (Equipment Engineer)
				Complete By Date:	10/03/00
				Approved Date:	12/03/00
				Approved By:	Names of approvers

Figure 5.16 Example showing part of a Level 2 risk assessment.

For equipment, a technician will make the work area safe before operators are permitted to work in it. He removes its tigers. Any potential hazards must be disabled (LOTO) or covered to prevent physical contact or damage. The Level 2 task must include instructions or refer to the safe working procedures that cover any cleaning methods, safety equipment, and inspection tasks.

The Level 3 is the risk assessment the technician follows when carrying out maintenance tasks (Fig. 5.17). In the example of the tigers, the task of moving the tiger from one cage to another would be a Level 3 task. (Removing a tooth from a tiger would be at least a Level 4!) Within equipment, Level 3 tasks are the complex tasks that need high skill levels: work behind protective panels that have to be removed for access; parts to be disconnected from the main tool, dismantled, and cleaned for routine maintenance; valves to be operated; gas bottles to be changed; or power supplies to be calibrated; and so on. The Level 3 will also include LOTO, safe working procedures, safety equipment, and special handling techniques.

Tool identity:	Implanter 3	Level #1:		Risk Assessment Date:	02/03/00
Tool Location:	Bay 19	Level #2:		Risk Assessment Team:	James Dean
Area within Tool:	Gas Box - Area 4	Level #3:	Yes		Dean Martin
					Martin Short

Task	Generic Hazard Type	Initial Risk Evaluation—No Countermeasures			Final Risk Evaluation—With Countermeasures	
		Worst Case Injury	Risk Calculation		Risk Calculation	
Replace gas bottle.	Electrical	Death	Severity	3	Severity	0
			Likelihood	3	Likelihood	0
			Initial Risk Evaluation	9	Final Risk Evaluation	0
		Countermeasures to Reduce Risk			Responsibility for Countermeasures	
		Follow Area 9 procedure. Follow Area 1 procedure. Ensure Gas Box is grounded. Two-man job (Buddy System)			Name:	James Dean
					Complete By Date:	10/03/00
					Approved Date:	12/03/00
					Approved By:	Names of approvers

Task	Generic Hazard Type	Initial Risk Evaluation—No Countermeasures			Final Risk Evaluation—With Countermeasures	
		Worst Case Injury	Risk Calculation		Risk Calculation	
Replace gas bottle.	Electrical	Death	Severity	3	Severity	3
			Likelihood	3	Likelihood	1
			Initial Risk Evaluation	9	Final Risk Evaluation	3
		Countermeasures to Reduce Risk			Responsibility for Countermeasures	
		Follow Area 9 procedure. Follow Area 1 procedure. Ground Gas Box. Follow Safe Working Procedure for bottle change. Close all of the the gas bottle valves. Restore power to the Gas Cabinet. Evacuate the lines up to the gas bottles. Ensure bottle valves not leaking. Two-man job (Buddy System) Both Wear Positive Pressure Breathing Apparatus. Both trained in use of BA. Understand Materials Safety Data Sheets (MSDS) Remove personnel from area near gas box and position barriers Remove empty bottle and place in sealed transport container. Fit new bottle. Leak Check.			Name:	James Dean (Equipment Engineer)
					Complete By Date:	10/03/00
					Approved Date:	12/03/00
					Approved By:	Names of approvers

Task	Generic Hazard Type	Initial Risk Evaluation—No Countermeasures			Final Risk Evaluation—With Countermeasures	
		Worst Case Injury	Risk Calculation		Risk Calculation	
Replace gas bottle.	Manual Handling	Death	Severity	0	Severity	0
			Likelihood	3	Likelihood	0
			Initial Risk Evaluation	0	Final Risk Evaluation	0
		Countermeasures to Reduce Risk			Responsibility for Countermeasures	
		Not Applicable.			Name:	James Dean (Equipment Engineer)
					Complete By Date:	10/03/00
					Approved Date:	12/03/00
					Approved By:	Names of approvers

Task	Generic Hazard Type	Initial Risk Evaluation—No Countermeasures			Final Risk Evaluation—With Countermeasures	
		Worst Case Injury	Risk Calculation		Risk Calculation	
Replace gas bottle.	Stored Energy (Gas Pressure)	Death	Severity	3	Severity	3
			Likelihood	3	Likelihood	1
			Initial Risk Evaluation	9	Final Risk Evaluation	3
		Countermeasures To Reduce Risk			Responsibility For Countermeasures	
		Follow Area 9 procedure. Follow Area 1 procedure. Ground Gas Box. Follow Safe Working Procedure for bottle change. Close all of the the gas bottle valves. Restore power to the Gas Cabinet. Evacuate the lines up to the gas bottles. Ensure bottle valves not leaking. Two-man job (Buddy System) Both Wear Positive Pressure Breathing Apparatus. Both trained in use of BA. Understand Materials Safety Data Sheets (MSDS) Remove personnel from area near gas box and position barriers Remove empty bottle and place in sealed transport container. Fit new bottle. Leak Check.			Name:	James Dean (Equipment Engineer)
					Complete By Date:	10/03/00
					Approved Date:	12/03/00
					Approved By:	Names of approvers

Figure 5.17 Example showing part of a Level 3 risk assessment.

TABLE 5.2 How Risk Varies for the Initial Assessment

Description of injury	Likelihood of injury	Severity of injury	Risk
Death	3	3	9
Loss of arm or eye	3	2	6
Cuts	3	1	3
No injury	3	0	0

Risk assessment categories. Any risk assessment must consider all the common hazard types to confirm whether or not they apply to the task being carried out. I have tried to include as many hazards as I could think of, but might have missed a few. The most common hazards to be considered would include the following:

- ❖ Electrical
  - ➢ Low and high voltages would have to be included.
  - ➢ Does the task expose the operator to electricity in any form or from any source?
  - ➢ Are there any power sources that could create an issue? Are there any with battery backups that could still be a hazard even when the tool is believed to be off and safe?
- ❖ Mechanical
  - ➢ Are there any moving parts that could trap the operator in the normal performance of his work?
  - ➢ What if he/she puts a finger or hand into an enclosed module, a hole, or through a grid?
  - ➢ Can any of the tools used for the task cause a short circuit, either through proper or improper use?
  - ➢ Are any special tools or test equipment required?
- ❖ Chemical
  - ➢ When connections to the unit are removed, will any material escape from the open ends?
  - ➢ During the performance of the task, either through proper or improper methods, is it possible to expose the operator to any harmful chemicals, liquids, or gases?
  - ➢ Does the operative use any chemicals or acids for cleaning?
    It is important to consider the risk of chemical reaction between the cleaning fluids and surrounding materials or surfaces?
  - ➢ Does the operator handle any chemicals, fluids, or gases?
    Check for operation of filling and draining systems.
  - ➢ Is it possible for the operator to reconnect lines incorrectly, open or close the wrong valve, or forget to close a valve?

- What if a pipe or connection leaks, or another valve in the system fails or is opened while working (either manually or automatically)?
- Do any pipes need to be blanked off and could any error or part failures cause a leak?

❖ Gas
- Check also the points for chemicals.
- When the connections to the unit are removed, is it possible for any gas or vapors to be released into the air?
  Remember even nontoxic gases can cause asphyxiation and death!
- Will any potential release of gas react with the air and become dangerous or corrosive?
- What if another valve in the system fails or is opened while working?

❖ Temperature
- Is any part of the module being worked on likely to be hot or cold?
- What about any nearby equipment?

❖ Manual handling
- Is the item heavy or awkward to lift or access?
- Is there any lifting gear that requires the operator to be trained.

❖ Stored energy
- When a part is removed or loosened, will any high-pressure air or liquid be released?
  Also consider vacuum as a negative pressure.
- If the part is removed, could any mechanisms topple, fall, or creep downwards or even upwards? Consider also gravity, sprung mechanisms, and pneumatic and hydraulic systems (compressed pistons and valve stems).
- Is there a capacitor or battery that could cause electric shock?
- Is there a chamber that is under vacuum either directly connected or as a reservoir to increase pumping efficiency?

❖ Slips and trips
- Will any oil or water spill onto the floor as the part is being removed?
- Are there any obstacles that might cause the operator to trip?
- Is the floor/ground/walkway in a safe state of repair?

❖ Height
- Does the operator need to use a ladder, steps, or a scaffold?
- Is there any medical reason that the operator should not be working at heights?

- ❖ Magnetism
  - ➢ Are there any magnets (permanent or electromagnets) that could affect a pacemaker?
      (Consider also the possibility of magnetic stripes being erased on ID, labels or credit cards.)
- ❖ Biological
  - ➢ Are there any bacteria that could live inside the equipment, cooling water, air filters, drainage systems, or extract ducting?
  - ➢ Is anything in the area, including product and cooking equipment, likely to be affected by airborne bacteria or viruses, sneezing, coughing, or touch?
- ❖ Radiation
  - ➢ Are there any x-rays, gamma rays, beta or alpha particles, radio frequency (RF) or infrared (IR) sources?
  - ➢ Consider high-voltage power supplies, smoke alarms, lasers.
- ❖ Moving vehicles
  - ➢ Are there any fork lift trucks within the area?
  - ➢ Is the area delineated to separate pathways from vehicles?
  - ➢ Are there any trucks or lorries to avoid?

There are many possibilities. It is advisable to list all of the possibilities and mark them Not Applicable (N/A) if they are not relevant. If marked N/A, it can be shown that an item has been considered but was found to have no likelihood of being an issue and has been excluded.

Quantifying risk. Risk is the product of the likelihood of an accident occurring and the severity of the injury caused by the accident.

$$\text{Risk} = \text{Likelihood} \times \text{Severity} \tag{5.1}$$

Taking the example of the bathroom switch, the worst case severity is death and the likelihood, when no cord is used, is high. High risk is not acceptable.

How did I evaluate the risk? Simple:

$$\begin{aligned}\text{Risk} &= \text{Likelihood} \times \text{Severity} \\ &= \text{High} \times \text{High} \\ &= \text{Higher}\end{aligned} \tag{5.2}$$

Somehow, the above calculation seems to lack something. What if we used numbers in place of words, a technique often used in statistical analysis to enable data to be analyzed? A system like this does exist and it works. It is based on the following logic. We must precisely define the various states of severity and likelihood that need to be allocated values. The numerical values need to range from low to high and have an incremental difference that is proportional to the different intensities of the written *standard* descriptions. But, how do we give likelihood a number?

The first way that springs to mind is the way we might interchange likelihood and the probability of the event. Naturally, "unlikely" would be zero. The maximum must be something like, "will definitely happen." Having established upper and lower limits, we now need to select the standard definitions for those in between and try to give them roughly equal increases in magnitude.

➢ Unlikely
➢ Might happen
➢ Very likely
➢ Will definitely happen

These definitions seem reasonable and well spaced, and so might be equated to the numerical values 0, 1, 2, and 3, respectively. For the sake of this explanation, the number system is realistic.

Equally, severity is simple at the two extremes: no injury must always be zero and death must be the maximum. We can now add intermediate definitions and allocate the appropriate numbers.

0. No injury
1. Slight injury
2. Severe injury
3. Death

The problem, as before, is how you categorize the injuries in between "no injury" and "death." I couldn't imagine an average injury, but it should be possible to list a range of real injuries, group them into levels, and convert them to numbers. This would, in fact, be a better system than the one used for our explanation and is a method that has been used in the past.

Let's assume Table 5.1 is used as a rough guide. (It can always be revised to reflect a different range of values.) The initial risk assessment must be evaluated assuming no safety precautions are in place. That means the likelihood would have the maximum value of 3 points. For the bathroom light switch example used above, the severity would be death and be allocated a value of 3 points. So,

$$\text{Risk} = 3 \times 3$$
$$= 9 \qquad\qquad (5.3)$$

This makes a lot more sense and, using this system, risk can be directly compared.

Now that we have established a method for quantifying risk, we now have to consider the countermeasures that reduce the risk.

Countermeasures. A hazard must be reduced or removed to achieve an acceptable level of risk. This is carried out by

1. Eliminating the risk.
   Just as is routinely asked in SMED, is the step actually required? The only way to eliminate a risk is to analyze the complete set of tasks and their order, and consider if any changes could be made that might change the need for the hazardous task. If we have a step where an engineer has to measure a high voltage using a handheld probe, could we have a permanent measurement system installed that would eliminate the need for the test?

   Another example might be the concern for a gas cylinder to leak and vent its contents. If the leak is uncontrolled, there can be a high pressure jet and the cylinder can empty very quickly. What about the recent availability of controlled gas flow cylinders? The gas cylinder has a ceramic "filter" at the neck of the cylinder that limits the speed at which the gas can escape.

   Use brainstorming techniques to look for alternatives. Check the Web and ask around for companies who might offer alternatives.

2. *Substitution.*
   Can a safe or less dangerous technique, gas, or chemical be used? For example, use helium and not hydrogen in airships.

   The Zeppelin airship had to use a gas that was lighter than air to give it the lift it needed to fly. The only suitable (?) gas available to the German designers was hydrogen, which is explosive. To eliminate the risk completely, another gas had to be used. There was another gas that the Americans were using, and for which the Germans were negotiating a purchase agreement. This new gas was called helium.

   Processes and procedures are usually the best practices available to the designers or engineers at the time, which might explain why some undesirable materials are used. It takes active intervention to identify safer materials and a willingness to invest in possible new processes being developed. Sometimes we have to wait for research to find a suitable substitute.

3. *Enclosure.*
   Can the system be modified so that the operator cannot come into contact with the hazard and the contents?

   The most basic example of enclosure is the use of panels to prevent anyone from physically coming into contact with exposed hazards.

Interlocked, software-controlled, automatic door panels can also protect the user from hazardous movements in areas where they have the need to manually load product. (Automatic loading might be a method for elimination.)

Back in Chap. 4, we used an illustration of a modification to a panel that made it possible to read a gauge without having to remove the panel (Fig. 4.2). A hole was cut in the panel and a Perspex sheet was used to seal the gap in the panel. The additional use of "visual controls," in the form of maximum and minimum levels marked on the gauge, eliminate the need for an accurate reading to be taken. Provided the gauge needle is within the two calibrated points, the setting is correct.

In industries where fork lifts drive through the factories, clearly defined pathways can be painted on the ground and barriers can be erected to prevent any possibility of impact with personnel and equipment.

4. *Using personal protection equipment (PPE).*
Particle masks, ear defenders, breathing apparatus, gloves, safety glasses, overalls, and so on ....
This is the last option to be applied in the safety manager's toolkit. It means that none of the other options has resolved the problem.

5. *Training.*
Training is one of the major factors in reducing risk. Skilled personnel and good procedures are invaluable.

The above five points assume that the equipment is operating properly and there are no dangerous defects. This is not necessarily the case. It is also the reason that inexperienced personnel would never be exposed to equipment unsupervised. Where a team is used, at least one person working in the area must have the skills to recognize an unforeseen danger arising and have the skills to either prevent it from happening or take steps to ensure the safety of himself and his colleagues. If possible, and without exposing himself to any danger, he should know how to prevent or limit the impact area of the incident. The "buddy system" relies on two people being able to support each other. This might be the use of an EMO, the use of a fire extinguisher, closing fire doors, warning others of the need for evacuation, and so on.

Just remember that where the buddy system is being used, both people must be capable of taking the corrective action. There is no point in one engineer wearing protective equipment when he is changing a gas bottle and the other "buddy" not. In the event of an incident where the gas leaks, the buddy without the PPE would be vulnerable to the gas and be effectively useless. Think about what can go wrong. Consider as many scenarios as possible. Gather information on similar situations from as many sources as possible.

✓ *Equipment Condition*
   This really goes without saying but the equipment must be in good condition and operate safely. No protective features should be missing, interlocks should not be bypassed, there should be no damaged wiring, and so on.

The target risk is always zero, but if that is not possible, then agreement should be made to decide what level of risk is acceptable. Previously, I have used a value of less than 4 from a maximum possibility of 9 as being safe. That system has now been refined. I do not feel qualified to decide on acceptable risk, having spent all my work experience working for people who try to put safety as their number one priority. I do know that not all industries consider safety as seriously as the ones I have worked for.

I must be a bit paranoid, but I do feel that deciding on an acceptable risk should always have the guidance of the safety department and what the courts decide is unacceptable. Current legislation can make the engineer responsible for injury if negligence or incompetence can be shown. I sometimes worry that despite best intentions, skill and experience, and the *advice* of health and safety departments, it is possible tasks will be accepted as safe and approved as safe by a committee. Then, if one task turns out to be unsafe for whatever reason, it is suddenly the fault of the engineer.

To get a better understanding of this explanation, refer to the area map (Fig. 5.12). The actual actions are only an illustration of the type of steps that would be taken. As discussed before, the three steps below are for making Area 9 safe:

1. Shut down machine, Control Power Off, and remove Control Power key. Retain the Control Power key on your person or initiate LOTO measures to prevent the system from being reactivated.

2. Open outer doors at gas box and listen for the bang the safety ground makes when the doors open. Ensure grounding bar functions correctly by checking the earth connection.
    Hold the grounding bar properly and ground the red terminal using earthing rod. Leave the rod hooked to the system.

3. Open outer doors at power distribution panel and ensure the drop down, grounding bar works OK as described in point 2.

So, if we are using a modular system, the above instructions required to work in Area 9 could be simplified to "Follow countermeasures for Area 9"—and then list the other steps required to make Area 1 safe.

We could simplify the risk assessment even further. The following steps make Area 1 safe at the gas box side. But let's assume, for simplification of the example, that they make the complete area safe.

1. Open inner red terminal doors at the gas box and ensure grounding bar functions correctly.
2. Ground gas box and source using earthing rod and leave rod grounding source.
3. Open inner red terminal doors at power distribution panel and ensure grounding bar is OK.
4. Turn "off" breakers: gas box and extraction electrode.
   Lock switches "off" and attach "Equipment Under Maintenance" signs.
5. The designated technician must work with the operator.
6. Follow the safe working procedures.
7. Train all users in procedures where required.
8. Wear PPE for cleaning.

The eight steps above could be summarized by the statement "Follow countermeasures for Area 1."

Now our modular assessment for working in Area 4 becomes

1. Follow countermeasures for Area 9.
2. Follow countermeasures for Area 1.
3. Follow the list of relevant steps for cleaning the exterior of the gas box with IPA.

Now, once Area 1 has been made safe, any number of tasks can be carried out and only the specific countermeasures to make each individual task safe need to be listed more fully. Naturally this is conditional on the assumption that we do not open any unauthorized doors or remove any covers and panels.

If it is decided that there is a piece of equipment in Area 1 that must not be turned off because it would have a negative effect on the process, or it would take too long to restart and stabilize, then steps must be taken to ensure that it cannot become a hazard to anyone working in the area. If it is hot/cold or has exposed electrical contacts, then it must be *enclosed* to prevent contact and be clearly labeled stating that there is a hazard and what it is. There is also the option of moving the hazard to a different area. Whatever is decided, follow the same kind of procedures that would have been applied for "hard-to-access" areas.

What would be a reasonable size for an area to be? There are two ways of looking at it. First, that the area should include only what you can actually see at any one time. This avoids the situation where two teams are doing something that might impact the other team. LOTO reduces this risk. Furthermore, if you cannot see into an area properly, it could be dangerous to put your arm into it, for fear of what might be touched. If the area can be made completely safe and the previous situation cannot arise, then the area can be as large as the team can realistically cope with.

Who creates the risk assessments? The technical members of the team will create the first drafts of the risk assessments since a comprehensive knowledge of the hazards and interlock system is required. The equipment engineer will provide support as will the health and safety department. Where different teams are working on the same type of tool, the option exists for one team to create all the risk assessments, with other teams approving or amending them. My preference is that each group should carry out their own assessments, or at least a significant batch of them. The different assessments can then be compared and the master developed from the best features. My logic in wanting all the teams to develop their own and then compare is to make sure that all the teams get as much practice as possible. It is only when it becomes second nature that the teams will start to think about risk assessments when they tackle any task.

Why is it that when different groups, working on the same tool model, are carrying out assessments, they will generate quite different risk assessments?

- *To minimize production impact, teams will probably be from different shifts.*
  This means that the teams tend to work in isolation and do not have the opportunity to standardize or even compare notes. Setting up an overlap for the teams at the beginning or end of a shift will help by enabling them to share ideas and progress.
- *Even tools of the same type can have different revisions of equipment.* Power supplies might be upgraded, some doors might have key locks on one tool but not another, different types of pumps or motors can be used. There can be in-house modifications or different revisions of software. These differences must be taken into consideration.

- *Different people see risk differently.*
  Consider the different sides of the arguments about speed cameras, driving without seat belts, or wearing crash helmets on motor bikes. Where they do agree on hazards, they might have different ideas on control measures.
- *Comparing the risks will generate the "best practice" method.*
  Each group carrying out their own assessments will take more labor-hours and might seem a bit over the top, but it also teaches every team member (including the operators) to think about safety as a matter of course and gives them a much better understanding of the tool. Another point worth remembering is that because TPM uses operators to carry out maintenance, it is important that no operator can ever be injured while working on a tool, and so all learning is beneficial.

I recommend that the approval of risk assessments has several stages. Initially the tool engineer and the lead hand/foreman would check the standards since they have the most experience on the equipment. For the final revision of risk assessments, it is advisable to have a safety committee. The members could be drawn from TPM steering group, the pilot team, the health and safety department, and equipment engineering. A controlling group is in the position to identify issues that one set of teams might have missed but a different project group did not.

Risk assessments must be routinely reevaluated or checked any time something has been changed. This will account for modifications in processes, procedures, or equipment. Even a new tool location can cause issues. The change could have introduced access hazards either to exits or other equipment. In one machine move, x-ray shielding had to be added to the roof of the tool to protect occupants who now worked in the area above the tool. This was not a consideration before the move as no one worked above the tool.

Safe working procedures: Using as standards

Over the years, I have seen hundreds of procedures being written but had rarely seen any being read. I attributed this to the formats used. Basically there were lines and lines and lines and lines of text. I don't think it is just me: I find that people do not appear to like reading this type of document. To overcome this, I developed my own technique. One which I believe is far more appropriate, creates fewer potential for errors, educates the reader, and does get read. Not only that, it gets used for training too.

Step	
44 Ensure the *Disk Chamber* area is clear. *(See Figure A)*	*Figure A* 
45 Enter *CMD 2150* at the *End Station Keypad* to *Close* the Chamber Door.	
46 When the door has closed, engage the *locking clamp* to prevent opening.	*The door weighs approximately half a ton and can cause serious injury if it opens when someone is working in the Disk Area.*
47 Check the *sliding seal height* is 0.125 in. *(See Figure B)* ➢ If it is *not,* proceed to *Step 48.* ➢ If it is Correct, Jump to *Step 52.*	*The sliding seal height is the distance between the bottom of the V3 valve casing and the sliding seal plate.* *Figure B* 

Figure 5.18 Example #1 of a "step-by-step" safe working procedure.

A safe working procedure should be as simple to understand and as accurate as possible. Pictures or drawings should be used as much as possible to illustrate the steps and remove any ambiguity. The procedure in Figs. 5.18 and 5.19 is made up of two main columns. The left-side column lists the actual *procedure steps* to be carried out. If the other column was missing, the procedure would still work by following only

148   Chapter Five

48 Loosen the *Lock Nut* on the *Index Switch*. *(See Figure 14.)*

49 Turn the *Adjustment Screw*:

➢ *Anticlockwise* to increase the gap size.

➢ *Clockwise* to decrease the gap size.

50 To reset the *Translator*
➢ **Turn the power *Off***

➢ **Manually rotate the lead screw to lower the Chamber height so that it does not operate the Index Switch.**

➢ **Turn the power *On***

Figure C

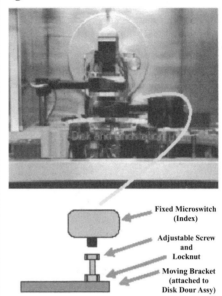

Note:
If an electron shower extension tube is installed, there should be a 1/8" gap between the bottom of the tube and the chamber.
After the gap is set, the lead screw should be clamped using the clamp ring at the bottom of the lead screw.

The translator "Auto Reset" will initiate a movement and drive the stage upwards slowly until it reaches the reset switch.

Figure 5.19 Example #2 of a "step-by-step" safe working procedure.

these steps. Even the steps are different from the way many procedures are normally written.

Since the figures are in black and white, we get the opportunity to see the advantages of using color in your procedures. People actually learn more from color documents than they do from black and white. This is important as the document style is a key tool in the documentation and training.

One major difference in my document is the use of words in different text, colors, and in the formatting of the sentence. I would use incorrect grammar if I thought it would make a document easier to understand. The steps in the left-hand column are basically written as a sentence but the words are broken up into short lines like a telegram (if you remember them) or like a text message. If we consider Step 44 in Fig. 5.18, we see the sentence "Ensure the Disk Chamber is clear" and "(See Figure A)." However, without color, the complete step would be shown in bold black text. Also the reference to the figure (which is extra information) would be in black italic.

If you look a bit closer you will notice the sentence is broken up into four lines, with the key words "Disk Chamber" on the second line. If there was a simple way to highlight the key words (like inverted commas) and if the technician knows the procedure, he can just read the key words as a memory jogger or to get dimensions, computer commands, switches to press, expected readbacks and responses, module names, and so on.

Now, imagine the words were in color. The main steps would still be in bold black, but we could use a different color, say brown, to make the key words "Disk Chamber" stand out. We could also decide that the extra information, the figure number, could be in a different color of italic, say green.

Now when we look at the sentence we have:

Line 1: "Ensure the" would be printed in bold black.
Line 2: "Disk Chamber" would be printed in bold brown.
Line 3: "is clear" would be printed in bold black.
Line 4: "(See Figure A)" would be printed in green italic.

We have, in effect, developed a color-coded sentence. In Step 45, we would have the key phrases "CMD 2150," "End Station Keypad," and "Close" also in brown; the rest of the words would be black.

The second column (the right-hand side) has a completely different purpose. It is for extra information. In a black and white document, the text is written in italic so it cannot be confused with the instructions, which are bold black and bold italic text. The right side contains reference diagrams, reasons for carrying out steps, explanations of what happens if the step is not carried out properly, and extra information that will make the user understand the task better or advise them on issues to look out for.

If we had used used color, the text on the right would be green italic. Why? Because we have just defined extra information in Step 44. Having said that, though, I am flexible in my use of colors. If something was potentially dangerous or must be avoided, I would use bold red italic or whichever color would make the point stand out to the user and not be missed.

The whole page is written in a tabular format, so that pictures and rows are always aligned. Only the vertical lines are shown in light gray, again to separate steps from information. New steps can be added simply as "insert rows" without affecting all the information below or above. The two examples used here are just to demonstrate a format that works really well. In reality, more precise drawings and more photographs would improve the examples shown in Figs. 5.18 and 5.19, but they are not true procedures, they were only assembled for illustration.

From experience, using this style of documentation has proved highly effective for increasing uptime. It is a regularly copied format.

It is worth pointing out that the intention of the documentation is to create procedures that will minimize the likelihood of anyone making a mistake. The document is not the end goal. The procedure should be the best way known to carry out the task. It will not remain the best practice for ever. In order to keep it at the top, it will have to be revised regularly. When creating a new procedure, bear in mind the following points:

- ☺ They should use the easiest language that says what the writer means and that the readers can follow.
  There should be no need to try and be clever, but neither should it be necessary to "dumb it down" too much. Write it for the audience who will be using it. Get advice if you need it.
- ☺ They are not intended to be works of art.
  PowerPoint is a really simple-to-use, but very powerful drawing tool. Use photographs or even sketches if needed. The artwork can be improved later if necessary.
- ☺ They are not intended to be perfect examples of grammar.
  Write it on a PC and it will correct the spelling and make grammar suggestions. Be careful though, sometimes it flattens the impact of the text. The user can choose to ignore either or both. It is also possible to write the procedure by hand. There is also the option to get a secretary to type out the master.

- ☺ They should not take ages to write.
  Take the time you need to get it workable. It need not be perfect; just have the correct information.
- ☺ To test the tasks for functionality and ease of understanding, recruit one or two people who do not know the equipment and ask them to read the procedures and review them.
- ☺ If "the reviewers" ask questions or do not understand parts of the document, then those parts should be rewritten and simplified.
- ☺ The format can be adapted to suit the needs of the procedure.
  Don't let the format detract the understanding. If a photograph, drawing, or table needs to be wider or larger, use both columns. (The columns should be merged.) Equally, if they need a whole page then do it.

The "semi" final document should be used to train any shifts on the new procedure. Improvements suggested by each shift, if agreed to be beneficial, should be added to the final document. After each shift has commented, we should have a "best practice" procedure and the added incentive of a decrease in maintenance issues.

	Actions
5-22	☺ Identify the tasks requiring attention.
5-23	☺ Create the risk assessments.
5-24	☺ Create the safe working procedures.
5-25	☺ Use each new procedure to train the shifts.
	☺ Record, discuss, and make any amendments to the procedures.
5-26	☺ Retrain shifts as necessary with any new procedures.

Chapter 6

5S: Organization and Improvements by Default

"Nice place...."
First impressions count. We all know that everyone likes nice, clean, bright surroundings—whether it's a shop, a restaurant, a house, or a factory. But sadly, the reverse happens too. I once had to work in a factory where the toilets were so bad I didn't want to go anywhere near them. Only in a situation of dire emergency would I use the toilets, otherwise I preferred to wait and find some place on my way home. The rest of the factory wasn't quite as bad, but you could tell that maintenance of the premises was not a routine topic at budget meetings (or anywhere else for that matter). At the time, I worked for a vendor company that supplied equipment and this was one of its customers. It was a place I *had* to go to work: I had no option. The company itself was pretty astute when it came to making profits; it just didn't appreciate that cleanliness and organization were sources of direct and indirect profit. The management had no idea that improving the place would have had real, positive benefits, even if only for the workers' morale.

Imagine that a potential customer has come to evaluate your factory. What will be his first impression: positive or not? Will he place an order? There is always a black and white option, but what about the grays? There is a common response by potential house buyers when they view dark, untidy, overpersonalized or cluttered homes. They can be turned off and they will have no idea why: it is a subliminal reaction.

In *all* workplaces there is a tendency for materials, documents, assets, brochures, spare parts, or stock to accumulate over time. Most of us are natural hoarders. I am too, but I justify it by convincing myself that I will need it for future training courses. But what you might not realize is that too much hoarding can have a real, negative effect on the working environment and on production. I am due carrying out a 5S exercise on my office as I am beginning to find that it is taking longer to find things. For me it is important. I have only one office and one "me" to worry about, but for a factory with a large production space and a group of workers all using the same materials, it is far more critical. Of all the business improvement books I have read, there is one line that sticks in my mind: "Factories, by definition, are places where something is made. Warehouses are places where something is stored."

Since "you only get one chance to make a first impression," the first step toward improving is to recognize that a change is needed.

5S: SSSSS—The Meaning

In Japan, the five S's represent Seiri, Seiton, Seiso, Seiketsu, and Shitsuke. Virtually everyone likes a catchy name: the two R's, the 4M's, the 5W's and 1H, DMAIC, and PDCA, but they become a nightmare when you try to remember them all. Even worse, depending on the industry you are in, the same letters will no doubt have different meanings.

When the 5S technique was adopted outside Japan, everyone had to find their own equivalent words, starting with an "S" (see Table 6.1). However, like poetry written in one language and then changed to another, it often loses something in the translation. Whatever substitutes are selected, the bottom line is that the five S's should combine to make a five-step formal program that introduces, implements, and maintains a clean, safe, clutter-free, and efficient site. There is even a 5C's, but I am not even going there!

The Benefits of 5S

If a factory is dirty, particularly the floors, there can be safety issues in the form of slips and trips. In today's workplace, poor safety standards will ultimately cost the company money. Since the onset of "no win, no fee" lawyers in the United Kingdom, many companies have seen a huge increase in litigation. It surely goes without saying that loose dirt can also contaminate production components and tools, which could lead to premature failure of finished product in the field. Your customers

TABLE 6.1 A Table of Currently Used "S" Equivalents

Seiri	Seiton	Seiso	Seiketsu	Shitsuke
Organize	Set	Scrub	Standardize	Sustain
Systemize	Systemize	Sweep	Regulate	Embed
Simplify	Neatness	Clean	Site-Wide	Self-Discipline
Sort	Set in Order	Shine	Standardization	Discipline

can easily be lost. They can lose their trust in your company, even if you only supply them with a few unreliable parts. Unless they have no option, customers will change suppliers if they lose confidence in their supplier's parts or even if they only suspect the parts could infect their own product, by making it unreliable. They will not, cannot, risk their own customers.

Equally, a production area with storage shelves covered with unidentifiable, assorted current and obsolete components is simply begging for an operator to lift and use the wrong part by mistake. I have been to companies that have suffered significant financial losses caused by this very problem. Using obsolete parts will also risk product failure or quality issues and, again, the loss of the customer. Remember this: once a customer has been lost, it takes about 5 years before you have *a chance* to win them back.

Excess storage space has often been found to get in the way of production line efficiency and improvements. It is pretty obvious when you think about it; all the storage gets in the way of simple reorganizational changes. The cupboards have to go somewhere on the floor and until the space is clear, it is hard to "see" a better layout. It is a bit of a Catch 22 situation. The new plan will always have to consider both the line *and* the storage. What would the layout be like if the storage areas were not needed or were smaller?

It is worth remembering that factory premises with stores need to be larger than premises without stores. (I know what you are thinking, "This guy's a genius....") Yet, if the stores spread across multiple buildings, after condensing the space they occupy, it might even be possible to reorganize and shut one building down. This action will reduce rent, rates, or taxes for the unnecessary (storage) space. At the very least, it will save on heating, maintaining, cleaning, and furnishing the space.

What can we do if we do choose to reduce the storage space and find that it causes problems? The answer is the same as with any new action you take. Find out what the new problem is, establish why it is happening, and find a way to solve it. When making *any* improvements, always follow the fundamental "Plan, Do, Check, Act" cycle.

1. *Plan*
   Decide what you want to do, what you expect to get out of it, how long you expect it to take, and how you intend to do it.
2. *Do* Proceed with the plan.
3. *Check*
   At all stages of action, monitor the changes and make sure they are working as intended. Look for trends that show deviation from the plan.

4. *Act*

   If any issues are discovered or are predicted, take action on the information to avert the situation or to restore it to the original, planned path.

5. *Go back to point 1*

   Repeat the cycle until all the changes work according to the plan.

The intent is simple: never make changes without monitoring if the change makes things better or worse.

The 5S process is widely recognized as a foundation step to continuous improvement techniques. It also provides a practical way to introduce employees to the concepts of waste and productivity. It promotes a clean, efficiently planned workplace and gives the employees the opportunity to improve the way they work. Because the 5S process will identify and remove *unnecessary* product, tools, and general clutter, it will further encourage the development of new layouts, better storage systems, and new, better methods for identifying the correct parts. By physically clearing and cleaning the area, 5S enables sources of contamination to be identified and removed, again improving product quality.

The five steps are represented in Fig. 6.1. The first two steps, Seiri and Seiton (Sort and Set in Order) are separate steps in their own right, but because of a natural overlap, they are often carried out together.

The Decision to Implement 5S

The *real* first step is when a company decides to implement 5S. But, before this is possible, the management has to hear about it in the first place. The initial source could be through an article in a magazine, a "How To" book, from an employee, from a colleague in another company, from a college course, or from an approach by a training organization or a business improvement consultant.

The senior management team has to weigh the benefits 5S has to offer and then make the decision to start the process. The next decisions will be the scale of the implementation and a time line for implementation. The management must evaluate the costs and allocate the funding. Figure 6.2 is a guide to the implementation cycle. It is important to appreciate the impact of the participation of the employees. The bulk of the cost will probably be the labor, which is more intensive at the beginning of a new project, although each team actually uses less time as it progresses. However, when the gains from the first project are seen, the team will likely be allocated a new project as the first one becomes embedded. Because 5S is a self-publicizing

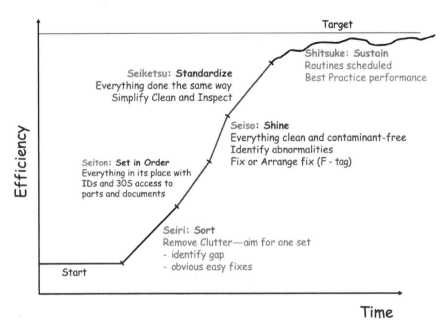

Figure 6.1 A schematic illustration of the 5S process.

venture for the teams, there will be investment in notice boards (activity boards), video and digital cameras, computer memory (hard disk space), and various stationary items. Then, as problems are uncovered, there will be the additional cost of the solutions: extra training, tools, designing and making modifications, allocating time to write procedures, time to train others, time to evaluate and test solutions, and so on.

Depending on a value assessment of the returns of the improvements, the degree of commitment can either be increased to complete a project sooner (and enable the benefits faster) or continue as scheduled in the original time slot, allowing extra time for the project to be extended until completion. The team members can experience a project firsthand and then become the plant steering group, which will promote the introduction of the initiative across the site.

Initial Management Implementation

The management team needs to collect the initial site data and decide whether a pilot team will be set up to test the procedure and promote its benefits. I like the idea of a pilot team. They are used in TPM and RCM as well as in 5S (Fig. 6.3).

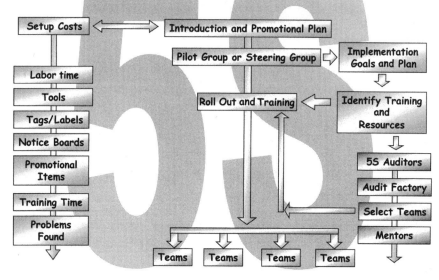

Figure 6.2 A summary of the stages of implementation.

Sample
5S Implementation Project Plan

Step	Responsibility	1-Jun-04	8-Jun-04	15-Jun-04	22-Jun-04	29-Jun-04	6-Jul-04	13-Jul-04	20-Jul-04	27-Jul-04	3-Aug-04	10-Aug-04	17-Aug-04	24-Aug-04	31-Aug-04
Establish Management Team	MD	■	■												
Budget	Brian		■	■											
Set Up Pilot/Steering Group	Steve			■											
Introductory Training				■											
Initial Implementation Plan				■											
5S Promotion					■	■	■								
Training Requirements					■	■									
Select Auditors					■										
Initial Factory Audit					■										
Define 5S Target Areas					■										
Select 5S Teams						■									
Select mentors						■									
Roll Out Training							■	■	■						
Start Teams								■	■						
Progess Reviews										■	■	■	■		

Figure 6.3 An example of an implementation plan. This one is for 5S.

Pilot team members are exposed to all the issues that a normal team will face and have to deal with them. It provides the opportunity to appreciate the issues the other teams will face. The plant managers can even be photographed participating in a 5S session and the photos be used for promoting the initiative. The pilot team should have the same membership content as any other team. Its members should understand the area they will be working in and all of the current issues. The only real difference will be that the members will be more senior.

The managers will need to pave the way for 5S, by selecting the auditors and directing them in their actions. From the budget allocated, the managers need to select the teams. Each one should consist of a small core, but have access to experts as required. The core team should contain about three or four members, consisting of one or two operators and a technician, an engineer, or someone from production. A small area could be limited to a two-person team. At least one of the team members should work in the area being 5S' d; if possible there should be more. There should also be access to a dedicated manager for support and assistance to clear roadblocks. The whole team should attend the meetings and participate in the initial clean.

Just as in TPM, the teams *must* be trained in general safety and all safety items that are needed for the areas they will be working in. This must include how to make a hazard map and an area map and how to carry out risk assessments. They should also understand similar safety procedures to those of the stores employees: for manual handling, sharp edges, warning labels, etc. Safety is covered in detail in Chap. 5.

Audit sheets

When auditing the site, a standard spreadsheet will be required that includes all the data needed to identify the location of the audit and all of the details needed to evaluate the condition of the area:

➢ The name/location of the area being reviewed.
  This information can be identified from the site map.
➢ The review date.
➢ The auditor/team names.
➢ An evaluation of the current condition of the area.
  This should include "before" photographs. The condition should include a simple overview of what might be needed to put it right; the final detail will be decided by the 5S teams.
➢ A numerical scoring system for the condition is preferred for comparing areas. "A bit dirty" or "some rubbish in a corner" can be less meaningful than, say

Poor = 0
Reasonable = 1
Good = 2
Excellent = 3

The score should be recorded for all five steps, although initially you might not expect to have any standards to compare with or, where standards do exist, they might have fallen by the wayside.

➤ The target condition or standard you expect to be achieved.

➤ The person who is (or will be) responsible for the area.

Some people have found a tendency for audits to be scored as average, and this makes improvement difficult to judge. It has been found useful to work to the upper and lower scores only: that means no averages. The score effectively becomes a "Yes" or a "No."

The red tag holding area

The management has to set aside a holding area. This is essential for 5S. Its purpose is to provide a buffer zone in the shape of a temporary store that should help avoid necessary or useful components from being accidentally scrapped.

Take care when deciding what to throw out. Even with the best intentions, there is likely to be accidental disposal of needed items. Often the scrapped parts are borrowed test equipment that has not been returned. It sits around for a while after its initial use (just in case it is needed); eventually it is adopted by the area owners, stored, and promptly forgotten about. The parts are only discovered to be missing when they are needed further down the line by their real owners, who then set off on a hunting expedition to try and find them.

The holding area is not intended to be used as a rubbish dump. It must be organized using basic 5S principles just as any storage area would be. All of the items stored must be easy to recover. Many of the parts that find themselves in the store will not be rubbish. They might be tools or jigs, essential for annual PMs, equipment installation, or alignment, but are rarely used. These are tools you don't want to bin, but return to their owner to be properly identified and stored. They might also be expensive, rarely used test equipment.

Most of the details for the *holding area log sheet* will be on the red tag log sheets, but should at least contain the following:

➤ The red tag ID number

➤ A description of the part

- Where it is stored
- How many there are
- A pound or dollar value
- When they were stored
- When they are scheduled to be disposed of
- How they will be disposed
  (Remember that some assets must be accounted for)
- Who/which section/which department the item belonged to at the time it was stored
- Confirmation that final disposal has been officially sanctioned
  Also
- Fit sheets to a clip board
- Attach a pen to the log sheet clip board

Management Actions

6-1 ➤ Maps need to be created for all the functional areas of the site.

6-2 ➤ Auditors will need to be selected and trained in 5S.

6-3 ➤ The auditors will carry out initial surveys of the factory. They should create a spreadsheet including the "headings" listed in the next section, the "initial audit sheet."

6-4 ➤ The auditors will identify all of the areas needing improvement and make recommendations of priorities to the management.
   ➤ The auditors should also identify possible locations for each area's activity board.
     They can be within the area, in a central location like the canteen, a rest area, or main corridor.

6-5 ➤ The management team has to decide on the teams. If they opt for a pilot team, the purpose of the team (in addition to improving the area) will be set up to test the procedure and promote its benefits.
   ➤ If no pilot team, they need to set up a steering group.

6-6 ➤ 5S teams and/or a pilot team will be provisionally selected, trained, and allocated their areas of responsibility.

6-7 ➤ The steering group should be formed.
   ➤ The group should plan the initial 5S training and draft an implementation plan.

> 6-8 ➢ Identify a location for a temporary holding area to store "red tag" items that are being relocated until their final fate is sorted.
> 6-9 ➢ The pilot team should create a layout diagram of the holding area.
> 6-10 ➢ The pilot team should create a holding area location log sheet.

Step 1: Seiri—Sort

The team should be allocated an area to be 5S'd. The size should be practical; go for "bite-sized" chunks. Small functional areas are easier to manage. The object of Sort is to get rid of anything that is not needed to make or maintain the making of the product.

There are a few records to be maintained in 5S and an "activity board" to maintain. It is a visual display of the details and status of the project. The activity board provides much of the formality. The discipline to record the details must be developed, which is one of the reasons for the fifth S, Shitsuke, which means discipline. Everything must be recorded in detail since some tagged parts might be in the holding area for long periods before they are either reclaimed or disposed of. The theory and layout of activity boards is covered in Chap. 12.

Red tag details

Not to be confused with a TPM F-tag, this "red tag" (which appears as black in Fig. 6.4) is key to 5S. It is the mechanism used to determine the functionality and usage of the components and parts on the site. It is an identification label that has all the information you need to know about the items.

The tag should record

➢ The tag creation date.

➢ The name of the person who created it.

➢ A unique ID number.

➢ A description of the part.
  If the part has a model number or an ID or part number, this should be identified.

➢ Where the part is located.
  This means that the storage areas and shelves must also have IDs to enable easy locating.

➢ A space to record any dates when the part has been used.

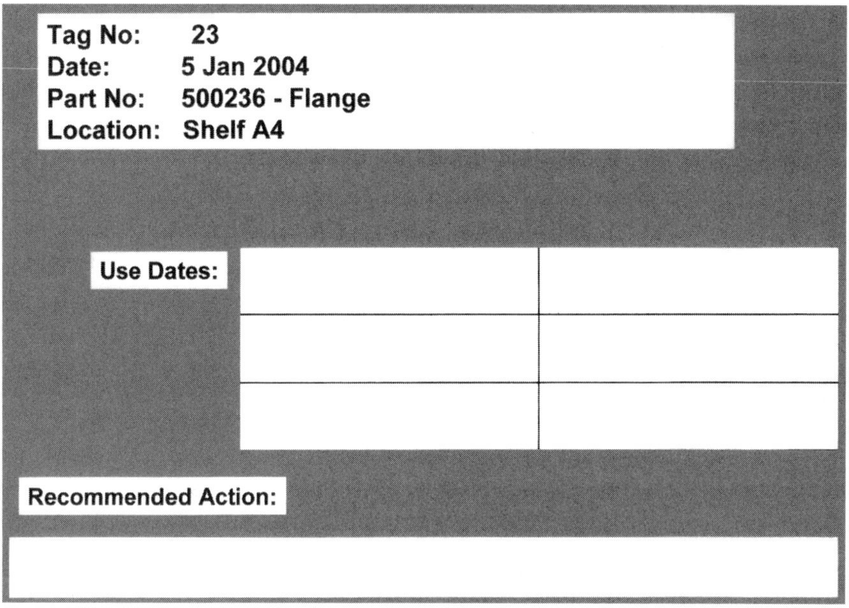

5S: Organization and Improvements by Default 163

Tag No:	23
Date:	5 Jan 2004
Part No:	500236 - Flange
Location:	Shelf A4

Use Dates:		

Recommended Action:

Figure 6.4 An example of a red tag (*which appears as black in the figure*).

- How many parts are present.
- When the status of the part will be reviewed.

Alternatively, and this is a big difference, just record the tag ID number and a description on the tag. The rest of the details can be retained on the spreadsheet. However, you must keep copies since you run the risk of losing the original sheet.

	Team Actions
6-11	Take photographs of the area before any work is carried out.
6-12	Create a number of red tags with unique ID numbers.
6-13	The first bit will be easy: take the stuff you know is not needed and for each item, complete a red tag and attach it to the item.
6-14	Create a red tag summary sheet and make copies.
6-15	Log the required details on the red tag summary sheet and place the items in the holding area.
6-16	Log the details on the holding area spreadsheet.

> 6-17 ≥ Send a memo, containing a list of all the items in the holding area, to all of the department managers. (This action will be repeated regularly.)
> 6-18 ≥ Red tag the remaining items and complete the log sheet.
> 6-19 ≥ Ensure *everyone* knows that each time the tool is used the recording/tracking system must be filled in.
> 6-20 ≥ Once it has been established that a part is used with a reasonable frequency, an appropriate position should be found for it and the tag removed.
> ≥ Parts that are not used in the chosen time should be removed to the holding area.
> ≥ Follow Actions 6-16 and 6-17 to identify the location where the part is stored and to inform the managers that new parts are being stored.

The next stage in the process is to identify if, and how often, the remaining parts are used. Two methods are used:

1. First, the team should ask around.
   Someone should know where the part lives. If the part has a local owner they should recognize it from the description or photograph.
2. Second, by attaching a red tag to the item and recording any dates the item is used.
   Provided everyone knows how to fill in the details, any use will be evident.

Over the next few months the usage rate of the parts and tools must be monitored. This will provide a guide to frequency of use, how many items should be stored, and where they should be stored. It will also highlight how often tools and equipment are used.

Step 2: Seiton—Set in Order

As previously mentioned, Steps 1 and 2 are often run in parallel. No time should ever need to be wasted looking for or picking parts. All storage spaces, including floors, should be labeled to enable quick identification of locations.

Using proper signs, shelf markings, and floor layouts will help the operator become more efficient. They will

≥ Reduce the time it takes to collect parts or tools.

≥ Reduce the time taken to find and select the part required by avoiding *searching time*.

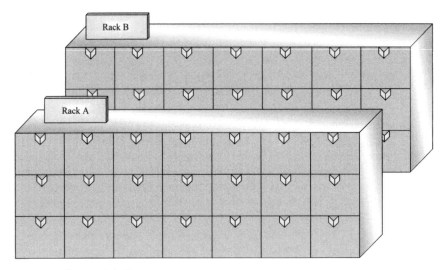

Figure 6.5 Storage labeling.

- Enable the operator to tell if the minimum number of parts has been reached. It will even help you decide what the minimum should be.
- Enable the operator to tell at a glance if a part is missing.

Removing the unnecessary components should have created space. Draw a layout map of the area and consider various locations for the remaining parts based on how they are used. Figures 6.5 and 6.6 are guides to storage locations. Time is lost when staff cannot find a part, if they do not know what part they should be using, or if the part they are looking for is out of stock. All storage spaces, walls, floors, drawers, and cupboards included should be labeled to enable quick identification of locations. It should be possible to tell at a glance not only what is there but also what is missing.

Step 2 is the step that sets the guidelines on how to label shelves and create visible maximum and minimum levels for parts. The key to all of this is "visible." It is important that the user is able to see the part he wants and, better still, see if the level is due to be increased, before it runs out. Pay close attention to the suitability of the labeling of the shelves. Ideally, it should be clearly visible and readable as it is approached and not have to be searched for when the operator reaches it. Figure 6.5 is a PowerPoint sketch of what shelf labeling could look like. Notice each rack is clearly labeled using large signs and each storage shelf also has a number that is visible from either side. Each time the layout is improved, the layout/position location map must be updated.

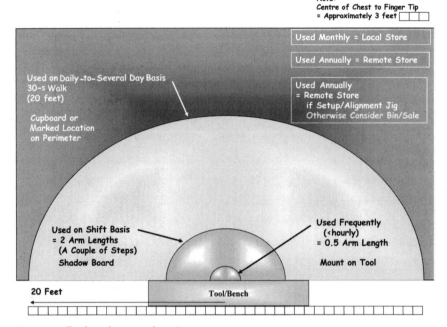

Figure 6.6 Preferred storage locations.

Once Step 1 has been completed, we are left with a more manageable amount of materials. Any items that have not been removed should be rearranged (Set in Order) as described above. They should be located in positions that improve the ease and speed of access by the operators.

Where parts are located should be the decision of the team, but weight should be given to the operators who work within the area. They are likely to have a better understanding of usage. Mistakes can easily be rectified: the choice is never cast in stone (plan-do-check-act). I might think a part should be stored right next to the operator. You might feel it should be further away or could even be a personal tool retained by the operator. It is a team decision, but should be based on accessibility. The proximity of a tool or part to the operator's work station will depend on how much it is used. If it is used virtually all the time, it should be positioned on the workbench, attached to the tool or on a rack within arm's length. It can even be suspended from above. If the chosen position does not work, change it. Figure 6.6 is a reasonable guide to locations for tools and parts.

Parts that are used only a couple of times a shift can be wall-mounted nearby, possibly on a shadow board or a shelf. As a guide, it should not take more than 30 s to retrieve it. Rarely used parts can be afforded to be stored furthest away, in a cupboard, or in a nearby storeroom. The advantages seen by gaining space near the working area are more important than the infrequent longer access time. Besides, using 5S systems, access to the parts should still be as quick as possible. For example, do not lock cupboards unless there are specific safety or security reasons. If a cupboard must have a door, could it be made of a see-through material like Perspex? If the door must be locked, could the operator have quick access to a key? Better still, a combination lock or swipe card access should be used.

The most contentious 5S feature is the use of shadow boards. Everyone, almost to a man, says they will never work "here." The reason is always the same: "All the tools will be stolen." My personal experience has been the opposite. In one case, I was losing tools from a laboratory in a university. They were stored in a series of drawers. Eventually, I made a bold, high contrast shadow board and, as far as I can remember, never lost another tool. This was in an area where anyone could be passing by outside the class and the class was frequently unlocked. The boards have worked well in a number of factories, but I imagine there has to be some factories where people will steal the tools. If the tools are not being stolen but disappear, then are they being used by other employees in the performance of their jobs? Is this a symptom of another issue? Perhaps an inadequate supply of tools?

What about issues with finding a part, selecting it, controlling the stock numbers, and if it is a tool will we have any difficulties when we return it? The objective is to store the components in an arrangement so that they can be easily seen by anyone. For example, where possible, do not use drawers: use a box with a cutaway front for viewing or have the box sitting on a sloping shelf rather than a flat shelf arrangement. The slope allows the user to see inside the box without having to peer into it. This is illustrated in Fig. 6.7.

The simple act of changing the shelf angle (front to rear) from level to a slope permits a visual display of the quantity available inside the box from a reasonable distance. It will be necessary to fit a lip to the front of the shelf to prevent the items from sliding off. Perhaps, although a less desirable alternative, mirrors could be positioned to see inside larger containers.

An improvement to the system would be a visual display on the base of the box that signals how many have been used or if new stock is required. Figure 6.7 shows that when the dark gray outline appears, it is time to replenish the stock. This will depend on the use of a first in first out system of use.

Imagine that the spare parts were located inside drawers or cupboards. Take a look around you. What is in the drawers and cupboards you see? How could you know what was stored in each one? Even if the drawers were labeled, how long would it take to find any given object, open the drawers, and count them? 5S is designed to make things more efficient, not less.

The box stored flat does not allow the contents to be visible without looking directly into it

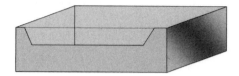

The box stored at an angle, higher at the rear makes viewing easier. Marking positions on the base clearly shows that there are four used

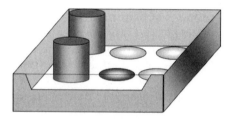

Figure 6.7 The best storage should display the contents.

If, as shown in Fig. 6.7, six components are used in one day, then this is the ideal number that should be available, not multiples of six. The only time more would be acceptable would be if topping up as required creates a significant issue. Buying excess numbers of components costs money and requires excess storage space. The parts should also be arranged so that they are used in order of replacement. Always use a first in first out sequence.

- Part numbers should be clearly visible.
  In Fig. 6.8, part numbers 00124 and 00125 have a minimum quantity of one. It is written below the item. Systems have to be set up to establish how the levels are controlled. Who organizes the reordering? How are the parts ordered? What about the other parts that have no minimums? The color of the outlined part could also be a warning that more parts are needed.
- The minimum stock level should be obvious.
  In Fig. 6.8, could the minimum be displayed in a clearer manner? Consider
    ❖ A shaded area outside of the part's shape
    ❖ A red outline

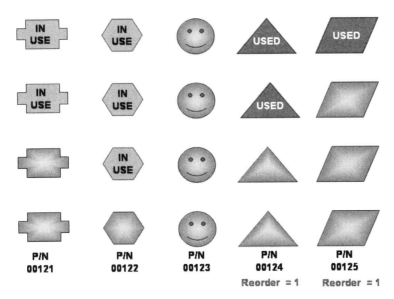

Figure 6.8 The shape of the parts can be drawn on the shelf so that it is visible when the part is removed.

- ❖ The word "Reorder" written within the outline.
- ❖ Or all of them together.
- ⚹ Only store enough parts on the production floor that are needed to carry out a fixed amount of work.
  If too many parts are out, this should also be obvious and avoided. Let the space limit the number. Where components are stacked on a shelf, there should be a marker that shows the minimum and maximum allowable quantities. This can be as simple as a red and a green line marked on the side of the shelf or compartment.
- ⚹ There should be a fail-safe mechanism for topping up.
  If they spill out of their allocated space or rise above the upper marker, it should be easily seen.
  A Japanese Kanban system could be used. If, say, two trays are used, when one becomes empty, it can be returned to stores. The arrival of the tray is the signal that more material is required. The tray is then refilled and returned to the floor.
- ⚹ Draw a spaghetti map (Figs. 6.9 and 6.10) to study the route(s) the operator takes when collecting parts. (Explained in the next paragraph.)
  - ❖ Do they collect one part at a time?
  - ❖ Could they realistically and safely collect more?

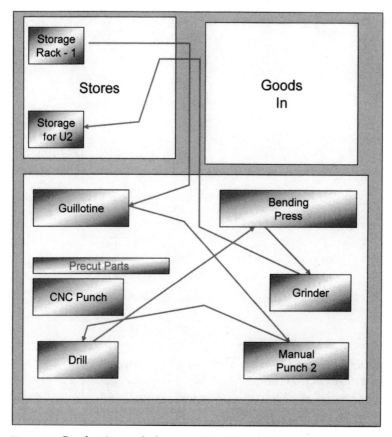

Figure 6.9 Spaghetti map—before organizing the tools in sequence.

❖ Can the parts be preassembled in any way—offline—before the operator collects them? Turnaround assemblies have positive advantages.

A spaghetti map is a floor layout plan that is overlaid with the routes taken by an operator or by the product. The intention is to identify unnecessary and wasted effort. This can take the form of walking extra distances, doubling back, picking up the biggest items first, and so on. Even supermarkets do not let the customer pick up big items too soon, but for different reasons: they fill up the trolley and subliminally limit how much the customer might buy. Watch the people who collect the parts. Does it flow or look awkward. Do the operators need to strain or bend to collect them? Are heavy parts collected from high shelves or are they sitting on transport trolleys. Consult with the people who collect the parts to find out how they would improve the system. Then,

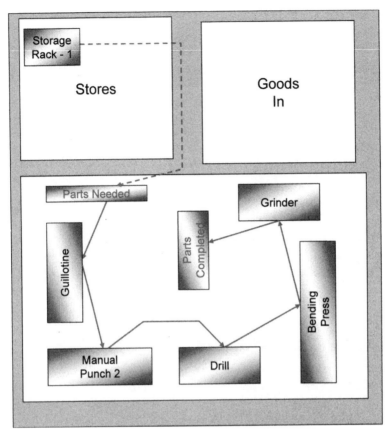

Figure 6.10 Spaghetti map—after rearranging the tools. (Notice the smooth flow.)

reorganize the layout on another sheet of paper trying to minimize the track lengths and make it more efficient. Keep It Safe and Simple (KISS). Then, test your ideas ... Plan-Do-Check-Act.

Using spaghetti maps can also highlight potential process failures. One possible consequence of a poor layout is the possibility of a part being taken to a wrong work station and, inadvertently, being put through the wrong process. One option to avoid costly processing errors is to use the colored line systems, common in American hospitals, to direct visitors to different departments. Paint color-coded lines on the floor to guide the process routes. Beware, too many tracks could become a problem in its own right! Other color-coding systems could also be considered.

5S uses floor marking to highlight a number of situations. Many items will be too large to locate in cupboards or on shelves. These usually end

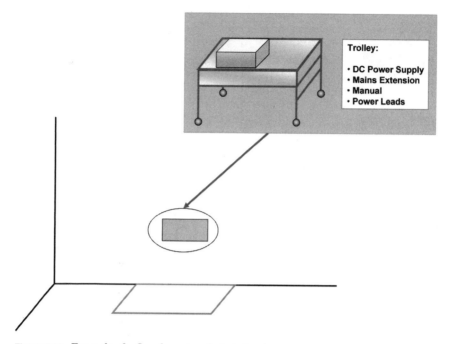

Figure 6.11 Example of a floor layout and photo locator.

up being stored on the floor, in a corner of the last place they were used. This is the best example of the opposite of the practice of 5S. It can often take ages to find these parts when you need them. (Unless, of course, you know where it was last used.) The solution is all too simple. Give it a place to live: a parking bay (see Fig. 6.11). Outline the floor area and clearly identify it with the part that now lives there. Now, it can be easily seen to be missing. If the part was, say, a vacuum cleaner or a portable test unit, the accessories could be "shadow board" mounted on the wall next to the unit. Then it can be seen if any parts have gone missing.

As discussed, a painted rectangle on the floor would have no meaning to anyone, other than to say *something is missing* or in use. So a photograph or drawing should be fixed to the wall to identify what the part should be. The detail should be enough to show any attachments, special tools, etc. and be supplemented by a list of items.

Floor markings can also be used to define

➢ Work areas
➢ Walkways
➢ Fork lift truck paths
➢ Keep Off areas

Seiketsu: Standardization

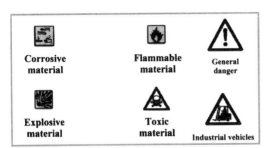

Hazardous Substances   Warnings

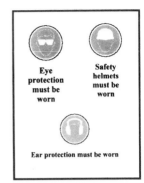

Mandatory

Prohibition

Dangerous Areas

Safe

Figure 6.12 Standard safety warnings that would have their own designated colors (e.g., green for safe and red for danger). It is advised to check the correct colors with your own safety department or (safety) government agency.

Initially, the floor layout markings should be made using sticky tape, as it is easy to reposition, should there turn out to be a better place for it. Also, before painting any permanent marks on the floor, it is a good idea to ensure the floor is suitable for the task and is in good condition. Take care with your color choices as many have accepted meanings like green for safe and red for danger. If you have multiple sites, standardize the color coding (Fig. 6.12).

	Team Actions
6-21	▰ Draw a map of the area subject to 5S improvement, showing the current positions of components and parts.
	▰ Find out how the parts are used by the operators.
6-22	▰ Draw a "spaghetti map" or maps of the operators' routes from their work areas to collect the parts and return.
6-23	▰ As a team, consider better functional arrangements. Discuss ideas with the operators, managers, engineers, and production staff.

> 6-24 🕮 Carry out a risk assessment of the hazards (if any) that might exist when rearranging the area or tools within the areas. Remember to include lifting weights, using ladders, cuts, slips, ingestion of materials, or eye damage through vapors or dust.
> 6-25 🕮 Test the selected new arrangements.
> See Fig. 6.13.
> 6-26 🕮 Improve positioning and visibility of the parts in their locations.
> 6-27 🕮 Check the condition of the floors
> 🕮 Cost and prioritize repairs
> 🕮 Test some storage markings/methods for large items.
> 6-28 🕮 Refine the setup.
> 6-29 🕮 Set up an activity board.
> (Chap. 12)

Step 3: Seiso—Shine

This step is the "spring cleaning" step normally associated with 5S by people who do not understand or appreciate the simple power of the process. As the application of 5S progresses, the time spent on physical cleaning will reduce. This is due to the fact that active plans for improving all sources of contamination found will be initiated and the problems resolved. This makes the starting point cleaner each time. Also, they will stay clean as permanent cleaning routines will be

Plan-Do-Check-Act Cycle

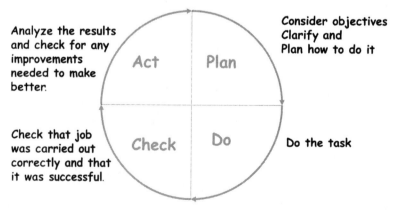

Figure 6.13 Plan-Do-Check-Act cycle.

developed by the teams. Graphing the cleaning times is a good measure of the overall improvement of the area.

Remember, the purpose of 5S is not to have to clean, but to stop the area from getting dirty in the first place by eliminating the reasons and sources of the dirt.

There are several benefits of routine cleaning cycle:

> To improve safety by reducing the possibility of accidents.
> To clean all of the area and make it suitable to be inspected for defects and show up new and recurring issues.
  All defects are recorded on a spreadsheet with the actions taken.
> To clean production tools, with a view to preventing damage to components that could be caused by rust, abrasion, or chemical contamination and affect the lifetime of the product.
> To create an environment that will give new and existing customers a positive impression of the high standard and quality of the product.
> To create an environment that the employees will be proud to work in.

The 5S cleaning map or assignment map

The responsibility for cleaning the areas will be evenly divided between the team members. In Fig. 6.14 the "blocks" have been allocated a person responsible for cleaning. In a real map, the areas would have equipment names or other location IDs to which the responsibilities would refer. These areas would also be the areas defined on the 5S "Clean and Inspect Checklist," which will be explained next.

Each tool can have its own "cleaning map" or assignment map that is broken down into functional areas. The Clean and Inspect Checklist will define all the tasks that need to be performed. Cleaning maps are also used by AM and PM teams. For more details refer to Chap. 2.

All of the tasks to be performed will be recorded on the Clean and Inspect Checklist illustrated in Fig. 6.15. The cleaning and the checks will be broken down into discrete tasks and listed. These checklists will be treated like maintenance schedules. The list will identify the areas, the tasks, the person who is responsible, and their frequency. Once they have been optimized, or are close, they will be added to the maintenance scheduling system.

Figure 6.16 is an audit sheet. Each "S" is given a score and then totaled to give an overall area mark. There is a task number column to identify areas that must be revisited for improvement.

The 5S teams must be trained in all aspects of any tasks they will encounter during the cleaning. They must be trained in simple risk assessment and safety procedures. If they are to clean equipment, they

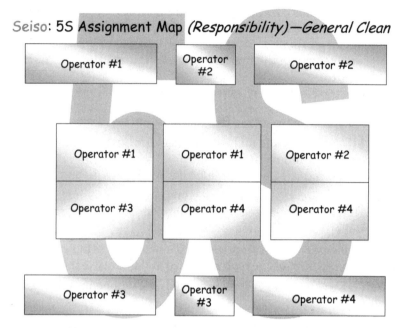

Figure 6.14 Alternative style cleaning map.

must know how to make it safe or have a procedure that checks it has been made safe for them. If equipment or chemicals are to be used for cleaning, again, they must know how to use the equipment and any hazards posed—no matter how slight. The team members must be tested to ensure they understand the hazards and equipment to be used. Training must be arranged in advance of the step.

The same 5S principles can be applied anywhere: to office areas, parts stores, roof spaces, utilities rooms, workshops, and even courtyards. Even if you have not applied them in a new area, like an office, consider what we are trying to do and look for ways of applying them. Every time improvements are made or graphs are updated, the information must be transferred to the activity boards to highlight the success of the team. The activity board is the means of promoting the team's progress, but remember, the purpose of 5S is not to create a board but to improve the efficiency of an area.

Step 4: Seiketsu—Standardization

One of the major failings in industry can be found everywhere, it has no borders. It applies to manufacturing, maintenance, design, facilities, purchasing, accounts, wages, stores, and process development.

5S: Organization and Improvements by Default 177

5S Task Number	Area	Task	Responsibility	As Required	Each Shift	Daily	Frequency in Days if <7	Weekly	Frequency in Weeks if <4	Period (Months)	Other
1	1	Dust terminal exterior	Jim					X			
2	2	Inspect oil tray	Ruth				3				
3	2	Check pump level and report if filling is required	Ruth				3				
4	1	Hoover floor	Mary	X							
5	4	Clean insulator	June					X			
6	4	Check/clean floor under source for signs of water leaks	June					X			
7	3	Record HCIG #1	Mary		X						
8	4	Check all screws are in place on rear panel	Mary					X			
9	2	Clean and inspect floor around Pump #1	Jim				4				
10	2	Clean and inspect floor around Pump #2	Jim				4				
11											
12											
13											
14											
15											
16											
17											
18											
19											
20											

Audit Date: TOOL ID: Auditor: ESTIMATED REPAIR FREQUENCY

Figure 6.15 Clean and Inspect Checklist.

Audit Date:			Auditor:						
TOOL ID:				AREA AUDIT SCORE					
5S Task Number	Area	Task	Responsibility	Sort	Set in Order	Shine	Standards	Sustain	Total Score
1									
2									
3									
4									
5									
6									
7									
8									
9									
10									
11									
12									
13									
14									
15									
16									
17									
18									
19									
20									

Figure 6.16 5S area audit sheet.

	Team Action
6-30	❧ Carry out a risk assessment of the hazards (if any) that might exist when cleaning the area or tools within the areas. (Chap. 5)
6-31	❧ Define the cleaning materials and method for the area. Include any safety implications and protective clothing.
6-32	❧ Arrange and carry out safety training and/or equipment training necessary to make the area/tool cleaning safe. ❧ Ensure the team members understand all the training.
6-33	❧ Create a 5S assignment map. (Fig. 6.14)
6-34	❧ Create a Clean and Inspect Checklist. (Fig. 6.15)
6-35	❧ Set up a cleaning schedule. Ideally make it formal like a maintenance period.
6-36	❧ Take "before" photographs of the area.
6-37	❧ Every time an area or section within an area is cleaned, time how long it takes.
6-38	❧ Chart the cleaning times.
6-39	❧ Estimate the cost of any repairs or improvements Ideally cost will not have a bearing, but that is not realistic. ❧ Get authority for repairs. ❧ Prioritize and arrange for repairs to any sources of contamination. ❧ Pursue any repairs not carried out.

This shortcoming is a lack of standardization. If everyone carries out the "same" job a different way, there will be no standard product at the end. I have seen 10 highly skilled technicians each doing the same task and most of them doing it their own way. The problem is, their way includes their own little tricks, picked up over the years. This leads us to problem number 2: that the completed part does not always work the same way, stay working for the same time, or, in some cases, work at all. The only way to guarantee a standard outcome and attain the expected reliability is to ensure that everyone carries out exactly the same steps, using the same tools and parts, assembling and testing them in the same order every time.

Let's take the simplified example of heating something to a temperature of 70°C in an oven. What could be easier? I mean, no one ever has problems with ready prepared meals ... do they? We are immediately faced with the questions, should the item to be heated be placed at the top, the middle, or the bottom of the oven? Does it make any difference where it is positioned? What if the oven has a heat circulating fan or is a vacuum oven? Would the temperature be the same if a different model of oven or manufacturer of oven was used? What if the product sat in a different tray?

Standardization requires a documented procedure that specifies all of the details so that the correct oven is specified. It should have a photograph. The actual oven will also be labeled with a unique ID, so it cannot be mistaken. The documentation complexity should reflect the complexity of the tasks. This does not mean it should be difficult to read or understand. It could be made up of two parts: have a simplified summary for those who need only use the procedures to remind them of the details but also have a more complete section for the less experienced to follow or the more experienced to reference when required. Pictures should always be used where possible to eliminate mistakes. Words are used to back up the pictures not the reverse. The correct trays will be illustrated; the actual trays should be marked. Any gauges will be modified to show minimum or maximum levels and set points. Any sources of variation that might potentially become the cause of problems will be identified and actions will then be taken to eliminate them or minimize their potential for damaging the product.

In 5S, even cleaning and inspection tasks must be carried out to the same standard. Hoping to make the task easier, people often ask, "How clean is clean?" (In case you are interested, the white glove test is the answer to that. Wipe the surface with a white glove, any dirt is too much.) The same initial cleaning frequencies should be used by everyone— unless there is a good reason to have a more frequent test! The same format should be used for forms, tags, labels, signs, floor markings, color codes, and activity boards. The standard used should be the "best practice" based on the results of *all* the teams.

The only way to guarantee that a task is carried out the same way is to document the method as it should be carried out and then train everyone in the method and police its adherence. The audit sheet in Fig. 6.16 is a perfect mechanism for this. The teams will carry out their own audits to the required standards but there should also be one or two general auditors who oversee all the areas and set the benchmark standards. Their purpose is to ensure that every team works to the same, "best practice," standard.

	Actions
6-40	≳ The auditors should review the results of their surveys and identify any teams that are performing better or more effectively than the others. ≳ The auditors should investigate the reasons for the improvements. ≳ They should also encourage the teams to find recommended improvements.
6-41	≳ The improvements should be analyzed and then disseminated as training to all the teams so that they can adopt the new standard method.
6-42	≳ A standard method for writing/creating procedures should be adopted. This must be used by all of the teams. (See Chap. 5)
6-43	≳ The improved method chosen in Step 6-42 should be documented as a procedure. The detail should be such that another team could be trained using the documents and safely carry out the tasks.
6-44	≳ The teams should update their documentation and monitor to ensure that the new standard is, in fact, an improvement.

Step 5: Shitsuke—Self-Discipline

If 5S, or any improvement program, is regarded as an "extra" or an "add-on," then it will only be carried out when all other duties are completed or when they are told to do it. When implementing 5S tasks, they must be treated as though they were everyday, routine production tasks. How do you do this?

Maintenance routines are developed as a schedule and displayed on a timetable or calendar. Often the PMs are planned at regular intervals or sometimes within a window in production, but they can be exact, for example, after a number of operating hours for the tool. PMs are not an extra. If they are not done, the equipment will deteriorate and eventually fail. So, maintenance must be carried out as scheduled and only be postponed when it is absolutely necessary. If a PM *has* to be postponed, the consequences must always be considered.

5S must be treated the same way, but possibly to a lesser degree. Production needs can indulge in the occasional postponement or cancelation, but it does not take many postponements for the teams to follow the "example" and start postponing routines on their own. Soon, the postponements become the norm and then the clean and inspect routines finally fade into oblivion.

The best way to overcome this problem is to prevent it from happening in the first place. The management team should set a reasonable protocol for cancelation that has to be followed. Imagine the situation where you were the manager responsible for production, but had no responsibility for 5S—Are you going to choose to fail in your responsibility or postpone 5S? Tricky decision—not! Perhaps a management team decision can be made, weighted on the number of times an inspection has been missed. It is important to remember that the interval between routines is based on the time it takes for the system to get dirty again. If you can afford to miss it two or three times in a row and show no issues, the frequency should be reviewed with a view to increasing the interval.

The 5S projects must be regularly audited to ensure that the same standard is maintained. This includes the standard, and regular updating, of the activity boards. These will show a failure of the 5S exercise and the cost will be an increased number of labor-hours to bring the system back to the way it should be or, in the worst case, we could move back to the potential for contaminated product. The concept of 5S is ongoing, so the team should be thinking about the next improvements or any steps that have failed and need to be improved.

The discipline extends to safety. Teams should always observe safety rules, wear any protective clothing, and check on risk assessments. They should also look for abuses in their own areas.

Chapter 7

SMED—Single Minute Exchange of Die

Where Did SMED Originate?

Taiichi Ohno, the famous Toyota President, was unhappy making cars when he knew he had no customers to buy them. He found it very difficult to reconcile spending money just so the cars could be driven from the factory and parked in fields. This reaction is a clue to one of his major attributes: he questioned everything. It was normal at the time to make cars and store them. Often companies made hundreds of cars. So, what was the logic behind this overproduction? When a car manufacturer changes from making one model of car to another, all the tooling and presses have to be changed to accommodate the parts required. This "changeover" eats into production time and is a drain on manpower. The mass(ive) production is carried out to minimize this loss of production time.

Let me explain.

Consider a company making only one product. If it can make one unit per hour, then in 4 days it can produce 96 units. The product pattern can be seen in Fig. 7.1. This production rate would continue until there is an issue that stops the line. To keep the example simple, we will now imagine that the same company has introduced a new product and now makes two products that it sells in approximately equal amounts. Following the car industry's old system, let us assume it makes lots of one product before changing over to the other one, say it makes 4 days worth. This way, the 4-h change happens only once every 4 days. The production loss is about 4 percent.

If the company now chooses to change after only 24 h and not 96 h, and the changeover still takes 4 h, the production time lost over the 4 days

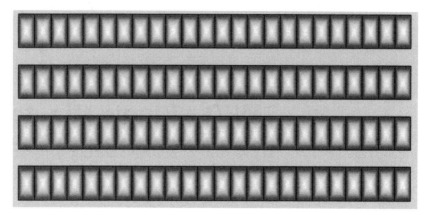

Figure 7.1 Four production days, each divided into 24 h.

increases to 12 h. The downtime pattern can be seen in Fig. 7.2. It seems that all we have done is create more downtime. So, the conclusion is obvious: there is no advantage in changing over sooner. Or is there...?

Let's revisit Taiichi Ohno's ideas. Basically, he studied what was being done from the point of view of the *customer* and set out to eliminate any procedures that did not have benefit to the customer. He started to look for *muda*. We would call it waste. This waste is not rubbish, or not only rubbish; it is anything that can be regarded as being "non-value-added" steps as would be seen from the customer's perspective. Table 7.1 could be such a review.

The immediately obvious part of Table 7.1 is that it doesn't seem to have a lot of points on the left side of the table: the positive side. So, if the smaller batches are better, what can we do to reduce the damage

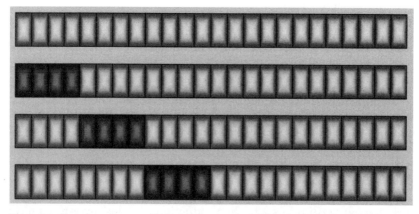

Figure 7.2 Four days, two products, 4-h change time. This time the changeover is 24 h each.

TABLE 7.1 The Pros and Cons of Large Batch Production

Positives (value-added)	Negatives (non-value-added)
You can save money on changeover times.	In order to minimize lost production time, he has to make extra cars that have no buyers—yet. Sales numbers have to be predicted.
You have lots of the cars already made.	Extra money is spent on parts and materials. Not recovered until the car is sold.
	Extra money is spent on facilities like electricity, water, heating, etc. Increases overheads.
	Extra storage space, which has to be rented, heated, and so on. Increases overheads.
	No income until the car is sold. Reduced bank balance.
	Perhaps have to sell off the car at a reduced price. Loss of profit.
	Possibly have to retrofit the car before selling. Extra expenditure on parts, facilities, manpower, workspaces, test equipment, and quality checks. Loss of profit.
	Increases the product "lead time." (That is how long the customer has to wait for the car he actually wants to buy.)
	Directly affects customer. Increases the payback time for making the car.

caused by the changes? What did Ohno San do? He selected one long changeover, one that lasted around 4 h, and decided he wanted the changeover time cut by 50%. The task was given to Shigeo Shingo, who went off and worked on the problem. When he had succeeded (not if), he was thanked and then asked by Taiichi to go and do it again ... and then again .... By the time he was finished, he had the time down to less than 10 min: that is in the range 1 to 9, or a *single* number of *minutes*. This became the *Single Minute Exchange of Die* or SMED as we all know it today. It is also known as Quick Changeover methodology.

Virtually every production system can be reviewed using SMED techniques, as can any maintenance task or series of operations. In addition to converting on-line tasks to off-line tasks, SMED looks for muda in the form of extra steps, wrong steps, steps that can be simplified using jigs, steps that can be prepared in advance, steps that can be combined, and, by using brainstorming, looking for better steps.

The analysis in Fig. 7.3 has seven basic steps. The first steps are based, like all continuous improvement techniques, on selecting a team and either a tool or an area to improve. The next steps are to study the existing procedure and gather data on the steps while carrying out the change. The final steps are applying the SMED techniques to reduce the total on-line changeover time. There is also a feedback loop. It is included to ensure that the SMED exercise is repeated regularly to continue to bring the changeover time down to a lower level.

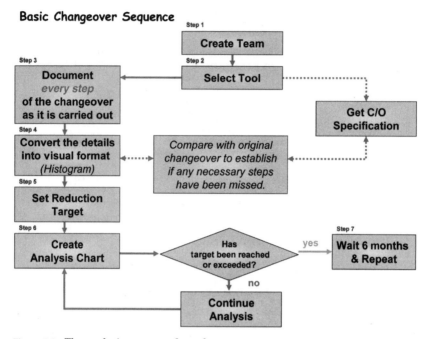

Figure 7.3 The analysis sequence for a changeover.

Step 1: Creating the SMED Team

Steps 1 and 2 are closely linked. The ideal team will be trained in SMED methodology before the team is formed or it will carry out the SMED routine as part of a training course. It will contain operators, equipment engineers/technicians, production engineers, and managers. At least some of the team members must have knowledge of the tool and be capable of carrying out the changeover.

Other useful team members are health and safety advisors, to ensure any more complex changes recommended are safe, equipment vendors, who might be aware of more up-to-date methods and parts, and trainers/facilitators, who know the changeover system and can maintain the momentum of the analysis.

The team members and their responsibilities

The team leader. The team leader will drive the progress of the team, keep the records, and share out the workload. He can be a working member of the team, but does not have to be. It is also the responsibility of the leader to organize meetings and equipment time for changes and modifications.

The facilitator. The facilitator must be SMED-trained and will lead the team through the various stages. He need not have any experience of the specific tool, but should be familiar with faultfinding and brainstorming techniques. The role can be rotated among all of the team members.

Operators/equipment engineers/technicians/process engineers. We need members to assign a range of essential tasks:

- Carry out the changeover.
  The changeover must be carried out exactly as it would any other time. Do not make any changes.
- Videotape the procedure.
  Once the video is started, do not turn it off until the procedure is over. If the battery or tale needs to be changed, stop the procedure, make the change, and restart. The video can be used as a time check.
- Take photographs—to be used in the new procedure.
  This might be difficult as the photographer must not inhibit the change.
- Record the process elements and microelements.
  This should be carried out in parallel by more than one person. This allows for steps being missed.
- Create the spreadsheets.
- Make parts for the solution.
  This will not be an immediate task. The change may have to be designed, justified on a cost/return basis, and tested.
- Write the new standard procedure.
  The procedure will be developed as described in Chap. 5.

Managers. The managers must understand the SMED process and its value to production. To be capable of this they will have to be trained. They will not require the full course, but it should be enough to convince them that the process really does work. It is also the responsibility of the manager to help clear roadblocks for the team.

Step 2: Select the Tool

There is no point in selecting a tool that is easy. The tool chosen should have a difficult changeover and stand to gain significant benefits from a revised procedure. It is likely that the production managers and operators will already know which tool would be the most suitable for selection. In the event that the tool is not a simple choice, to help with the selection, look for

- The changeover that takes the longest time.
- A bottleneck route.
- A single route that needs to be on-line more than it is.
- A tool that is changed frequently, even if the time to carry out the changeover is not excessive, it could provide cumulative time benefits.
- A tool that has a recognized, wide variation in the time taken to carry out the task.
  The variation suggests either lack of skill, lack of organization, or unusual difficulties in the task structure.
- If the tool is to be selected for an SMED training course, it might be prudent to either select a tool that is scheduled for a changeover during the course or, even better, plan it well in advance to ensure that the best changeover is available at the time.

Reminder: If the tool selected has any hazardous steps, it is imperative that all safety considerations are made and that any changes do not introduce new safety hazards.

	Management Actions
7-1	☺ Decide on the implementation plan.
7-2	☺ List the changeovers and prioritize them in terms of benefits.
	☺ Select the managers to be trained.
7-3	☺ Identify the team.
7-4	☺ Organize the training.
7-5	☺ Purchase a digital camera and a video camera.

Step 3: Document Every Step of the Changeover

Table 7.2 lists some specific functions that must be carried out for the analysis. Every task has a series of actions or *elements* that need to be carried out in order to reach completion. Just as every word is made up from a series of letters, each element is made up from a series of smaller elements, known as sub- or microelements. Some elements take a long time and others are short. SMED requires that *every* action be recorded,

TABLE 7.2 The Elements* of Making a Cup of Tea

				Column #4	
Column #1	Column #2	Column #3	Parallel	Preparation	
Make a cup of tea		Get box of tea bags	Step 3-2	Yes	Get box of tea bags
		Remove two tea bags	Step 3-2	Yes	Remove two tea bags
		Place next to teapot	Step 3-2	Yes	Place next to teapot
		Empty teapot	Step 3-3	Yes	Empty teapot
		Turn on water tap	Step 3-4	Yes	Get milk from refrigerator
		Fill kettle	Step 3-4	Yes	*Add milk to cup*
	Heat water	*Boil water*	Step 3-5	Yes	Get sugar bowl from cupboard
		Heat pot with hot water	Step 3-6	Yes	Get spoon from drawer
	Insert two tea bags	*Insert two tea bags*	Step 3-6	Yes	*Add two spoons of sugar to cup*
	Fill teapot	*Fill teapot with water*	Step 1	Yes	Turn on water tap
		Leave for 5 min to brew	Step 2	Yes	Fill kettle
		Remove tea bags	Step 3-1	No	*Boil water*
	Pour tea into cup	*Pour tea into cup*	Step 4	No	Heat pot with hot water
		Replace teapot	Step 5	No	*Insert two tea bags*
		Put on tea cosy	Step 6	No	*Fill teapot with water*
	Add milk	Get milk from refrigerator	Step 7	No	Leave for 5 min to brew
		Add milk to cup	Step 8	No	Remove tea bags
		Get sugar bowl from cupboard	Step 9	No	*Pour tea into cup*
		Get spoon from drawer	Step 10	No	Replace teapot
	Add sugar	*Add two spoons of sugar to cup*	Step 11	No	Put on tea cosy
	Stir tea	*Stir tea*	Step 12	No	*Stir tea*
Drink tea	*Drink tea*	*Drink tea*	Step 13	No	*Drink tea*

*The elements are in italic; the remaining are the microelements.

even if that action is waiting or looking for something. No, it is more than that: *particularly* if the action is waiting or looking for something!

I would imagine that a group of computer programmers would make the best team to carry out this kind of analysis. After all, a computer program is only a series of actions. Programmers understand the need to tell a computer absolutely every step it has to do, because if they don't the program will stop running. Sadly, most teams initially have difficulty with the concept of "every action," possibly because they are not aware that many of the actions actually exist, even when they have carried out the task before. The elements are easy, after all they are documented in the procedure (unless of course there is no procedure), but the microelements are less obvious. These are often the automatic and enabling actions carried out to help achieve the main steps—like using the clutch while changing gear. Microelements also overcome minor obstacles that get in the way.

If I asked you to "turn on the television," automatically you would locate its position in the room. You would stand up and move across the room, automatically avoiding all the furniture, find the switch, turn it on, and return to your seat, again avoiding all the furniture. But what would happen if you were blindfolded? Suddenly, the task becomes much more complex. The single instruction is totally inadequate. Even when you are allowed to receive instructions during the task, the detail required can be vast. We rely on our ability, or instinct, to fill the gaps by ourselves. SMED requires us to identify and record all actions, even the automatic ones.

If we were going to boil some water, what would we need to know? Here are a few questions that need to be asked:

- Where is the kettle?
- How do I get to it?
- How long does it take to get there?
- Do I have to search or wait for access?
- Where is the water tap?
- How do I get to it?
- How long does it take to get there?
- Do I have to search or wait for access?
- How much water do I need?
  (More water than necessary is "waste" in water cost, filling time, electricity charges, and heating time.)
- How long does it take to turn on the water tap and fill the kettle to the chosen level?

- Where is the socket for the kettle?
- How do I get to it?
- How long does it take to get there?
- Do I have to search or wait to get access to the socket?
- How long does it take to plug it in and switch the heater on?
- How long do I have to wait until the kettle boils?

The goal of SMED is to reduce time, effort, and waste; we must question the necessity of every step. Is "this" the best place to store the whatsit? Is the kettle always stored in this place? Why do we store it here? What if the tap was closer to the area the tea is made or if it had a better, faster flow? Is there a kettle that heats faster? Could I replace the kettle with a temperature controlled boiler? Could the boiler be self-filling? Basically we are looking for ways to make time work for us, not against us. (Table 7.3 includes more detail.)

Have you ever made a cup of tea in a friend's house? Did you notice how often it was necessary to ask where things were stored: the cups, tea spoons, sugar, tea bags, and so on? All of these points are mirrored in a factory environment, especially at tea breaks, and also on the production line. Before the days of SMED and 5S, changeover parts and tools were stored in the storeroom or wherever there was a space. Storage in a place appropriate to where and how they were used was never a consideration. The consequences of which include a *time* cost. SMED teaches the user to become aware of this cost. It also counters that if there is not a place close to where the parts are used, is there another better way? Could they, perhaps, sit on a cart, in a dedicated location, and be brought to the production tool *before* the change?

Another good example for identifying elements and microelements can be seen in the example of how to make a cup of tea. Figure 7.2 illustrates the tea-making task. It is a table format, divided into four main columns, used to compare the various levels of detail. Column 1 has very little information. Some people will only record to this level of detail until they have been trained to look deeper. Column 2 has only the *elements* listed. The elements would equate to the instructions in a simple procedure. Column 3 has more details: it includes many of the microelements. Column 4 has the same details as column 3, but the step order has been rearranged. It has also been further divided to include extra information for the purpose of illustration. "Preparation" steps are external and can be carried out at any time. The column "Parallel" refers to whether or not a task can be carried out at the same time as another. This might mean that extra manpower is required for some steps, but for SMED it is important that the primary objective is not lost—to keep the production lineup. One hour of labor will cost around $50. Compare this to the value of one hour of lost production downtime? More ... or less? Figure 7.4 shows the benefits of the optimum use of manpower in an imaginary nine-step change.

Did you notice that in Table 7.2 some steps required to make the cup of tea are missing. Either that or David Copperfield was making the tea and made the cup appear by magic? Also, there is no mention of where the tea maker had to go to collect the components needed to carry out the task. However, it is only an example. One other point to notice is that in Step 3-1, when the water is being boiled, we have a range of steps that can be carried out while we are waiting (Steps 3-2 to 3-6). These steps are the preparation for the cups and the teapot. There is, obviously, no need to wait for the water to boil before preparing them. It is interesting though that in a changeover, many people choose to wait for things to finish before proceeding.

In a real, first-time, changeover it might be necessary to wait for access to equipment, tools will need to be found, wrong tools will be selected, parts will be stored too far away, and nothing will be obvious or color-coded to simplify selection and assembly. In short, the 5S-type fixes that interact to speed up the changeover will not yet exist. Microelements will also be missed, but they can always be added later, when identified, or they will surface during the next analysis.

Figure 7.4 shows the advantage of using multiple personnel. The purpose could simply be to pass tools to an engineer or to work on another section at the same time. The costs can be evaluated to confirm that it is a gain, but it is very unlikely it will not be. Also consider the possibility of a semiskilled person providing the support. If the tasks were not carried out as shown in Fig. 7.4, the overall time would increase by the sum of the times for Steps 4, 6, 7, and 8.

There was a situation where the operator had to replace about 30 parts. The trolley had to be positioned about 15 ft from the workstation where they were needed. To carry out the exchange, the operator had to

1. Remove the current part,
2. Carry it to the trolley,
3. Find the new part from the pile on the trolley,
4. Walk the return distance while carrying the part,
5. Fit the new part,
6. Repeat Steps 1 to 5 at least 30 more times.

It was actually even more inefficient than it sounds. The operator also had to climb inside the tool to reach parts and crawl under barriers to

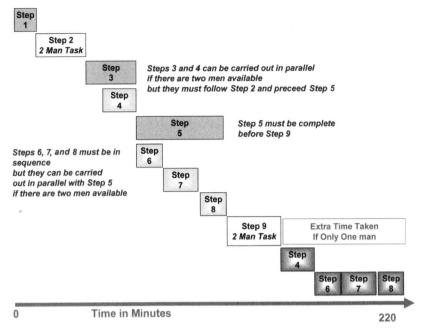

Figure 7.4 Parallel task allocation in SMED—shown as a project-plan.

get the job done. Had the parts been organized and there been a second person to hand over the parts, there could have been substantial time savings. But the saving does not end there; the actual job the operator had to do was very exhausting work. Getting help also made the job easier. What about the cost of the second person? Let's assume the production line makes around £5000 in an hour and we could have saved 15 min of the downtime. The saving would have been around £1250. The helper earns only £100 in an hour .... All right, so what if I am trying to make a point? The operator earns about £20 in an hour. So we win £1250 and lose £20 on every changeover. Even if there is only one changeover in a week, the saving is still more than £60,000 in a year. However, there will be more changeovers each week and the profit will likely be more than £5000, so the extra income is potentially pretty good. Oh yes, and the job was easier for the operator.

In the real analysis, we will use four methods for recording information:

❖ *Post-it Notes or a Notebook*
  If we use a notebook, the data can be recorded and transferred later. The detail can be improved as it is transferred or it can be reduced, if the notes are not clear enough. If Post-it Notes are used, which colour

is the preferred option?—I have been using yellow. This is the color used by Dave Hale of Proven Training, a company I have worked with as an associate. Although the color is not really significant, it is prudent to select one and standardize on it. Using a standard means that anyone looking at a chart will know what the Post-it means. The purpose of the Post-it Notes is to record *every action* as it is carried out.

- Each microelement is recorded on a separate Post-it sheet.
- The elements are recorded independently on the Observation Sheet (Fig. 7.5) before the analysis starts. If there is a standard, the elements can be taken from that. If there is no standard, the elements will have to be written down and checked.
- At the end of an element, signaled by the person completing the Observation Sheet, an "X" or an "E" for end is recorded on the corresponding Post-it. This breaks up the chain of microelements into elements in the same way a period breaks up a series of words into a sentence.

❖ *The Observation Sheet (Fig. 7.5)*
This is either a table that doubles as a horizontal bar chart or a horizontal bar chart with extra columns for *x*-axis descriptions. Either way, it is a very effective chart.

- Column 1 is the element (step number).
- Column 2 is a brief description of the purpose of the step and what is it trying to achieve.
- Column 3 is the cumulative time to the END of the task, recorded in minutes and seconds. A stop watch is used to record the time. The time is recorded at a point initiated by the operator carrying out the changeover.
- Column 4 is the actual time the task takes, calculated from the data and converted to seconds. This time is plotted on a horizontal bar graph. (See Step 4 and Fig. 7.5.)

❖ **Photographs**
To be used for details in the new standard procedure. These might have to be taken over time to avoid getting in the way during the SMED analysis. The spare parts can be photographed off-line.

❖ *Video*
The video camera is the most useful recording tool. It is used to record *every step* and allow playback to refine any microelements listed. It also gives a course time backup—so it should *never* be stopped or paused.
Consider an SMED analysis to speed up the battery or videotape change—almost a joke. Plan for the change and have all the parts.

SMED—Single Minute Exchange of Die 195

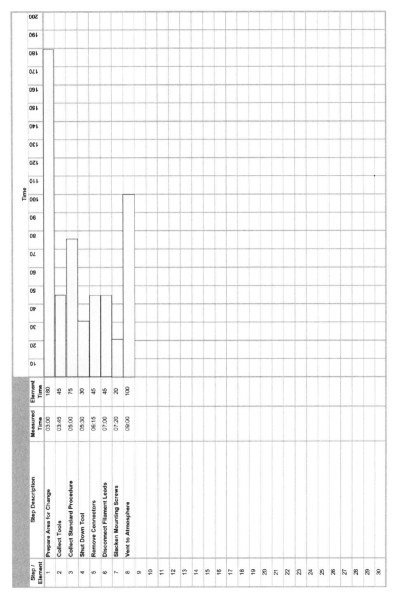

Figure 7.5 The Observation Sheet.

- The camera operator is not intended to be Spielberg, so don't worry about accurate framing.
  There might be a need to consider lighting, though. It would not be a bad idea to test it out before the real analysis.
- Never use the pause as it stops the timer.
- Try and get as much detail on the changeover as possible—not just the back of the engineer. Unless, of course, it is Angelina Jolie.
- Make sure the batteries are charged, or set up a power unit.

From the plotted element times you can create a visual display of the areas most likely to produce large savings. (The method is shown in Step 4.) Big bars equal big potential for savings. It is unlikely that the smallest bar (Element 4) will provide much room for improvement.

	Team Actions
7-6	☺ Order a selection of colored Post-it Notes. Yellow, pink, and blue.
7-7	☺ Order a stop watch.
7-8	☺ Create an Observation Sheet. (Fig. 7.5)
7-9	☺ Organize the production tool for analysis.
7-10	☺ Get the specification that defines the changeover.
7-11	☺ Compare the specification with the way the procedure is carried out in normal practice and define when to start and end recording.
7-12	☺ List the elements, in order, on the Observation Sheet.
7-13	☺ Assign team responsibilities for the changeover.
7-14	☺ As a team, carry out the changeover.

Step 4: Viewing the Changeover as a Bar Graph

The Observation Sheet (Fig. 7.5) not only records the data, but it also doubles as a sheet of graph paper and is used to display the changeover times. If desired, it can also be used to show the reductions as the analysis progresses. The element times are found by subtracting one element time from the next. For example, Element 3 minus Element 2 gives the time for Element 3.

Select a scale for the axis that allows for the longest element time.

Team Actions
7-15 ☺ Calculate the element times in seconds.
7-16 ☺ Select a scale for the Observation Sheet bar graph.
7-17 ☺ Plot the data.
7-18 ☺ Create an analysis chart. (Fig. 7.7)

Step 5: Define the Target Time for the Changeover

I am a great believer in targets; they increase a team's momentum. The reduction time for the analysis will normally be set by management. It should be at least 50 percent. If the reduction is too low it will not stretch the team.

Notice the reduction asks for 50 percent: but 50 percent of what? Is it the time the tool is off-line or when the engineers start preparing? Is it from the planned start time—should the time spent waiting by the team be counted? Remember, this is the first changeover and no allowances should be made. The routine should be carried out exactly the same way as it is always carried out.

How long should the team take to carry out the analysis, to define any changes, to design modifications, to write the new standard, and to evaluate costs and gains and implement any improvements? Excluding the changeover, it should take about 8 to 12 continuous hours to carry out an SMED analysis. Creating the standard could take around 4 h for the team to create a skeleton and up to 2 days for one member to turn it into a document, depending on the complexity involved. If the analysis time is split over a number of days, extra time will be required for refamiliarization, setting up, and getting the momentum back. Modifications that have to be machined could take a few weeks. It is difficult to decide how much to ask from the team. Be realistic, there is no point in setting up the team to fail. However, having said that, they will probably need your support and that of other senior managers to ensure time is made available for the team and funds are set aside for any changes that are valid.

Figure 7.6 is an illustration of a reduction plan. Basically the reduction assumes a linear reduction over time, from the initial time to, in this example, 50 percent of the initial time. If the analysis is carried out over one day, it is not really needed, but if over a number of weeks, it can be used to ensure the momentum is not being lost.

198   Chapter Seven

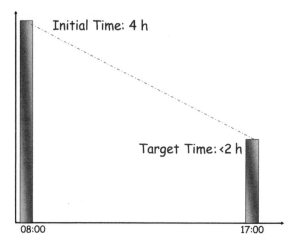

Figure 7.6 Example of a reduction plan: A standard bar graph with start and end points.

The bar graph is the only way to represent the plan. No other graph has as much impact. A target of 50 percent is a reasonable reduction for a first-time change. It is possible to get even more if the teams are motivated and have a successful brainstorming session. However, as the number of repeat analyses increases, the savings will reduce unless someone has an original idea that does not cost too much money to manufacture. While maintaining continuity of team membership, there are benefits to adding fresh minds and ideas to any repeat analyses.

	Team Actions
7-19	☺ Define the reduction target.
7-20	☺ Define the start and stop points for the analysis. It should be the line down time i.e. lost production time.
	☺ Define the time for the project.
7-21	☺ Plot a reduction plan.

Step 6: Analysis of the Elements

The analysis is carried out by the whole team with the facilitator as defined in Action 7-1, so a large visual display is desired which the whole team can view easily. The room should ideally be set up in a semicircle to provide an informal atmosphere. The analysis needs to define several parameters that are either printed on as large a sheet of paper as you can get, say, 5 ft by 4 ft or 6 ft by 5 ft, or they can be drawn on a large white board, provided it is clean enough to let the Post-it

Notes stick to it. Alternatively, the spreadsheet can be marked out on a table top using paper and pens or tape. The width of the cells should be sufficient to take one Post-it. Post-it Notes are favored in many analyses because they are easy to add, remove, and move about.

Figure 7.7 is a suggested spreadsheet format. The categories required are

➢ The original element/action.
  These are the main steps.
➢ The time taken to carry out the element/action.
➢ The microelements that make up the element.
  These are recorded on the Post-it Notes. The "X" or "E" defines the end of each element.
➢ External task times.
  We need to have a row to enable tracking the external task times. These can be carried out while the tool is running product. Tasks that are external can be organized in advance and have the capability of saving time.
➢ Internal task times.
  We need to have a row to enable tracking the internal task times. If it is an internal task, the tool must be off-line before the task can be carried out? These are the tasks to shorten or to try and convert into external or partially external tasks.
➢ What improvements and changes can be made to reduce the time?
➢ How much time the improvements have saved or will save when the modification is completed.
  This acts as a motivator.
➢ The new, reduced time for the element.
➢ The total time for the changeover at its current state of analysis.
  The final column identifies the time totals.

The elements can be identified on the top row. On the row below that we enter the element time. The time can be copied from the Observation Sheet. These are the times we are trying to reduce. Next, separate the microelement Post-it Notes into the chains that make up each element (they are separated by an "X" or an "E") and attach them as shown in Fig. 7.7.

If Post-it Notes are not used, the elements can be listed on flip charts that are divided into columns (see Table 7.3). This method is not as elegant as the analysis chart but it is workable. Have a look at the example. Notice that italic text has been used to highlight changes.

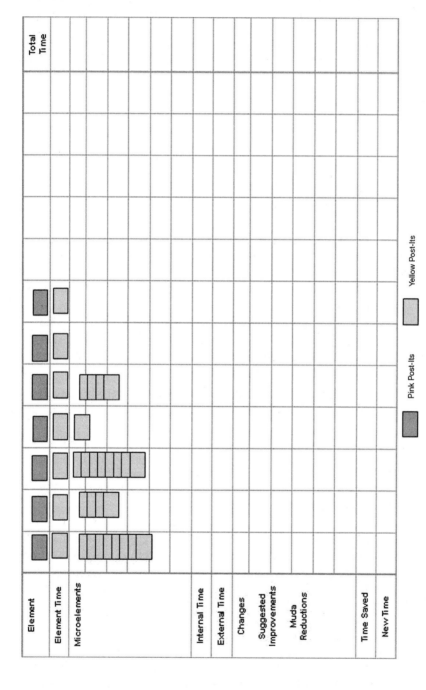

Figure 7.7 The SMED analysis chart.

TABLE 7.3 Flip Chart Analysis Example

Microelement	Change 1	Change 2	Change 3
Walk to kettle location –30 s	Walk to kettle location	Walk to kettle location	Walk to boiler location –30 s
Unplug kettle –3 s	Unplug kettle	*Use kettle that sits on autoconnect base – save 3 s*	
Carry kettle to water tap –10 s	*Locate kettle closer to tap – save 4 s*	Carry kettle to water tap	
Remove kettle lid –2 s	*Save 2 s*		*Use temperature-controlled water heater with tap*
Add water to desired level –7 s	*Use kettle with filling spout –7 s*	Add water to desired level	
Turn off tap –4 s	Turn off tap *save 2 s*	Turn off tap	
Replace kettle lid –2 s			
Return to kettle location –10 s	Return to kettle location	Return to kettle location	
Plug kettle into socket –3 s	Plug kettle into socket	Place kettle on base – *save 3 s*	
Turn electricity on/switch on kettle –1 s	Turn electricity on/switch on kettle	*Save 1 s*	
Wait for water to boil –180 s	Wait for water to boil	Wait for water to boil	
Turn off electricity/kettle –1 s	Turn off electricity/kettle	*Auto switch off – save 1 s*	
Remove teapot lid –2 s	Remove teapot lid	Remove teapot lid	Remove teapot lid –2 s
Pick up kettle –1 s	Pick up kettle	Pick up kettle	*Pick up teapot – 1 s*
Carry to teapot location –2 s	Carry to teapot location	Carry to teapot location	*Carry teapot to boiler –2 s*
Add hot water to teapot –8 s	Add hot water to teapot	Add hot water to teapot	Add hot water to teapot –8 s
Total Time = 266 s	**Total Time = 258 s**	**Total Time = 250 s**	**Total Time = 43 s**

Using this method leads to flip chart sheets being attached (in sequence) around the walls of the meeting room. If you have the option, use an analysis chart.

It is generally recognized that the Formula One pit teams are the best examples of SMED. They have changes down to fine art and still try to improve. How do they do it? Well, it is a bit like choreography but more of a ballet than a line dance. Everyone and everything is located his/its own designated place and the order the tasks are carried out is practiced until perfect and then practiced more. The tools are specially modified to minimize physical operations: jacks that rise to height in one operation. The wheels have only one nut to remove and a hub that speeds up repositioning. Parts are taken from the operator as he removes them and the new parts fed to him just as he needs them. The team members can communicate via headsets, but to maximize efficiency, they only talk when it is absolutely necessary. I am buying a similar helmet for my wife.

In the workplace, people rarely plan a changeover in advance. In fact, many companies do not even commit to a time when the tool will be returned to production. Figure 7.8 shows the pre- and post-SMED patterns of tasks. Compare the post-SMED pattern to the factory engineer changing the oil in his own car. He will buy the oil and a filter

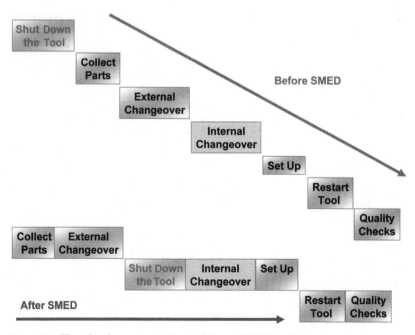

Figure 7.8 How the elements are changed by an SMED analysis.

before starting the job. He will even make sure he has the tools and drip tray to hand and, if not, he will get them too. He knows that after the job is started and the oil is drained, he will not be able to move the car until the job is complete. So there will be no easy way to get to the shop for anything he has missed. Besides, it would take much longer to get there and back by bus and he *is going* to use the car to go to the football in a couple of hours. Notice the "is going" is not a "hopes" to go.... This job has been planned and will be completed on time.

On completion, our engineer cleans any spilled oil from the inside of the engine and starts the car. The plan is to run it for 5 min and make sure there are no obvious leaks. After a couple of minutes he will put the tools away, check the engine once more for drips, and, if there are none, remove the drip tray and empty it into a waste drum. After the 5 min are up, he moves the car to a clean area, locks the doors, and has a cup of tea while he washes up. After all, he is confident all will be well. Finally, before he gets dressed, he nips out and has a quick look for oil on the road. No oil pool leads to a quick change of clothes and the application of his football supporter's kit—normally a top, a hat, and a scarf. Finally, he jumps into the car and drives forward 12 ft. Engine still running, he makes a quick, final check for oil on the road. Still no oil ... CD player on, inserts team CD and sings, "Here we go ... Here we go ... Here we go ...."

If we go back and look in more detail at how the engineer works in real life, in an average factory changeover, it is a wee bit different from changing his oil. It has pretty much the same issues as uncontrolled maintenance. This is a true changeover.

1. Step 1: shut down the tool.
2. The operator does not get the tools until he needs them.
3. Around 50 percent of the time the tools are missing and have to be found.
4. The changeover parts are stored, pretty much anywhere and they, too, can have missing parts that will not be noticed until they are needed.
5. Some parts might be damaged which will not be noticed until they are picked up and ready to be positioned.
6. The operator does not always follow the instructions because his way is better.
7. He cannot carry out any other tasks while he waits because the auto functions are faulty.
   The interlocks are broken, the software has bugs, or it is too slow.

Most drives and setup commands must be entered manually, so the operator is tied to the controller.
8. The quick release systems have deteriorated over time and many of the parts are not appropriate for the task.
A good example of this is saddle washers replaced by ordinary washers—requiring two hands to release the clamps.
9. A spare is needed, but he decides to wait and collect it just before or after the break—to save time.
10. The equipment has not been designed with fast changeovers as a priority.
11. Most of the bolts, nuts, screws, and fittings are a different size: no standardization.
12. The threads on the bolts and screws are longer than they need to be.

The SMED Analysis

When we review the microelements on the analysis chart, we consider each one in sequence to establish, as a team, if it is a valid step. Always ask the same questions.

List 7.1: Element Conditions

1. Why do we do this step?
2. Is the step actually needed?
3. Why do we do it at this point in the changeover?
4. Does it need to have this done now?
   What would happen if we did it at a different stage in the sequence? Would it be an improvement?
5. Why do we do it the way it is currently done?
6. Is there an easier way to do it?
7. Is there anything we can do to make it become an external element either completely or in part.
   (Can we precondition it? Preheat, precool, preclean, or preassemble it? Incorporate some sort of turnaround part?)
8. Can the time be reduced in any way?
9. Could this task be coupled with another task or be carried out in parallel with any other tasks?
10. Would two or more people doing the job make it faster?

SMED—Single Minute Exchange of Die 205

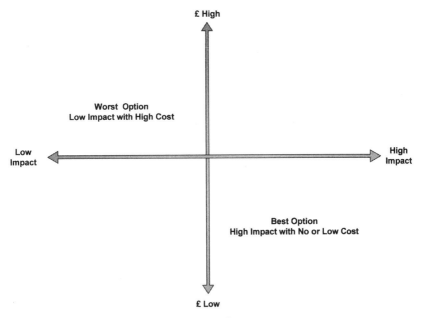

Figure 7.9 The cost vs. improvement impact chart.

There is a vast amount of time that can be saved simply by developing a system that utilizes advance planning. Just think about the oil change example. In fact, many of the techniques used in 5S (Chap. 6) will speed up a change. Brainstorm each element for improvements. Each idea must be recorded on a Post-it. Again, the color is not critical, but recently I have been using blue (Dave Hale again!) and changing the color too often confuses me.

The cost vs. improvement chart in Fig.7.9 will help prioritize which ideas to follow up on. The money pot has only a limited volume and it has to cover a lot of changeovers, so make them count. The best option, other than free, is a high impact with a low cost; the worst option is a low impact with a high cost. Mark the Post-it with a Priority 1 or Priority 4 depending on the outcome. I am more flexible with the middle levels, but quantify the costs and look at the ratios before committing or rejecting.

Brainstorming rules:

1. Every idea is a good idea.
   —Record it on a blue Post-it and stick it on the chart.

2. Each person makes one suggestion in rotation.
   This keeps everyone involved and sustains the momentum. "Pass" if you have no suggestion.

3. Never criticize.
   Criticism kills the mood and destroys the free flow of ideas.
4. Get as many suggestions/ideas as possible.
5. Do not interrupt the flow of suggestions.
6. Do not try and analyze, evaluate, or discuss ideas.
   This is to keep the idea stream flowing. Discuss the ideas later.
7. No comments—ideas only.
8. "Yes, but" is forbidden.
   This is a mood busting phrase.

Select a problem to brainstorm and be sure it is understood by everyone in the team. Often, writing the problem in words makes it easier to understand. There are scores of bullet points in this chapter that should act as triggers, so use them. Consider printing some of them on a chart and fix them to the wall as a guide. The very act of doing so will prime the team's imagination. Let the ideas flow and follow the rules. A negative environment inhibits the flow of ideas. When the ideas have run out, they can then be analyzed and prioritized into viable options.

A selection of improvement ideas are listed below:

➢ Use shadow boards for tools and parts.
➢ Color-code tools and fittings.
➢ Use trolleys to make kits with the parts arranged so that they are removed in the order of fitting and that can be wheeled to where they will be needed.
  Consider having the essential tools with the parts.
➢ Locate the parts at user-friendly heights to suit the task.
➢ Can any of the external elements be assembled in advance of the tool going off-line? Think of turnaround parts.
➢ Can any of the parts be tested/heated/purged before fitting to reduce the requalification time?
➢ Can any of the fittings be standardized or modified to reduce the number of different tools or fittings used?
➢ Are the threads on screws and bolts the optimal length.
➢ Are tools essential? Can wing nuts, clamps, twist locks, or quick connects be used?
➢ Can any adapters, bases, or flanges be designed to simplify the change?
➢ Can precision measurements be replaced by a setup jig or fitting?

- Can assemblies be modified to lock into the correct position without the need to check? For example, the use of spring-loaded pins or latching stops.
- Can an internal setup be reduced by having a turnaround unit on a special adapter frame? Think reel-to-reel tapes versus cassettes.
- Does the product need to be a different size? Is it possible that the design people can modify the size to make it the same—from a changeover point of view?
- Ensure there is a standard procedure.

Remember the other common failing we all exhibit from time to time. Make sure our improvements actually are better than the original systems. Follow the cycle:

- *Plan:* what you want to do and how to do it.
- *Do:* carry out your plan.
- *Check:* monitor the changes and make sure they work.
- *Act:* on the information gathered.
- Repeat the cycle until the changes work.

Never make changes without monitoring if the change makes things better or worse.

Figure 7.10 illustrates how to record ideas, denote internal and external elements, and display the time savings as the improvements are made.

Note any changes you make to the elements, particularly if they affect the sequence of the changeover. This will be necessary for the new procedure.

	Team Actions
7-22	☺ Populate the analysis chart. Elements, Element Times, and Microelements as shown in Fig. 7.7.
7-23	☺ Revise this chapter with the team. Print out a list of improvement ideas and fix on the wall. Print out the list of questions for reference and fix on the wall. (List 7.1)
7-24	☺ Split the Elements into Internal or External by moving the Element Times to the appropriate position as in Fig. 7.10.

7-25 ☺ Convert the Internal Elements to External Elements. Use brainstorming for ideas.
☺ Record all ideas on blue Post-it Notes and attach to analysis chart.
Fig. 7.10.
☺ Estimate the time saved (or will be saved).
Note on a Post-it and attach to the Time Saved and New Time positions on the analysis chart.
7-26 ☺ Reduce the time for all remaining Elements, including External.
Apply 5S methodology to improve the storage location, storage method, component identification, and component delivery.
☺ Estimate the time saved (or will be saved) for each element.
Note on a Post-it and attach to the Time Saved and New Time positions on the analysis chart.
7-27 ☺ Create a list of the final structure of the elements.

Implementing ideas

Once an idea has been approved, carry out a W3 (Who, What, and When) and record the information.

1. What is the improvement?
2. Who is going to do it?
3. When will it be completed?

Include cost of the improvement, the expected saving in time, and, if possible, as a cash equivalent based on production values.

Do not discard the rejected ideas. They might be helpful for future analyses as the SMED cycle is repeated.

Create the new procedure

When Action 7-26, reducing the time for all remaining elements, has been completed, it will be possible to create a new list containing all of the elements, as they will be used, with a summary of the ideas and improvements. This will become the skeleton of the new procedure. The team should collaborate in the creation of the new procedure.

We have already discussed how to write a comprehensive procedure that uses a table format to separate the steps from the illustrations and

SMED—Single Minute Exchange of Die 209

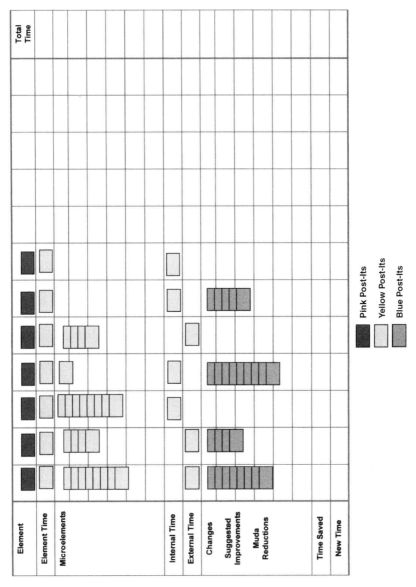

Figure 7.10 The SMED analysis chart with improvements and ideas.

secondary information. This can be found in Chap. 5. To some this might be too complicated, but I have seen procedures that are very Spartan, with totally inadequate information. The worst, when I was more into fixing cars, I remember finding while reading through a manual. It was describing how to remove a linkage. Step 1 was "Remove engine" and Step 2 was "Loosen the 15mm nut at the top of the spline." Not enough detail, would you not agree?

For people who prefer minimalist instructions, I would still recommend the procedure as written and explained in Chap. 5, but having a cover page that has the summary steps or a flowchart that can be used by the more experienced engineer. I say this with one condition: that adherence to the procedure is monitored. There should be no "skilled judgments" used in place of absolute measurements or jigs. If the changeover takes much longer than expected or fails after completion, the root cause of the delay or failure must be found. It is very important to remember that one of the most common reasons for failure is not following procedures and the most common offenders are those who *know* the procedure best.

The procedure can use the photographs taken during the changeover. Drawings can be used where no suitable shots are available and better photos can be taken at the next opportunity. Once the new procedure is complete it must be tested. The team can decide whether or not to circulate the procedure to the other engineers for review and comment until after testing. If it is decided to send it for review, take note of any suggestions and when all the suggestions have been returned, have a team meeting to discuss them. Not all suggestions will be improvements, so minute the meeting and keep a record of any reasons for inclusion or rejection. Never make the comments personal. If any amendments are made to the procedure, call it Revision 2 and send it back out for review with a copy of the meeting minutes. Test the procedure on a real changeover as soon as possible. Plan-Do-Check-Act.

	Team Actions
7-28	☺ Create a W3 for all approved ideas.
7-29	☺ Create the new procedure—Revision 1.
7-30	☺ Decide whether to circulate the procedure for review or wait until after it has been tested on a changeover.
7-31	☺ At the test changeover, take any necessary photos and account for any time used.
7-32	☺ Revise the procedure from the lessons learned during the test changeover—Revision 2.

Step 7: Repeating the Exercise

The final step in the SMED analysis is never really reached in theory. This is because the analysis should be repeated at fixed intervals of about 6 months, for an indefinite period. In reality though, it is likely that the changeover priority list created in Action 7-2 will be used to control the utilization of the SMED teams, more so in the early days. Each new procedure should be reassessed in relation to the list and reprioritized in accordance with the anticipated benefits. This should give you the best return for your efforts. Besides, if you have managed to get this far, you can use your new tea-making skill to brew a pot, sit back, relax... and prepare for the next emergency to arrive.

Applying SMED to Maintenance and the Use of Turnaround Parts

Here is my dilemma. I have a fixed maintenance budget and have to keep my spending below it. This PM task is the issue. I can do the maintenance in 2 h for £100 or I can do it in half the time for £500, provided I exchange the whole module. Should I spend the extra £400? The sweetener is that production can make £1000 worth of extra production in the 1 h saved. The obvious answer is I should do it the quick way, and get the maintenance budget increased from the cash saved. If only it was that simple.

We do have the option of applying SMED techniques to the PM task and tweaking tasks to suit basic equipment needs. We can also use RCM to refine the PM schedules developed and ensure their content is appropriate to failure cost and we also have 5S. We want to be sure that we have been balancing profit as well as minimizing costs and maximizing production.

Scheduled *restoration* and scheduled *discard* are often selected for reasons other than the cost of the repair. Discard, which replaces entire modules, is more expensive than stripping apart and servicing. Often it is the only option, when the part is not easily serviceable on site. But, even where the part can be serviced, modular replacement is usually the most reliable and is the quickest way to get the tool—and production— back on line. It is relatively easy to calculate whether the *profit* value of the extra product made during the recovered time is greater than the increased cost of using scheduled discard. The higher is the value of the product (like microprocessors or beer), the higher the profit. Even better, the benefits of using discard in preference to restoration do not end here. TPM and RCM have both shown the number of failures following a PM is linked to procedures and skill level. Do you remember the TPM failure categories?

212 Chapter Seven

- ⊗ Basic condition neglect
- ⊗ Operating standards not followed
- ⊗ Unchecked deterioration
- ⊗ Inadequate skill level
- ⊗ Design weakness

Modular replacement usually provides increased reliability because it limits the impact of *operating standards not followed* and *inadequate skill level*.

SMED is a favorite ingredient of lean manufacturing. It is also a part of TPM in another guise. The main objective of an analysis is to minimize the time a tool is *not capable* of running production, usually during a setup change from one product to another, but it works in other procedures too. It uses two primary techniques:

1. By manipulating the task "categories" it identifies which tasks actually prevent the machine from running product and which ones can be carried out while production is running.

2. By reviewing every action at the smallest, practical level, it identifies which are essential, which can be improved, and which ones can be removed.

SMED defines the two categories referred to in bullet "1" above as *external* and *internal*. External tasks are the ones that can be carried out while a machine is running production. Internal tasks can only be carried out while the machine is shut down or unavailable for production.

In the first few paragraphs of this chapter, we reintroduced the advantages of scheduled *discard* over scheduled *restoration* from the perspective of minimizing lost production time. Now that we appreciate that the purpose of SMED is basically the same, we can see how the discard option might be useful. If we could only find a way to reduce the extra cost in using discard . . . .

Well . . . we already have. In Chap. 10 we compared the off-line time savings made by an *electron shower* using a third option, turnaround parts. They are excellent examples of the SMED technique of transferring *internal* work that shuts down production into *external* work that does not.

Chapter

8

Deciding on a Maintenance Strategy

The PM Analysis: TPM or RCM?

I would bet that your current maintenance system was developed from the original schedules recommended by the equipment manufacturer. It is probably a reasonable good first step in setting up a maintenance program, but the thing is the manufacturer's schedule was not developed for you. It is a generic plan that will probably cover only the essential, more obvious components and the consumable parts. It was developed to cover all the tools of the same type and probably has too many checks. When analyzing PMs for content, bear in mind two points: the primary function of maintenance is to keep the tool making as much *usable* product as it can for as long as it can and that it is the product that pays the wages. We need to appreciate the obvious conflict: time spent on maintenance is time not spent on production and yet we still must not neglect maintenance. Impossible? Not when you consider that wrong maintenance eats production time and bad maintenance has a banquet.

What happens to a tool if we simply ignore maintenance? Deterioration would get a foothold. Performance would become increasingly more unreliable. The deterioration would gradually take over and, in about a year, it would eventually force reliability and quality to plummet. The answer to the maintenance dilemma is to do only the maintenance that we definitely need to do; to ensure it is the correct *type* of maintenance; and to make sure that we do it properly, without ever having to repeat or correct a bad job. And you thought it was going to be difficult. . . .

Well it can be done. It only requires commitment, determination, and time. The amount of effort required to put a maintenance program right is directly proportional to the state of the entire maintenance organization. Fortunately we have access to a toolkit in the form of a series of procedures that we can apply to transform the situation. *Total Productive Maintenance* (TPM) and *Reliability Centered Maintenance* (RCM) are two of the best tools in the box. TPM will restore the maintenance infrastructure, the operator and engineering skill levels, and the equipment performance. RCM will be used to refine and tackle specific areas and guide us through some of the cost evaluations. Although their methods differ, they are the right and left hands of a pair of gloves.

Did you notice that the equipment performance was third in the order of improvements? As we follow the procedures, we will identify issues, prioritize them for resolution, identify their causes, and solve them. Once the groundwork has been prepared, we can start to focus on the problems, which will be solved in a way that brings returns as we move forward. Another worthwhile point to remember is that, in most cases, it is not the actual downtime that is unacceptable; it is the *unplanned* downtime that causes the biggest problems, because it just happens at random, can kill product, and it messes up everyone's day.

One point that has not yet been considered is how bad the current maintenance state actually is. For some, all they have is minimal routine maintenance with only DIY (do-it-yourself) repairs to keep the equipment running. For others, there might be no, or a very limited, system for recording the breakdown histories. There will be no system for identifying if a problem has been resolved and no record of how the problem was resolved, how effective the support group is, or how well the equipment operates. There might even be factory "experts" in specific faults: a testament to their failure to resolve problems. With the introduction of TPM, the turning point has been reached. It is time to become a bit more professional and efficient in the approach to maintenance.

The TPM PM Analysis

The first step in any plan is to establish what exactly the problem is: to find out its magnitude and what the TPM teams will be up against. This is a combination of hands-on investigation, equipment restoration, and an analysis of the past performance of the equipment. Ideally we would go back as far as possible and analyze the data, but that is not practical. What we will do is choose a time period, probably 3 months, and begin with that. A number of techniques will run in parallel as we make our way to establishing a bespoke maintenance schedule, on the basis of the successes and failures of the existing maintenance system.

Once the teams have worked their way through the equipment history and the minor stops analysis and they have the initial clean routines under control, they will have enough of the performance data required to create the *malfunction map* and all of the *PM maps*. TPM has a broad audience, so it favors the simplicity of visual imagery. It is much simpler to put a dot on a map that can be seen by everyone than it is to describe a precise position in words. The malfunction map uses this technique; it is a representation of the tool displaying the position and type of every failure identified during the analysis. The PM maps are a similar concept, except that they display the failures in the areas of the tool that are subject to routine maintenance—or should be. The reasoning for the PM map is simple: if the part is maintained correctly, it should not break down before its next scheduled service. In this case there will be highlighted areas to confirm the effectiveness of the PM.

The malfunction and PM maps

The malfunction map is the same basic diagram as the area map except that it is populated with the locations of *all* of the failure modes: F-tags, machine history, and minor losses. This is the record of the tool at its current—some would argue worst—state of deterioration. The completed map should be displayed on the activity board. In Fig. 8.1,

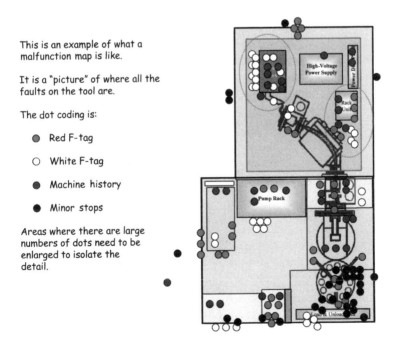

Figure 8.1 Malfunction map—recognize the base diagram?

it can be seen that the minor stops (black dots) are mostly on the periphery of the tool or the areas that handle the product. They must be; apart from a very few exceptions, they cannot really be found anywhere else. If the tool had to be shut down to repair them, the time taken to repair the fault would be too long and would need to be logged. Historical faults tend to be found within the tool. The white dots (white F-tags) show the areas where there are tasks that could be handled by *Autonomous Maintenance* (AM) teams. The malfunction map helps the team identify these areas and, *provided the areas can be made safe*, the AM teams can be trained to carry out the tasks.

If more detail is required to spread out a tight cluster of dots, extra drawings and photographs can be added.

PM maps have two objectives: to check the effectiveness of the current maintenance and to identify new areas that fail or might fail but have no current maintenance to prevent it. Apart from its simplicity, the beauty of TPM is that any maintenance data collected is based on how the equipment has actually performed under its current *use* conditions. This means that any two identical tools, if they run different processes or are located in different environments, might have variations in their required maintenance schedules. Use conditions are also covered in this book.

The basis of the PM map is a drawing or photograph of the maintained part. An example is shown in Fig. 8.2. Naturally any dot color can be used, but this example uses blue and green dots to mark the positions. The first step is to analyze the data needed to create the maps for every area/module that has a failure and one for every area that is maintained. The dot colors are used to represent PM steps and failure positions.

Blue dots (which appear as gray in the figure) are stuck to the "map" to identify which parts are maintained.

Green dots (which appear as white in the figure) are stuck where failures have been found.

A PM that does not fail would show blue dots only. Any other combination of dots means a PM needs to be improved or developed. Although the green dots are included to confirm successful PMs, even a map with no green dots (failures) should be reviewed to check the PM frequency and content.

We need a method to review the data we have collected. The Green Dot Allocation Table (*not everything gets a clever sounding name in TPM* ...) is the starting point. This should be similar to Table 8.1 and is compiled from the failure data. Every failure that occurred in the analysis time period must be recorded. It is worth remembering that the purpose here is to improve the standard of the PMs, so if any other failed part is known about, it should be added to the table. If the table is

Deciding on a Maintenance Strategy 217

Figure 8.2 A PM map of a mechanical "electrode" assembly. This assembly is listed in Table 8.1 as "PM 9/Electrode."

created as a spreadsheet, in an application like Excel, the information can be easily sorted. If the Green Dot Allocation Table is sorted by PM ID, it is possible to prioritize an order for their analysis. The team can also prioritize on either the number of failures, the downtime hours, or the frequency of the PM.

TABLE 8.1 Green Dot Allocation Table

Failure source	Failure number	PM ID number/ module	Failure description	Green dot number
F	R1	Pump 1	Leaking oil seal	144
M	87	PM 9/Electrode	Electrode "x" alignment wrong	33
M	154	PM 21/Sliding Seals	Uniformity issue due to vacuum leak. Shoulder screw not fitted.	76
F	R97	PM7/AT4	Loose indexer screw affecting positioning.	82
M	43	PM 9/Electrode	Electrode arcing. Set too close to face.	16

Failure Source Key: F: F-tag; M: Machine history files.

TABLE 8.2 PM Analysis Sheet

Step number	PM action	Blue dot number	Green dot number
24	Set "Z" adjust plate to 192 mm from rear face and adjustment	24	16
25	Set "X" adjust to 47 mm from RHS edge	25	
26	Set "Y" adjust screw to center position	26	
27	Reconnect the power cables	27	33
28	Refit the outer door	28	
29	Check the door latch is tight	29	
30	Remove the grounding rod and replace in its holder	30	

Note: PMID: 9; Module: Electrode.

However, we need to select a PM to analyze. In our example we chose PM9—The Electrode, a monthly task that was discovered to be a regular cause of major downtime because of either failure or improper setup, but we could have chosen anything—a labeler, a drill, or anything else. To carry out the PM analysis, we need to be able to compare the failure positions with the positions of the parts that are maintained. To do this, we need to construct a second table, this one will be similar to Table 8.2.

Collect a copy of the most recent PM specification (Standard). Then, from the PM specification or instructions, list every maintenance step in the order the tasks are carried out. Allocate each task a numbered blue dot. Stick each dot on the map as close as possible to the position on the part that the maintenance step is describing.

The flow diagram in Fig. 8.3 shows the steps to follow during the analysis. There is a more detailed explanation of the possible dot combinations on the pages following the flowchart.

The PM map is created using data from the machine history, the F-tagging, and the minor stops if they are able to help.

Technical support is essential in the creation of the maps; an operator could never carry out this task on his/her own. Notice there are both blue and green dots (which appear as gray and white dots respectively) in Fig. 8.2. Green dots represent failures. This means that there are three parts of the PM, we currently know about, that do not prevent problems. There must be something remiss with the procedure. Possibly the steps are not explained properly in the procedure and the job is, therefore, likely to fail. It could also be difficult to carry out these particular steps because of equipment design or there could be accessibility issues. It is also possible that the technicians do not have the correct skills needed to carry out the maintenance.

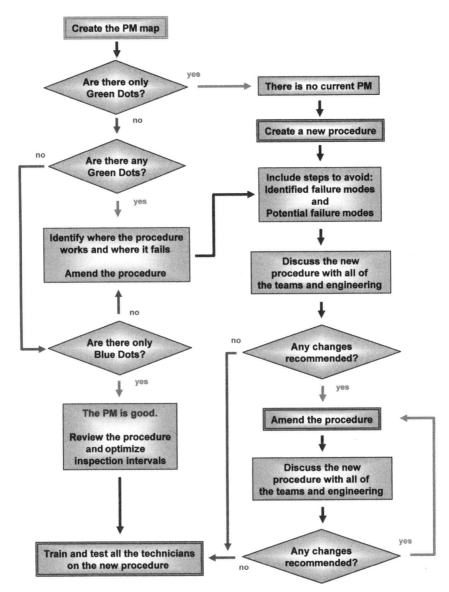

Figure 8.3 The flowchart is a simplified set of steps for analyzing the PM map. The steps are also listed below.

	Actions
8-1	☺ List all the F-tags, history, and any known minor stop failures.
8-2	☺ Create a green dot allocation spreadsheet similar to the one in Table 8.1. Using a spreadsheet allows easy sorting of the PM identification or specification number.
8-3	☺ Ensure that the maintenance reference (PM ID) or specification number and the name of the module are listed.
8-4	☺ Allocate each failure listed, a numbered green dot.

Interpreting PM maps

The steps for analyzing the PM map are listed below (see also Fig. 8.3):

Blue dots only	☺ The PM works. ☺ The PM can now be reviewed with a view to optimizing the time interval between the inspections. ☺ Consider removing any tasks or steps within a PM that are not necessary.
Green dots only	☺ There is no PM. ☺ Create a full PM procedure for the module. ☺ Include steps to prevent the failures already identified. ☺ Include any potential failure modes that have shown up during the creation of the procedure. ☺ Meet with the teams and engineers to discuss the new procedure. Record their comments. ☺ Update the procedure. ☺ Meet with the teams and engineers to discuss the updated procedure. ☺ Record their comments. ☺ Update the procedure. ☺ Circulate the procedure highlighting the changes. ☺ Create the Final rewrite. ☺ Use the best practice procedure to train all of the technicians.

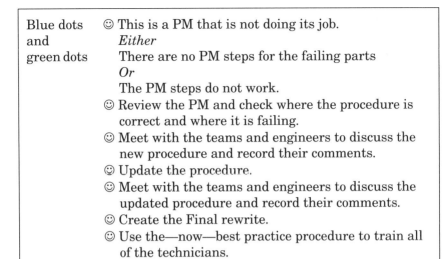

In Fig. 8.4 the assembly still has issues. The equipment history records the following faults:

- Vacuum leaks caused by contaminated o-rings.
- Shoulder screws found in the wrong positions causing mechanical damage to the plate surface.
  This will cause particles and can lead to vacuum leaks that will affect the process.
- Seal plate overgreased and causing particles.

This is a complex procedure to carry out, but not for a trained engineer. The PM maps clearly show that the procedures had to be reviewed as it should not be possible for an engineer to put screws into the wrong holes. The technicians also needed to be trained on the changes. The amount of greasing was a more interesting issue. This problem had to be analyzed to evaluate the mechanism and the use conditions and perhaps increase the PM frequency in addition to optimizing the amount of grease. In this situation a good place to start is with the equipment vendor, who in this case would be only too pleased to support the customer.

All the PMs need to be analyzed. Give priority to those with both colors of dots on the images. It is also worth carrying out a quick condition-based analysis of the PM frequencies applied to all the modules: are they being carried out too often? If they are, the interval can be extended. If they are not often enough, the 3-month time interval for this analysis should have picked it up. If the inspection interval is much greater, say 6 or 12 months, then a longer term review is needed. Bear in mind

222 Chapter Eight

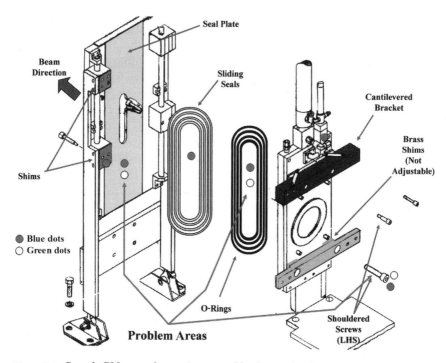

Figure 8.4 Sample PM map of a moving assembly that maintains a vacuum as it moves.

that, should any new failures occur, they will be F-tagged and flagged up during the team analysis. The PM team will get a chance to put the situation right.

	Actions
8-5	☺ Sort the data listed on the Green Dot Allocation spreadsheet based on the column "PM ID Number/ Module." All the failures that have the same PM number (or module if there is no PM) will be listed together.
8-6	☺ Select a PM to analyze. Example PM 9/Electrode.
8-7	☺ Take the photographs or collect the drawings for the PM map.
8-8	☺ Stick the numbered green dots on the PM map at the position nearest to the failure points.
8-9	☺ Collect a copy of the maintenance instructions for the PM.

8-10 ☺ Create a PM analysis spreadsheet. Reference Table 8.2.
8-11 ☺ List every step of the PM procedure and allocate it a numbered blue dot.
8-12 ☺ Stick the blue dots on the PM map at the position nearest to the step.
8-13 ☺ Interpret the photographs and take action on the basis of the instructions.

Scheduled maintenance or scheduled restoration

If you have a maintenance system in your company and it has not yet been through any kind of continuous improvement or content analysis, it is likely to consist of only scheduled maintenance tasks, which are all time-based. At the beginning of this chapter there was a short discussion on how PMs are likely to have developed. This section is more concerned with the content and practicality of scheduled maintenance.

Although this section was initially written for an RCM analysis, any maintenance system should consider both of the procedures. Both TPM and RCM target maximum reliability, but TPM believes it can do it and have zero failures. I understand why and even believe it is an excellent target to aim for but, although we might get close, it is not really practical. Perhaps it will be when the improvement exercise has been running for a while. By that time, the technical infrastructure will have improved and the major issues should be under control.

I am certain that RCM would like to target zero failures too, wouldn't we all? However, RCM recognizes that, even with the best of intentions, failures will still happen. This being the case, it takes the view that every failure has a cost to fix and that the failure event initiates a train of consequences, all of which have their own $cost that must be added to the initial failure cost. It is from this perspective that RCM is also interested in the true cost of maintenance and, for that matter, the cost of no maintenance.

For a technique like RCM to be applied, the user must understand the mechanism behind all of the different types of failure.

Scheduled maintenance is most suited to tools that follow a standard bell curve pattern of deterioration. Figure 8.6 represents the bell curve of failures. There are many different types of equipment

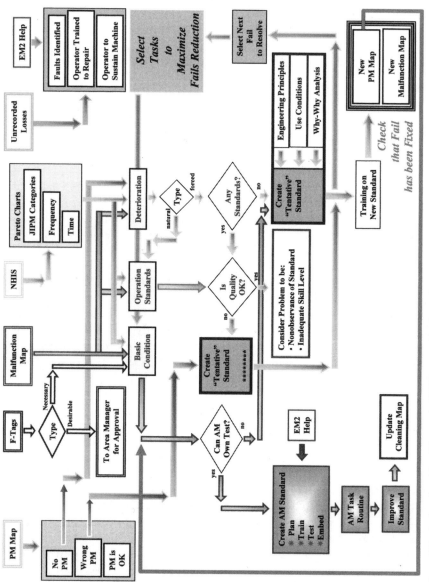

Figure 8.5 A visual guide of the main TPM steps that brought us here.

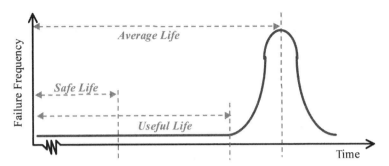

Figure 8.6 The statistical pattern of failures of a part over time.

pattern and parts that will follow this distribution and will deteriorate at around the same rate, provided the equipment is used under the same or similar conditions. This repeatability enables the average lifetime of a part to be established with a reasonable statistical accuracy. This lifetime is known as the Mean Time Between Failures or MTBF for the device. Once we have established that a part follows this pattern of deterioration, we can optimize the servicing interval to get the maximum use from the part and still minimize the risk of failure while the equipment is in use. Hence the name *time-based maintenance* (TBM).

Notice in Fig. 8.6 that by the time a part reaches its mean or average life, half of the parts have already failed. These are not very good odds for machine reliability. In practice, we need to base our maintenance on something a tad more predictable. To solve this problem, the "useful life" was established. It characterizes the operating time during which the part has a low probability of failure. Referring again to Fig. 8.6, you will see the point chosen as the end of the useful life is just before the graph starts increasing, that is, before the probability of failure starts to increase rapidly. There is also a "safe life" limit, used for protective devices. It is based on a fraction of the useful life. It is used for extra reliability where a failure has the potential to have safety or environmental consequences. The useful life does not guarantee there will be no failures, but that the probability of the part failing has been substantially reduced.

Parts with predictable lifetimes are suitable for TBM. It is worth noting, however, that according to the American civil aircraft study (from which RCM was developed) actual TBM patterns accounted for only 11 percent of all fails analyzed. I am prepared to bet this is a much lower percentage than you would have expected. I believe that the percentage of tasks being maintained *using* time-based procedures was much higher. It is important to remember, though, that the study was based on aircraft failures and so might have a different failure distribution pattern than other types of equipment. An RCM analysis would be necessary to establish the variations in different tools. Even so, 11 percent is not a lot!

The options for TBM are either restoration or replacement, with restoration being less expensive to carry out than replacing an entire part. If a part can be restored, it *must* be returned to its original condition and reliability. The end state must be exactly the same as it would have been had it been a new part. Scheduled discard or scheduled replacement relates to the exchange of the complete unit.

Not all components within a module deteriorate at the same rate. It is the ones that deteriorate fastest that define the useful life. For example, in a handheld flashlight or torch, the battery will fail before the bulb fails, the bulb will fail before the switch fails, and so on. It is the battery that defines the useful life of the flashlight. More often than not, scheduled maintenance procedures change only the most frequently failing consumable parts and few of the longer life or random failure components. So, after a number of routine restoration cycles, some of the longer life parts are getting close to failure and are preparing to mess up reliability. Eventually, barring accidental damage, the bulb in a flashlight will be ready to fail, but only after a high number of battery changes. It is important to ensure that the components that do fail at lower frequencies are also replaced as required. This effectively means that the same module will need to have its own, different time-based PM routines. The flashlight would have a regular PM to change the battery and after every "X" battery changes, the bulb will be replaced. We must have two different PM frequencies for the flashlight.

A complex assembly can require any combination of scheduled maintenance, scheduled replacement, and even "on-condition" monitoring. The on-condition monitoring is ideal to predict the longer lifetimes and the random failures, provided its use is "cost-effective" (see Chap. 9).

Scheduled replacement or scheduled discard

If scheduled restoration is not a realistic option, then the entire part must be replaced and the original part either thrown away (discarded) or returned to the manufacturer for refurbishing to the *as new* condition. Many large companies prefer this modular option. It is a costly way to do maintenance, but everything is relative. Discard minimizes the tool downtime, can make the PM task simpler, and, provided the value of the extra product produced makes it worthwhile, it can be a practical alternative. In short, if we can trade off the extra maintenance costs with the increased production then scheduled discard is the way to go.

Turnaround parts are an excellent compromise. This is a unique combination of restoration and replacement because the replaced part is still restored by the on-site maintenance team, but not while the tool is down. It is similar to an internal task in an SMED exercise being changed into an external task (see Chap. 7). The replacement part should be tested before fitting and must be restored to its *as new* condition. Turnaround parts minimize the tool downtime but do not reduce the total manpower use. It can reduce the workload on skilled technicians, however, if suitable semiskilled employees can be trained to do the work. If we consider the flashlight example, the turnaround unit might be a complete replacement flashlight. If we simply exchange them, we have no downtime. Now the original torch can be serviced at the engineers' convenience.

It is not a bad idea to carry out a quick RCM analysis on complex turnaround parts. In this case, complex also meaning parts that have a large number of components. This would help identify which parts are most likely to wear and which, if any, might have different failure patterns. It would also be an opportunity to evaluate whether or not a maintenance step for the less frequent or random failures might be more expensive than the cost of the failure's consequences.

The RCM PM Analysis

Harry Houdini had one, blockbuster, impossible trick. In fact it was the last trick he didn't do. We, the maintenance groups, also have our own Pagoda Torture Chamber: we are tasked to make real improvements to maintenance and still maximize production. At least our task is less impossible. We do it by

- Introducing proper procedures.
- Eliminating bad maintenance methods.
- Upskilling the maintenance groups.
- Adopting the most suitable maintenance techniques.
- Carrying out only the maintenance we need to do to ensure safety and provide the performance we need.
- We ensure that we use the best manpower-efficient methods we can, while still controlling the costs.

What could be easier! I believe that RCM provides the best method for deciding on the PM task. Downtime is unavoidable; the best we can do is to make it predictable and stay in control. RCM provides the user with a method to establish the complete cost and consequence of breakdowns. If it is decided not to do a specific maintenance task, then at least the decision to do so will be based on data and a realization that the function will eventually fail with a given set of costs.

A maintenance schedule developed using TPM is based on real fault data and it evolves as the experience of the tool increases. To say that RCM is a theoretical analysis does the procedure a huge injustice. Although it is true to say, a fair degree of *potential* breakdown situations are based on probability or, more accurately, educated guesswork as to the likelihood of a failure. RCM can be applied to

- Equipment as it is being manufactured.
- To an old machine that has earned its living in the field.
- To analyze a complete tool or only a selected, perhaps problematic, area of the tool.
- To current PM modules.

The RCM decision diagram

The key to devising PM routines in RCM is the decision diagram. It is a flowchart through which each *failure mode* is processed and whose progress is recorded on the decision worksheet. The outcome is a recommended PM task. Chapter 9 details how RCM initially treats equipment as *functions*, not parts, and then defines how these functions can fail. Only then does the RCM analysis look for the components or modules that could cause the failures and need to be considered for maintenance. Also in Chap. 9 is the method for identifying these parts and evaluating the cost and consequences of their failure.

This chapter compares RCM to TPM and starts from the assumption that the failure mode is known. You will discover it is a tad more complex than TPM, but you will get much more out of it.

The columns are trying to identify the most reliable and cost-effective method for maintenance. The decision blocks are considered in the following order:

1. On-Condition Monitoring (HC)
2. Scheduled Replacement (HR)
3. Scheduled Discard (HD)
4. Failure Finding (HF)
   Used in the case of hidden functions only.

The complete decision diagram, shown in two halves, can be seen in Figs. 8.7 and 8.8. More details on the overall structure of the complete diagram can be found in Fig. 8.13.

Deciding on a Maintenance Strategy 229

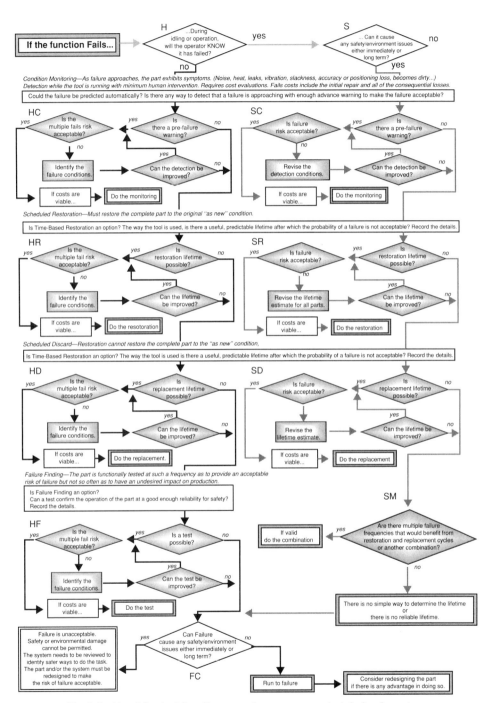

Figure 8.7 The left side of the decision diagram column sequence in tabular format.

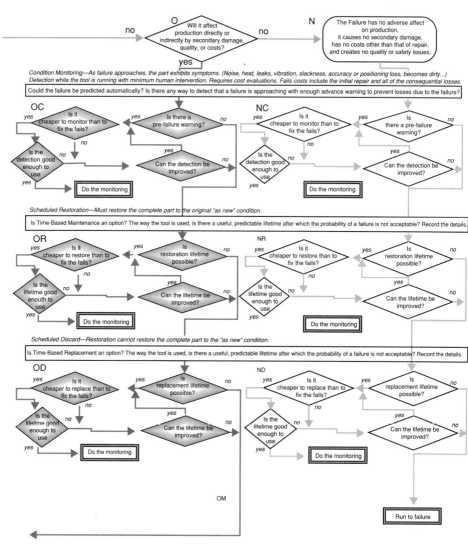

Figure 8.8 The right side of the decision diagram.

Hidden Failures	Safety	Operational	Nonoperational
H	S	O	N

Figure 8.9 The decision diagram column sequence in tabular format.

Figure 8.9 identifies the headings for the four-column format. The failure mode enters from the top left side of the diagram. Depending on the answer to a gating question, the decision diagram will direct the user either to the right, across the row and to the next gate, or down the columns through the various maintenance check blocks, to identify the maintenance action that is most suitable for that failure mode. The columns are divided into Hidden Functions (H), Safety (which includes Environmental) (S), Operational (O), and Nonoperational (N) fails.

Initially we will take a close look at the first column to explain what each of the decision block objectives are. The first column identifies hidden failures. Normally only protection devices are subject to hidden failures. The "hidden" part means that they can fail without anyone being aware of the failures. If the answer to the gating decision diamond is "No," the failure falls into the *Hidden Failure* category and we travel down the first column. The gating decision diamond in Fig. 8.10 asks, "... During idling or operation, will the operator know if (the function) has failed?"

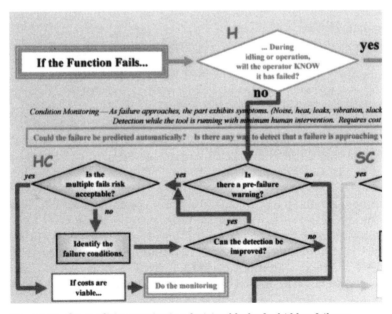

Figure 8.10 On-condition monitoring decision blocks for hidden failures.

The decision diamond in the first row (HC, Fig. 8.10) is trying to identify whether or not the part follows the pattern of a P-F curve. It asks,

"Is there a prefailure warning?"
If there is, it must be reliable enough to be used to predict the need for maintenance. The team need to consider the following:

➢ Does the failure pattern or wear of the part follow a repeatable P-F curve?
➢ What is the potential failure warning?
➢ Is there only one prefailure warning?
  (Perhaps there is heat, followed by an oil leak, then by noise...)
➢ Does the warning give enough time to detect it and prevent the failure?
➢ Is the warning good enough and repeatable enough to use.
➢ Is the warning reliable?
➢ Is there a better way to detect the warning?
➢ If we can use on-condition maintenance, can we modify the system at a cost that will enable us to use it to our benefit?

If the part does not follow a P-F curve, the "No" exit flows to the right and down to the next row.

If the answer is "Yes" we move left to the next question in the decision block:

"Is the multiple failure risk acceptable?"

This is asking if the on-condition test is reliable enough to be depended on. A protection device will have failed and we, the users, will not know anything about it. Failure means we have no safety interlock and the module is free to fail with all its consequences. The protected device might not fail, but it could. It is a risk, so we have to be concerned with the probability of both failing at the same time: the sensor *and* then the protected device. This is the multiple failure the question asks about. If the detection is not repeatable, relying on the prewarning might not be good enough to avoid the failure consequences. What do we do if the system gives a warning and we miss it or if we don't have time to take action to avoid the failure? If we are not satisfied with the reliability, the "No" exit will lead us to reconsider the detection system.

"Identify the failure conditions."

What is the current system? Check that we have understood it, its capability, and its reliability. When we know the answers, we move on.

"Can the detection be improved?"

Is there a better way to detect the failure or can the method we are currently analyzing be improved, possibly by using better detectors, better amplifiers, or any other part or technique that makes it better? If "No," we move down to the next row.

If "Yes," we move back into the loop until we either leave to enter the box considering if the monitoring "... costs are viable" and making the changes or, if not cost-effective, we move down to the next row.

For the condition "If the costs are viable..." the solution has to be cost-effective and the warning monitoring must be practical and possible. We would not spend $20,000 to save $400 unless there were other considerations. If they are not viable, we move down the row to the next option. The risks are all considered against the cost.

The second and third rows are asking if the part follows a repeatable deterioration time. Is it possible to predict when maintenance will be needed? If not, we move downwards. Row 2 offers scheduled maintenance and row 3 offers modular replacement, both covered earlier in this chapter. Figure 8.11 shows the complete decision blocks. The team must ask the following questions:

⇒ Does the deterioration of the part follow the bell curve as shown in Fig. 8.6?
⇒ Does the failure pattern of the part follow a repeatable, useful lifetime that can be identified?
⇒ Can we restore the part to its original standard of reliability? If not, do we need to replace it?
⇒ What must the maintenance frequency be to ensure the reliability we want?
⇒ In the case of operational failures, does this maintenance frequency—considered over a time interval—cost more than the cost of the failures and their consequences over the same time?

We know from the American aircraft study that a vast number of components did not follow time-based failures, but remember the data was based on aircraft. In the case of restoration we must establish whether the unit does follow a time-based cycle and, if it does, can it be restored to the original reliability? This applies to all of the components in the module and not only the ones just replaced in a PM. If it is not as reliable we must consider why and what can be done to make it better. Discard is looking for the same basic criteria.

In the situation for column 1, where we can have a failure that no one is aware of, we must consider a fourth option: failure finding (see Fig. 8.12). If the system cannot be reliably tested to minimize the chance of failure to an acceptable degree, then the function *must* be redesigned

234   Chapter Eight

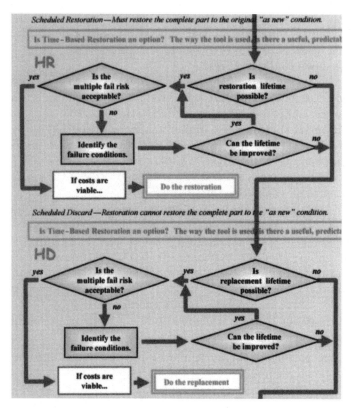

Figure 8.11   The scheduled restoration and scheduled discard decision blocks for hidden failures.

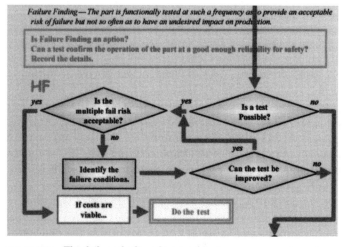

Figure 8.12   The failure-finding decision blocks.

to make it reliable and safe. Failure leading to a safety or an environmental incident is unacceptable. Even if there is a detection task, the frequency of the task is a consideration. If it will be required too often to be practical, then redesign should be considered.

The first question of the decision block HF is

"Is a test possible?"

Are we able to carry out a practical failure-finding test that will confirm the part is still operational? If we can, then

➤ How often would we need to carry out the test to give us the necessary acceptable reliability?

➤ If this frequency is not financially viable, we need to redesign the system. It will probably be necessary to shut down the tool to carry out the check. If this is the case, we must calculate the $cost of the checks—including the downtime losses. If they are too high, and because the failure will affect safety or the environment, we will need to consider redesign.

If none of the above options lead to acceptable ways to avoid a failure, we end up at

➤ Redesign
If the consequences are not acceptable and could lead to a safety or environmental issue.

➤ Run to Failure
If nothing would have helped *to find, predict, or prevent* the fault, it is not a time-based failure and PMs plus failure finding do no good.

Let's take a look at the rest of the decision diagram. Figure 8.13 is the diagram drawn at the same scale as Table 8.3, which is a simplified decision diagram, with the decision blocks and decision diamonds condensed to text that fits the cells of the table. Each cell contains the test and which outcome is under consideration. Notice that the columns in Table 8.3 and Fig. 8.13 align. The rows also correspond.

The second column of the decision diagram is designed to identify and maintain safety and environmental failures. The gating decision diamond asks

"... Can it cause any safety/environmental issues either immediately or in the long term?"

"Yes" leads us down the *Safety* category.

Safety applies to personnel, environmental safety, and contamination. New standards and legislation make breaches in health and safety or environmental releases potentially very costly and, sometimes, even a criminal offence punishable by imprisonment. The risk of failure

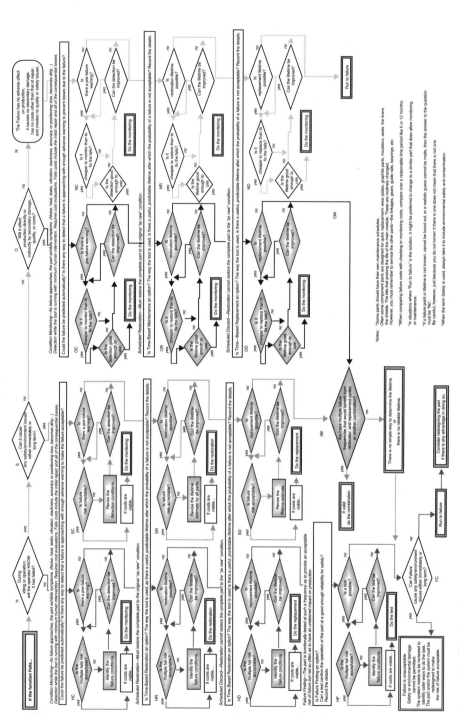

Figure 8.13 Decision diagram.

TABLE 8.3 The Decision Diagram Sequence in Table Format

Hidden failures H	Safety S	Operational O	Non-operational N
On condition Multiple fails risk	On condition Failure risk	On condition $Consequences/$Checks	On condition $Repair/$Checks
Restoration Multiple fails risk	Restoration Failure risk	Restoration $Consequences/$Checks	Restoration $Repair$Checks
Replacement Multiple fails risk	Replacement Failure risk	Replacement $Consequences/$Checks	Replacement $Repair$Checks
Failure finding Multiple fails risk	More complex PM Failure risk	More complex PM $Consequences/$Checks	
Final safety Multiple fails risk	Final safety Failure outcome	Final safety Failure outcome	
Redesign Safety fail unacceptable	Redesign Safety fail unacceptable	Redesign Safety fail unacceptable	
Run to failure	Run to failure	Run to failure	

must be as low as possible and, where possible, more than satisfy the legislation.

The third column is designed to identify and maintain operational failures. Figure 8.14 shows the first three decision blocks. An operational failure is one where the failure has an impact on production or product quality. The gating decision diamond asks

"... Will it affect production directly or indirectly by secondary damage, quality, or costs?"

If the answer to this decision diamond is "Yes," then the failure falls into the operational category and we travel down the third column.

Virtually every company exists to make money. Therefore it must take steps to understand the magnitude of any losses and limit the cost or damage to product. In addition to the cost of repair, consequential losses include production time, time on tools waiting for product, labor-hours, scrap, wasted facilities, remaking product, tool downtime, and the impact on the customer. The decision blocks are different from the Hidden and the Safety columns. Although they are really looking for the same things, using the detection capability or lifetime stability is compared to cost before any time is spent considering if it is good enough.

The fourth column is designed to identify and maintain nonoperational failures. It uses the same gating diamond as column 3, except the "No" answer leads you out a different exit. Nonoperational issues have a cost to rectify but they do not have any negative impact on the product, its manufacturing, or the running of the facility. Nonoperational costs are simply the cost of the repair against the cost of the failure.

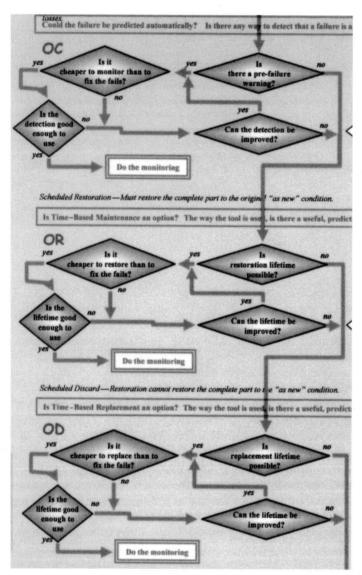

Figure 8.14 Three rows of operational decision blocks.

The outcome is all down to the consequences of the failure. In the operational case, making a mistake will only cost money or affect product. This does not mean we can afford to be careless, but the questions are testing the reliability of the pre-fail warning against financial and customer costs only. Is it good enough to use? Could the signal as it stands be capable of warning of a failure coming and give enough notice to be

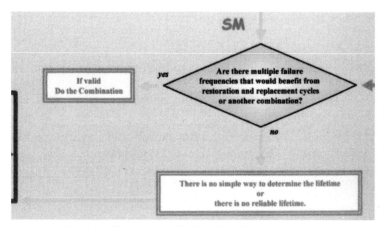

Figure 8.15 Decision diagram—multiple failure frequencies.

able to avoid the failure by taking immediate action or allow time to plan or to take action? Is it reliable? Will it give the same warning every time? Where safety is impacted by failure, stricter limits might have to be used, so we establish a "safe life" limit—at a fraction of the useful life.

Care must be taken to ensure that components that do fail at different frequencies are also replaced as required. It is the ones that deteriorate fastest that define the useful life. Section SM in the decision diagram considers this point (see Fig. 8.15). After a few routine restoration cycles, some of the longer life parts could be getting ready to fail and mess up reliability. This effectively means that the same module will need to have its own, different time-based PM routines. They can also have a combination of scheduled maintenance, scheduled replacement, and condition monitoring. Remember, the checks must also be cost-effective.

We now appreciate that the decision diagram directs the user in different directions on the basis of answers, which in turn are based on the best knowledge available. *Beware:* a lack of knowledge can lead the user down the wrong path. When considering on-condition, restoration, or replacement tasks the user is questioned about P-F curves and useful lifetimes. If the user does not know the lifetime, the answer must be "No," which will ultimately lead to "No Scheduled Maintenance."

Consider the example of a new drive motor that you don't know much about and follow the questions on the decision diagram. "No Scheduled Maintenance" will be the outcome since the user does not know the answers—even when you know that the answer is wrong. This point has been reached because the team does not yet know what the useful life is or how it will deteriorate in your system. In this case "No Scheduled Maintenance" means "We Don't Know."

If we do not have a precise answer, we can try several different sources to find one. We can

- Find out the information from the part manufacturer, the equipment vendor, or other users.
  Finding out the correct answer is a lot of work so it is often not pursued—or even started.
- Make a "best guess" of an initial TBM period and modify the frequency each time the unit is checked until the answer is found.
  While learning, it is better to make the checks too often rather than not enough.
- Make a "best guess" as to what the replacement lifetime will be. The guess can be evaluated by iteration.
  - Is the lifetime more than 1 year?
  - Is the lifetime more than 20 years?
  - If the lifetime is not more than 20 years, could it be more than 10 years?
  - If it is more than 10 years, is it likely to be more than 15 years?
  - If it is less than 15 years, could it be more than 12 years...?

What we are doing here is using the experience of the entire team to work out a best guess. Everyone might think it will be 5 years, but they might also lack the confidence to be decisive. They might be positive that it is more than 2 years, so check it at two or maybe three and see how it looks. Err on the safe side—always. Then go back to the manufacturer and confirm your estimate. Your current estimate can be revised any time you want.

In a place I once worked, I found a sign marked "Not Drinking Water." It was lying against a wall at the top of the stairs. It was true; in fact there was no water at all.... Looking around, I noticed there was a sink nearby that had some holes in the wall above it. I checked around and found there were quite a few sinks that did have notices above them, all with the same message. So it was safe for me to assume that the sign belonged to the sink, but it was not fulfilling its purpose. The sign was a protective device intended to stop people from drinking the water. The missing sign was a *hidden failure*, since no one was aware that it was missing. In addition, if someone had drunk the water, we would have had a *multiple failure*.

Signs were never checked, just replaced when reported to be damaged. Besides, who would think about checking a sign anyway? Interestingly enough, RCM would. If someone had carried out a failure-finding task—walked around the building and checked that the signs were in place—then we could have reduced the likelihood of a multiple failure. 5S might have found the same issue.

The sign was in a hospital ward and failure was not acceptable. The problem was analyzed and the outcome was to improve the system. The "Not Drinking Water" signs were all removed and the sinks where the water was suitable for drinking were allocated a "Drinking Water" sign. A general rule was issued and supported by notices throughout the hospital that no one could drink from any sink unless it had a sign saying it was safe—otherwise, to assume it was not safe. This way a missing sign would imply the water was not safe to drink. A pretty clever solution, I thought. I wonder who's idea it was?

RCM Example 1: The Missing Sign. What would RCM have done if it had been given the same failure mode "Warning Sign Missing"? Follow the decision diagram as shown in Fig. 8.16.

What do we do about the missing sign? Follow the bold decision diagram arrows in Fig 8.16:

Question H (Hidden)
Would the operator, or user in this case, know the sign was missing?
No.
Move down the column.

Question HC (On-Condition)
Is it possible to use on-condition monitoring to tell you it was missing?
A micro-switch, pressure pad, or other sensor behind the mirror, linked to an alarm and a lamp? These are possibilities but not a good idea.
Too expensive a solution—Not cost-effective.
No.
Move down the column.

Question HR (Restoration)
Is it possible to use scheduled restoration?
Yes, but a sign never really deteriorates, so the answer must be
No.
Move down the column.

Question HD (Replacement)
Is it possible to use scheduled replacement?
Not a practical option, so the answer must be
No.
Move down the column.

Question HF (Failure Finding)
Is it possible to use failure finding?
Yes.
Move to the left.
Is the risk of a multiple failure acceptable?

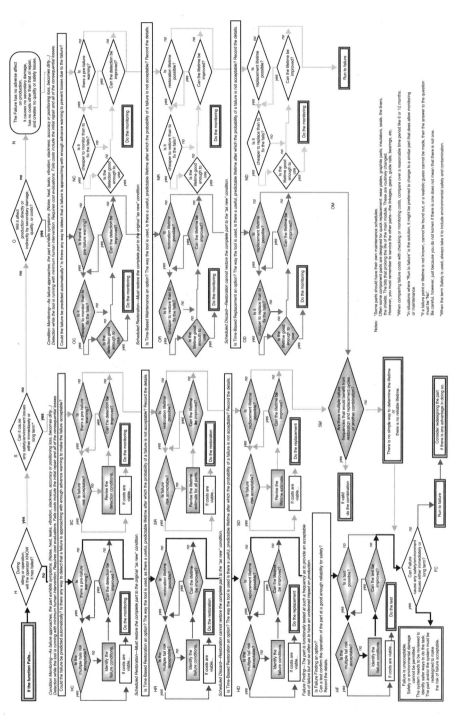

Figure 8.16 The missing sign.

Someone could drink the water the same day the sign disappeared. So the risk is not acceptable.
Move down.

Identify the failure conditions. Can the test be improved?
No.
Move down the column.

Question FC (Final Safety Check)
Can failure (the missing sign) cause any safety issues either immediately or later?
Yes.
Move to the left.

Failure is unacceptable: Redesign the system

The outcome was the same as the hospital engineer's decision. The only difference would have been that the problem would have been an imaginary, anticipated one. The issue would have been identified before the sign was missing and the risk of anyone drinking the water would have been avoided.

RCM Example 2: Heating plate thermocouple failure. This check is not black and white. Figure 8.17 highlights the path that would be followed for the analysis of the failure of a thermocouple used to monitor the temperature of a heating plate. Follow the bold decision diagram arrows in Fig. 8.17.

Question H (Hidden)
Would the operator know the thermocouple had failed?
There would be no heating plate control, so they would discover something was wrong.
Yes.
Move to the right to the safety gating question.

Question S (Safety)
Can the failure cause safety or environmental problems?
Yes.
Move down to the on-condition decision diamond.

Question SC (On-Condition)
Is it possible to use on-condition monitoring to predict failure was coming?
There is a difficulty here. There might be a couple of options that would allow the problem to be predicted.

1. *Measure the "Set" temperature and compare to the "Actual" temperature and look for control instability.*

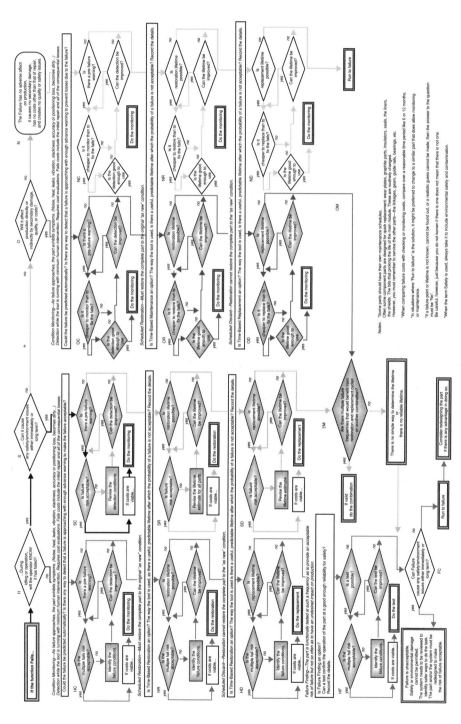

Figure 8.17 Failure of a thermocouple measuring the temperature of a heating plate.

2. *Measure the resistance of the element and look for changes caused by potential failure.*
3. *Look for increasing electrical noise levels as the end of the life approaches.*

Yes, the fault can be detected.
Move to the left.
Is the failure risk acceptable?
Yes.
Move left and down, within the decision block
Is it viable to set up the monitoring?
Evaluate the cost of experimentation and apply the tests.
(If it does not work, go back to the decision diagram and go through it again. This time, redesign coupled with on-condition might be the outcome. For example, a dual heating element might be used with current monitoring. The element could be replaced on the failure of any section of the element.)
Yes.
Move to the right: Flow stops at:
Do the monitoring.

RCM Example 3: The blocked exhaust line. Figure 8.18 highlights the path that would be followed for the analysis of a blocked exhaust line in a production system. Follow the bold decision diagram arrows in Fig. 8.18.

Question H (Hidden)
Would the operator know the exhaust line was blocked?
There is an audible alarm and a warning lamp. It would only be missed if no operator was present.
Yes.
Move to the right—to Safety.

Question S (Safety)
Can the failure cause any safety or environmental issues immediately or in the long term?
The system shuts down the pump.
No.
Move to the right—Operational

Question O (Operational)
Will it affect production directly or indirectly?
There will be equipment downtime, probable product loss, and manpower to clean the pipework and repair and requalify the tool and the vacuum system around the pump.
Yes.
Move down the column.

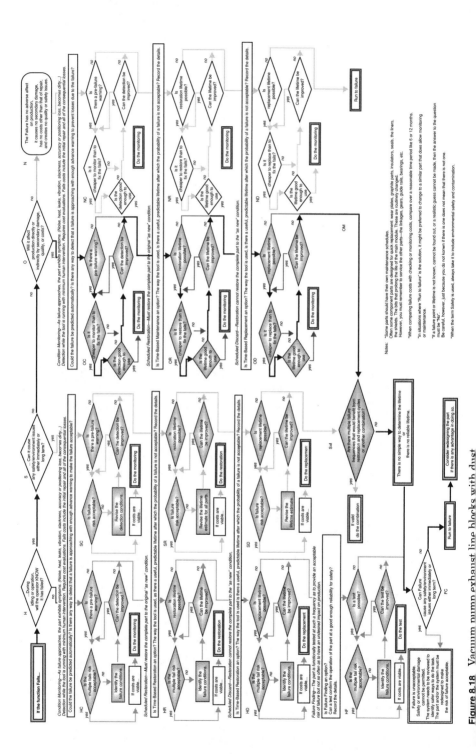

Figure 8.18 Vacuum pump exhaust line blocks with dust

Question OC (On-Condition)
Is it possible to use on-condition monitoring to predict failure?
Yes.
Move to the left
Is it cheaper to monitor than to fix the fails?
Yes.
Move to the left.

Is the detection good enough to use?
The vacuum in the pump exhaust line is already monitored but the transition from operating to fail is very quick.
The definition says, "Is there any way to detect that a failure is approaching with enough advance warning to prevent losses due to the failure?"
It works sometimes and is better than having no test.
No.
Move to the right.
Can the detection be improved?
No.
Move left and down to the next row.

Question OR (Restoration)
Is it possible to use scheduled restoration to prevent the exhaust from blocking?
Yes.
Move left.
Is it cheaper to restore than fix?
It is restored but still fails frequently.
Yes.
Move left and down.

Is the lifetime good enough to use?
No.
Move right.

Can the lifetime be improved?
Yes, but it would be too frequent. So
No.
Move right and down to next row.

Question OD (Discard)
Is it possible to use scheduled replacement to prevent the exhaust from blocking?
Yes.
Move left.

Is it cheaper to replace than repair?
Yes.

Move left and down.

Is the lifetime good enough to use?
It is replaced but still fails?
No.
Move right.

Can the frequency be improved?
No.
Move right and down.

Question SM (Multiple-Task Restoration)
Are there multiple failure frequencies that would benefit from a combination of tasks?
A combination is already used, but it can still fail. The lifetime can be shortened, but would be too short.
No.
Move down and left.

Question FC (Final Check)
Can the failure cause any safety issues, etc.?
No.
Move right and down.

"Run to Failure"
This is what happens on occasions. We have a degree of predictability and control but not enough.
Move right.

"Consider redesign"
Weigh up the costs of the number of failures versus the repair costs and decide if redesign is a worthwhile option. We might try changing the pump to one that reduces the amount of dust it exhausts, perhaps fit a filter or some kind of dust trap, modify the exhaust line (heat trace) to reduce the chance of dust settling, or change the routing of the pipes to eliminate susceptible bends and restrictions.

In the meantime use the best standard we have for maintenance and on-condition monitoring. It works sometimes, which is better than never. Use and improve.

Actions	
8-14	☺ Learn and understand P-F curves and useful life.
8-15	☺ Learn and understand the different types of maintenance tasks.
8-16	☺ Learn and understand the decision diagram. Check that you understand the meanings of the simplified questions in the flowcharts.

The blocked exhaust is a good example of using a system plus your own intelligence. RCM leads you to redesign, which is the correct option. But, rather than wait for a design and new solution, use what you have now to maximize the possibility of avoiding a failure in the meantime.

This problem also highlights that an RCM analysis before the installation could have saved oodles of money. No one considered it was a problem, let alone appreciated the extent of the problem. All the downtime was blamed on the first piece of equipment following the vacuum pump, when, in fact, it was the design of the pipework connecting the unit to the pump and the lack of heat tracing. Even a small RCM team (possibly even one man) would have detected these two issues from just reading the manual. He could have made an evaluation of the cost of the failures and the consequences. If there was no manual, a telephone call to the vendor would have identified the issues—they were well-known problems.

We must be in control of the desire to drive for faster equipment installations. Managers need to see reduced costs—engineers often only see potential issues. An RCM style analysis would have provided the $numbers for most managers to appreciate the real gain in doing the job right the first time.

Recording the process on the decision worksheet

If we are going to use a flowchart as complex as the decision diagram, we need a way to record our options. It must have been pretty obvious that the previous few pages were difficult to follow. Would it not be nice if there was a simpler way to record the answers? Enter the decision worksheet.

The design of the decision worksheet enables us to look back and trace how the failure modes were evaluated and the maintenance task was selected. Although RCM is based on cost versus consequences, it is important to remember that the costs are not only financial, but also legal and moral. Team analysis is essential for applying the decision diagram and completing the decision worksheet. We need the teams to ensure that the knowledge input is as high as possible and to maintain a broader perspective. Keeping everyone involved, particularly the maintenance group, will prevent the views of any one person becoming dominant.

Unless previously agreed, when following the decision diagram, where on-condition monitoring or a redesign is selected, do not use the RCM analysis time to create a solution. That is not the purpose of RCM or the responsibility of the teams. The teams need to limit themselves to their own tasks only. It is reasonable, however, to note any ideas that

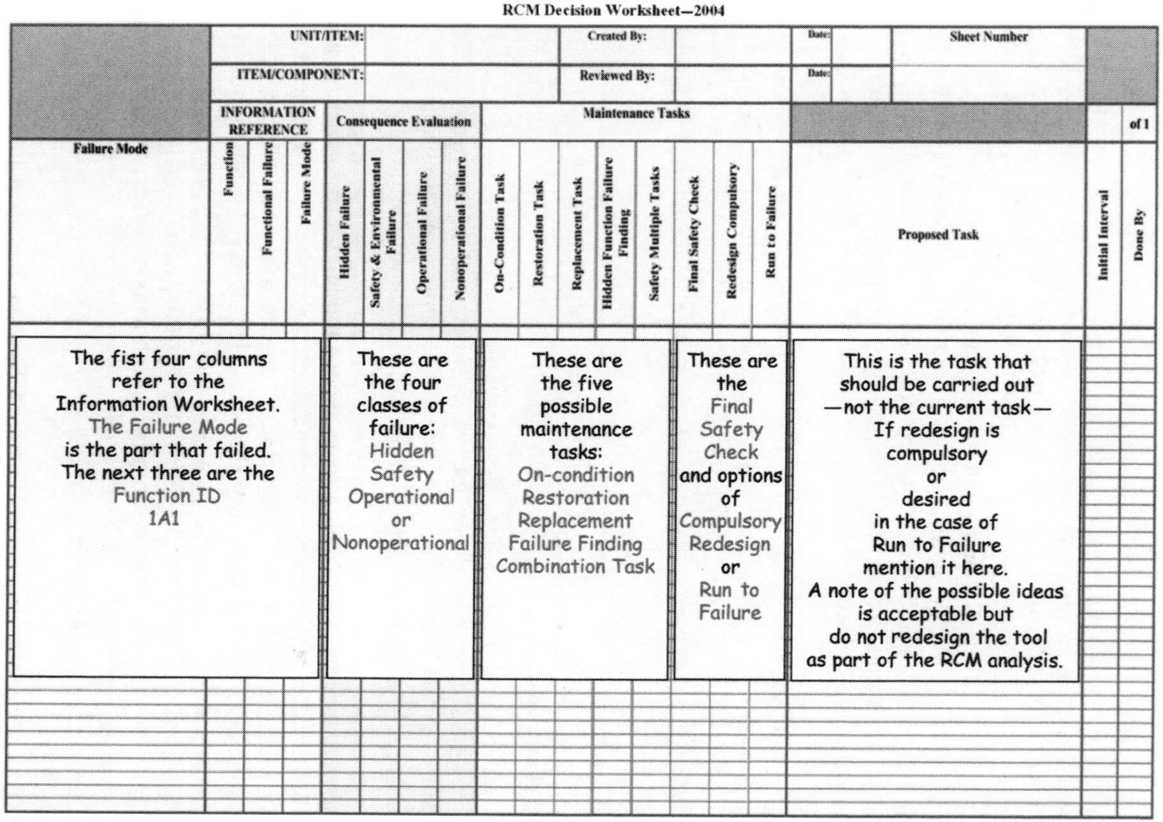

Figure 8.19 Decision worksheet layout.

RCM Decision Worksheet—2004

UNIT/ITEM:			Examples		Created By:					SB			Sheet Number		
ITEM/COMPONENT:			Examples		Reviewed By:								1 of 1		
													Date: 03-Nov-04		
													Date:		

INFORMATION REFERENCE			Consequence Evaluation				Maintenance Tasks							Proposed Task	Initial Interval	Completed By
Function	Functional Failure	Failure Mode	Hidden Failure	Safety & Environmental Failure	Operational Failure	Nonoperational Failure	On-Condition Task	Restoration Task	Replacement Task	Hidden Function Failure Finding	Safety Multiple Tasks	Final Safety Check	Redesign Compulsory	Run to Failure		
			H	S&E	O	N	HC / SC / OC / NC	HR / SR / OR / NR	HD / SD / OD / ND	HF	SM	FC				

Failure Mode

"NOT DRINKING WATER" SIGN IS MISSING.	1	A	1	N			N	N	N	N	N			Y	FAILURE FINDING IS POSSIBLE BUT WOULD NEED TO BE TOO FREQUENT. REDESIGN HAS TO BE THE SOLUTION.	TBD	
HOT PLATE THERMOCOUPLE FAILURE.	1	A	1	Y	Y			Y					Y		COMPARITOR CIRCUIT CHECK SET AND READBACK TEMPERATURES.	TBD	
EXAUST LINE BLOCKED	1	A	1	Y	N	Y		N	N	N	N	N	N		Y	SYSTEM IS MONITORED BUT FAULT HAPPENS TOO FREQUENTLY. RECOMMEND RETAIN ON-CONDITION MONITORING AND TIME-BASED PM, BUT CONSIDER REDESIGN TO TRY AND AVOID THE UNPREDICTED FAILS.	TBD	

Figure 8.20 Completed decision worksheet using previous examples.

surface during the brainstorming sessions. They will serve as guides later for the design team, who have to implement recommendations.

Figure 8.19 shows how to complete a decision worksheet. The first column has been added to make it easier to use. It was suggested by an RCM team who felt that having a real failure to look at and not just a number would make things easier. It does. Enter the answers to the decision flowchart in the corresponding worksheet column. The "Proposed Task" column is for the maintenance task and suggestions.

In Fig. 8.20, following the example of the missing "Not Drinking Water" sign:

➢ The answer to the Hidden Functions gating question
 "... During idling or operation, will the operator know it (the function) has failed?" is "No," so enter an *N* in column *H*.
➢ Following the flowchart sequence down the column gives four more negatives in a row. Enter an *N* in columns *HC, HR, HD,* and *HF*.
➢ The final safety check gate, however, asks if there will be any potential safety issues. This is a "Yes," so a *Y* is entered in both column *FC* and in the *Redesign Essential* column.
➢ The Proposed Task column is as described.
➢ The Initial Interval for maintenance or failure finding will have to be decided.

Action
8-17 ☺ Complete the decision worksheet.

Failure finding and calculating acceptable risk

Failure finding is used to verify that a protection device is working. The best option is to plan in advance and use a device with a readback of its state, so its operation can be confirmed during the normal operation of the tool. If this is not possible, then the next option is to check it operates. Where the device is inaccessible, making a check is not possible. Examples could be: the lower limit switch in an acid tank; an oil level switch inside a high-voltage transformer; a flow switch inside a radioactive core; an amplifier power detection switch underground or under sea; a door entry switch in a satellite. It could also be checking the presence of a thermal tile on the underside of the space shuttle while in orbit. There are also some switches and safety devices that are only capable of single operation (a fuse or a shock sensor).

If the failure of a device is not acceptable, double or triple redundancy is an option. This means that backup units must be set up to take over automatically, in the event that the main unit fails. The backup then effectively becomes the protective device and a signal should be generated to warn the users that a changeover has occurred. Backup units are more critical than the main functions. Should the function fail, the backup must take over and, if all operates as designed, there should be no impact on the system. Aircraft have several backup hydraulic systems. They have to. Imagine your surprise should the pilot say, "This is your captain speaking. We are diverting to Bouncy Castle Airport, as we have just developed a failure in our primary hydraulic system. But please do not be concerned as the backup system takes over automatically...OOPS!"

Failure finding. How often should we need to check a protection device? If the device is not working, it simply means that there is no protection and, if we are lucky, the protected device keeps running. However, the system is now ready, poised, and waiting to fail and we all know Murphy's Law. (If it can fail it will and it will fail at the worst possible time....) We need to find a checking interval that ensures an acceptable level of probability of the function failing and causing problems. We have to choose an acceptable risk. How often we test is known as the *failure-finding interval* (FFI).

Everyone says the yearly car MOT test "...is only valid until it leaves the testing station." True, it is possible that a part can fail immediately after it has been tested, so the part can be dead for the whole year. Equally, it can last until the day before it is tested. So the average time the part *could* be in a failure state is 6 months—half of the test interval. This means that for any one device, the average time it could be in a failed state is half of the FFI. The hard bit is deciding what the acceptable unavailability should be. This has to be a company-guided decision and involve the safety department, especially in situations where there is a risk to safety or the environment. It is much simpler if the risk is purely financial; then all you need to do is compare the costs. Previously, we tended to opt for the highest level of availability—99 percent plus.

Figure 8.21 is the mathematical proof of the derivation of the formula for calculating the faultfinding interval. Table 8.4 is a table showing some of the more common intervals and required performance levels.

a Protective Devices have failed in *b* years

The Mean Time Between Failures $= \dfrac{b}{a} =$ MTBF

The Failure Finding Frequency $=$ FFI

The Maximum Unavailability, knowing that only *a* failed in time *b* is:

The Maximum Unavailability $= a \times$ FFI

The Minimum Unavailability has to assume zero time was lost in each of *a* FFIs

The Average Unavailability $= \dfrac{(a \times \text{FFI}) - 0}{2}$

Total Unavailability $= \dfrac{\text{Average Unavailability}}{\text{Total Time Period}}$

$= \dfrac{\dfrac{a \times \text{FFI}}{2}}{a \times \text{MTBF}}$

$= \dfrac{a \times \text{FFI}}{2a \times \text{MTBF}}$

Unavailability $= \dfrac{\text{FFI}}{2 \times \text{MTBF}}$

$\boxed{2 \times \text{MTBF} \times \text{Unavailability} = \text{FFI}}$

Figure 8.21 Proving the testing frequency formula.

If it was important to guarantee 90 percent availability on a protected device that has a mean time between failures of 50 years, then the FFI would need to be

$$\begin{aligned}
\text{FFI} &= 2 \times \text{MTBF} \times \text{Unavailability} \\
\text{FFI} &= 2 \times 50 \times (100 - 90)/100 \\
\text{FFI} &= 100 \times 10/100 \\
\text{FFI} &= 10 \text{ years} \\
&= 20\% \text{ of MTBF}
\end{aligned}$$

Deciding on a Maintenance Strategy 255

TABLE 8.4 The Most Common FFI Values as a Percentage of MTBF

Required availability of hidden function	99.99%	99.95%	99.9%	99.5%	99%	98%	95%	90%
FFI as % of MTBF	0.02%	0.1%	0.2%	1%	2%	4%	10%	20%

If we wanted 95 percent availability:

$$\text{FFI} = 2 \times \text{MTBF} \times \text{Unavailability}$$
$$\text{FFI} = 2 \times 50 \times (100 - 95)/100$$
$$\text{FFI} = 100 \times 5/100$$
$$\text{FFI} = 5 \text{ years}$$
$$= 10\% \text{ of MTBF}$$

The further away you get from 100 percent, the greater the error in the calculation. However, where safety is concerned, you should never be that far away.

The manufacturer of the part should provide information on the MTBF of the protection device. If he cannot, check how many have been replaced over a time period. Beware of using the stores department for the information. Ensure the part has only been used in the analysis tool: different tools could have different use conditions. A good option is changing the device to another of the same or better quality that does have all of the required data. In any event, if you are not certain and have to make an educated guess, work out a range of options to either side of what you think the MTBF is and err to the safe side. Always involve the safety department in the analysis of the data and when making the final decision. They might even have a list of approved safety devices.

Action

8-18 ☺ Calculate the FFIs for the hidden failures and update the decision worksheet.

Chapter 9

RCM—Reliability Centered Maintenance

Reliability Centered Maintenance (RCM) differs from *Total Productive Maintenance* (TPM) in the way it sees the equipment: it looks for the functions of the tool and not a list of its modules. This technique opens up the analysis to such a degree that it enables the team to consider a broader spectrum of *potential* failure conditions that could lead to more maintenance options. At the very least, a properly completed analysis will let you discover any PMs that have consequences likely to come back and bite you on the butt, could cost you a lot of money to fix, or could have safety or environmental impact. Fortunately, RCM's cost versus consequences methodology enables the user to choose not to include maintenance, based on data, if the return is not justifiable as a $cost. Figure 9.1 is a block diagram of the stages of an RCM analysis.

The First Stage in an RCM Analysis: The Operating Context

RCM works within a set of guidelines known as the *operating context*: it is a cross between a contract, an operating manual, and a maintenance manual. Essentially it is intended to provide the team with the information needed to carry out the analysis. Figure 9.2 gives a reasonable summary of what should be included in an operating context. A perfect one would make completing all the spreadsheets similar to carrying out an "interpretation" question in an English exam. I don't know if they still have interpretations in exams today; there does not seem to be the same emphasis on the less important stuff like spelling, counting, and understanding any more, but basically the tester provides the student

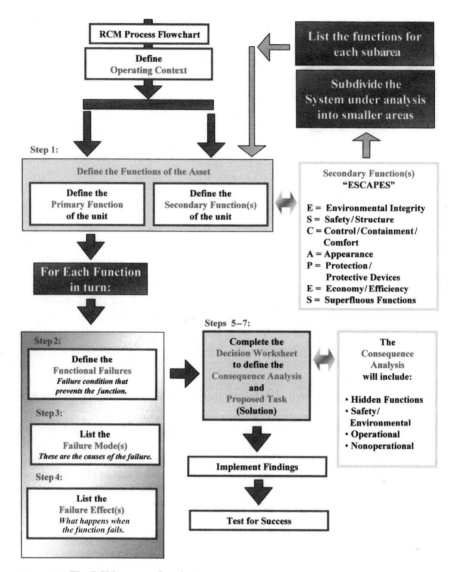

Figure 9.1 The RCM process flowchart.

with an essay to read and then tests his understanding of the contents. The answers to every question the team would want to know should be within the operating context, so I guess it is also a training document for the tool. In reality, it will not have all the answers. I suspect it will not even be close. It will be more general. This is not a real problem, it simply means the team will need to make regular investigations at the tool to find the information they want. This will become most

Management Level	Section Level	Tool/Analysis Level	Support Equipment Level
A description of the factory			
The products made at the site			
How much product			
The turnover			
The profitability			
The materials that factory uses			
Company ambitions			
Cost of failure of the asset	Cost of failure of the asset		
What the tool does in the process	What the tool does in the process		
Safety/environmental considerations	Safety/environmental considerations		
	Which department the tool is in		
	What happens to the product in the department, i.e., what processes does it see		
	Department environment, i.e., open factory, clean room, dark room, temperature/humidity controlled, etc.		
	The number of tools		
	The number of backup tools		
	Department throughput		
	The product quality/yield	Tool availability	
	The uptime/availability	Maintenance cost per hour	
	Maintenance cost per hour	Operator cost per hour	
	Operator cost per hour		
	How long can the tool be down for before causing problems		
	Time lost to PMs	Time lost to breakdowns	
	Time lost to breakdowns		
	Where are the raw materials stored		
	How is the material transported		
	Operator structure		
		Safety features on the tool	Safety features on the tool
		How the tool works & the process	How the tool works & the process
		How the tool is controlled	How the tool is controlled
		The actual tool throughput	
		The tool limitations & effect on product quality	
		The materials the tool uses	The materials the tool uses
		Air/nitrogen/water/extract/process cooling/toxic exhausts and other facilities	Air/nitrogen/water/extract/process cooling/toxic exhausts and other facilities
		Safe working procedures, risk assessments, and MSDS	Safe working procedures, risk assessments, and MSDS

Figure 9.2 Examples of the contents of an operating context.

TABLE 9.1 The Ideal Team Composition

Team member	Attributes
Technicians	Knows how the tool is maintained, problem areas, and the shortcuts taken.
Equipment engineers	Normally writes the PM task instructions and knows any omissions. Knows the tool capabilities, what it should be able to do as opposed to what is asked of it—required for understanding chronic issues that might be caused by use conditions. Writes the initial operating context.
Process engineers	Writes the recipes for the tool and defines how precise and accurate the process needs to be. May not appreciate the tool's limitations.
Operators	Knows how the machine operates, fails, and responds to failures.
Facilitator	Knows RCM. Need not have any experience of the tool.
Team leader	Drives the progress of the team, keeps the records, and shares out the workload. He can be a working member of the team.
Equipment vendor	
Experts	

apparent during the brainstorming and analysis stages. It will come as a (big) surprise just how much knowledge is missing in the team, mostly to the engineers and technicians who thought they were the experts—until now. This is not a serious flaw; in fact it keeps the technical staff motivated and prevents the task from becoming a purely theoretical exercise.

To carry out the analysis the team needs access to all of the tool's manuals, drawings, technical specifications, production and manning requirements, maintenance schedules, loss information (costs per hour), raw materials used, and so on. For companies that do not have manuals for their equipment, this would be a good time to contact the vendors. The RCM team itself needs members with knowledge of not only how the tool should work and its potential, but also how the tool is actually used today, since this is the basic criterion for an RCM analysis (see Table 9.1).

Basing the analysis on how the tool is actually used NOW, not what people would like or imagine it to be, does not limit the standard that can be set for the machine, but merely provides a realistic picture of the way it is and has been previously accepted. The next three questions define what RCM calls the Primary Function.

1. What does the tool actually do?
2. What does the tool do to the product or raw materials?
3. Why was it bought?

When Fig. 9.2 is studied closely, the reader will see that there is an overlap between operating context levels, but it really does not matter; if something is included in the wrong section, it can be taken out or vice versa. The team will decide on any changes needed as the analysis progresses. The operating context will probably take several weeks to write and requires someone with comprehensive knowledge of the tool, perhaps an equipment engineer.

Example of a Furnace Boatloader Operating Context: Tool Analysis Level

The original context was more than 12 pages long and was written by a colleague and friend of mine, Marshall North. Marshall is a strong proponent of RCM and was instrumental in its introduction to the organization. This is an edited example in which many of the technical details and drawings have been changed to make them more suitable as a teaching document for writing procedures as opposed to being an operating context to follow in an analysis. It is not important that the reader understands how a furnace works, but that he understands the level of detail required for the analysis.

- When the equipment is being described, the details of the tool should be comprehensive and include operation and functionality. Chapter 5 explains how to train on the subject of equipment. Use photographs and drawings to simplify the explanation and enhance understanding. Even though most team members will have experience of the tool, their knowledge will be centered around different areas like maintenance or operation.

  "The furnace has four separate tubes, stacked one on top of the other. Each tube works virtually independently. Their control systems are also independent, acting as a complete sub-system that has its own gas feed, loading system, heating element with control and monitoring systems. Figure 9.3 is a rough schematic of a furnace stack. Some facilities are common and fed to all four tubes (nitrogen, compressed dry air and hydrogen) but the process is independently run for each tube."

When you are writing your own procedure, you will zoom in on important details. "Figure 9.4 is a close-up of the quartz boats sitting on a loading Paddle. The operator positions the boats on the silicon carbide Paddle, which can support a load of up to twenty pounds or six boats, each containing twenty-five wafers. An adjustable arrow on the rear panel points to the position of the middle wafer of the centre boat—when the paddle and arrow is set up properly. The position of the arrow can be a source of nonuniformity if it is incorrectly set up.

Dummy wafers are positioned at both ends of the process wafers. The dummy wafers (known as baffles) act like heat buffers and help stabilise the temperature in the production volume of the tube to $\pm 1°C$."

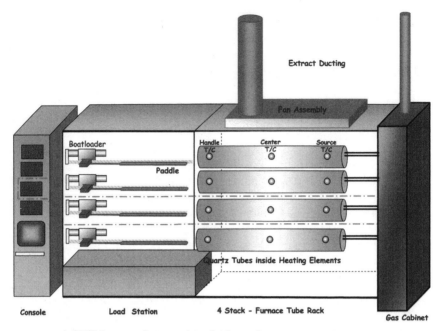

Figure 9.3 A TMX furnace photograph and schematic.

 Describe in detail what happens when the tool is operated. How does the operator interface with the tool? Is the operator always present? How much time does it take to load and unload? Is the machine capable of being left unattended? How long do process runs take: do the times vary? Are there any commands or operations that are known to cause problems? What about the surrounding layout; does it simplify issues and make the job easier or could it be improved? What are the audible and visual warnings the operator sees when a failure occurs?

"The operator downloads the process recipe and controls the Start and Stop routines at the Control Console, shown in Fig. 9.3. When

Figure 9.4 Wafers, boats, and a paddle.

"Start" is pressed, the Paddle moves into the furnace tube at a controlled speed. This is to protect the wafers from thermal shock. Too fast a speed will damage the wafers. The process for the tool under analysis limits the speed to 9 ± 0.5 cm/minute as it moves into and out of the furnace tube. The Tube Computer constantly monitors the speed and position of the Paddle, getting its data from the motor's encoder pulses. If the encoder fails or the lead screw sticks or stops, an "Event 0" will be initiated. This is known as a Boat Stall alarm."

- Explain how problems affect the tool at an appropriate level. The issue needs to be understood enough to enable the team to take action to avoid the failure or to understand the magnitude of the repair. Include known problems, repair times, and costs.

   "Slip is a problem that breaks wafers at a molecular level. The wafer is a three dimensional crystal lattice, like a cube of bricks. Slip has the ability to crack the joins between the atoms in its crystal structure (bricks). The damage happens when the wafers enter the furnace with their front face towards the heat. The edges of the wafers heat up (or cool on the way out) at a greater rate than the centre of the wafer. It is kind of like the way an ice cube cracks when dropped into a glass of liquid. The thermal stress...."

   Another example could be, "... The Paddle drives a distance of 195 ± 1 cm into the tube until it reaches the calibrated Zero Position (0 ± 1 cm). If the Paddle travels beyond Zero by more than 1 cm it can crash and cause a boat stall and create impact damage....

   The damage can be severe and has the potential to affect the other stacks in the tube. If the paddle stops 1cm before the zero position, the tube will not seal at the entrance. This might allow gas and heat to escape which will affect the flow dynamics of the process and cause the product to be affected."

- Explain the actual process with specific values and tolerances. It is important to be precise.

   "The wafers are not only warmed by the heating element and the time the wafers are exposed to the tube temperatures, but are also cooled by the flow of gas across the wafers. (In the same way tea is cooled by blowing.) This means that wafers can be under-processed if the Boatloader moves in too slowly or the gas flow increases or is unstable...."

- Explain in detail any control logic or circuits within the boundaries that might help the analysis. If a description is omitted, it can be added later when it is needed.

   "... the drive assembly is checked at three points along the length of the lead screw. (See Fig. 9.5) The "real" position is confirmed by the location of the microswitches. There is a cam mounted on the

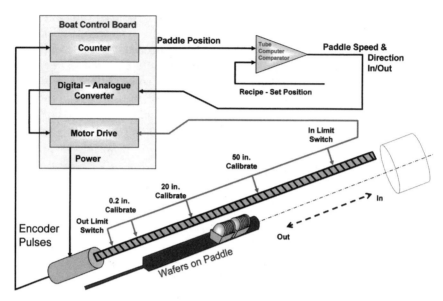

Figure 9.5 The paddle drive and positioning.

paddle drive assembly that operates the switch as it passes. It resets the Tube Computer to the correct position."

Or, "The lead screw has limit switches positioned at each end of the travel. Activating either switch will generate an 'Event 4' which will alarm and inhibit the Boatloader motor. The IN limit is just beyond the Zero position and, if the system is properly set up the limit switch will protect the tube—and product—from damage should the Paddle drive too far. If it is not correct, an overrun can cause physical damage." These examples relate to switches, but any essential electronics should be included.

Notice that Fig. 9.5 has details on the motor and the board that controls it. Use other diagrams, perhaps from the vendor manuals, that show how the systems interconnect. As I mentioned before, the operating context is like a training document that helps boost the knowledge of the team. Figure 9.6 also shows the board that controls the motor, but this one has a line to the central computer, which feeds out to the furnace controls. The problem we are dealing with here links the speed of the motor as it enters the furnace with the process temperatures and the gas/vapor residue. The previous two sentences do not have adequate detail for a real operating context. As a rough guide, if you have to ask what it means, it is wrong.

- Are there any manmade or design problems? The tool we are using for this analysis is very well designed. In actual fact, its uptime

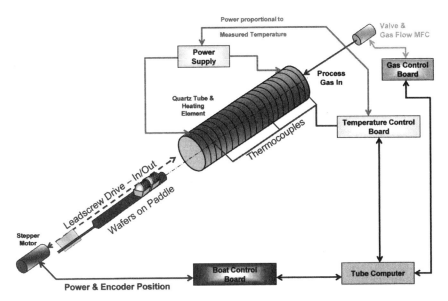

Figure 9.6 Use of overlapping diagrams to show linkage.

is regularly only a few points away from 100%. This tool has been around for many years, but is still in demand by companies. Any limitations it has tend to be due to its age and how its design specifications conflict with process engineering desires. Even so, the tool still performs very well.

"... initiates an audible alarm at the tube computer. This alarm has shown itself to be unreliable since, if no operator hears it, the product can be heated for too long i.e. over processed. The system is not interlocked, the audible alarm can be halted by resetting even when no action is taken to correct the problem.... There have been reports of alarms being accidentally reset by personnel unaware...."

📖 Describe all the safety circuits and modules. Include all the specifications and "whole system" functions. If I describe the functionality of my kitchen, I would have to include the alarm system, including the control box, even though it is not in the area. I would also have to include the "toast ready" alarm, which is located in the hall. You might recognize this function under its more common name: the smoke detector. The alarm is a "whole system" function: it serves the whole house.

"An extract is located at the entrance of the tube. Its function is to remove any toxic and process gases as they leave the process tube. Each furnace has its own, controllable Scavenger which has a monitored extract, fed from.... The gas flow rate can affect the quality of the product, so the setting of the Scavenger, which effectively "pulls" the gas through the tube, has a direct impact on the flow and will affect the uniformity of the product. The main extract duct supports all four scavenger units and sits at an extract rate between 145 and 155 ft^3/minute.

The scavenger is also affected if the loading door does not close properly as it draws air in from outside. This has the effect of reducing the "pull" on the process gas. Also…"

The above points reference ways that the product quality can be affected. RCM needs to know what these issues are to enable it to evaluate the cost of failure or missetting.

☐ Explain how to set up production and what the maintenance schedules are. RCM is concerned with costs that are generated by machine failures. It needs to know how much it costs to run a line: the time it takes the operator to run the machine and how much it costs the technician/engineer to maintain the tool. It also needs to know the cost of the materials, the amount of waste, how often runs are abandoned, the yield of the product, etc. In short, it needs to know where the money goes. Not only the obvious money that has to spent, but also the internal, invisible cash flow that is paid whether actions happen or not—like heating, operators waiting with no work, or materials being used while there is no production.

☐ What are the limits of the operating context? Define any boundaries. The operating context sets the area the analysis will cover. It will be defined coarsely by the manager, "I want to review the loading system…." The engineer will have to decide which areas to include. If he excludes the electrical supply to the tool, we would have no functional loading system. However, he does not have to consider the faultfinding systems that do not deal specifically with loading.

The bottom line is that if the boundaries are set and found to be in need of modification, this can be changed if approved by the team. Limiting the area to only the relevant parts is deliberate and planned to save time on the analysis and avoid unnecessary work for the team.

Actions

9-1 ☺ Select the tool to be analyzed.
9-2 ☺ Select the team members.
9-3 ☺ Train the team members.
9-4 ☺ Collect all the relevant technical data and manuals.
9-5 ☺ Select the appropriate author, and write the operating context.

Equipment Defined as Functions

RCM, like TPM, is a manpower-hungry system, requiring a high level of equipment knowledge. To get the best return on manpower I recommend using TPM to train the teams, restore basic condition, develop the initial maintenance schedule and infrastructure, and then apply RCM to refine the maintenance and eliminate chronic issues. RCM requires a deeper knowledge and experience of the equipment from all areas: operators, engineers, production, and facilities. The one RCM feature that really appeals to managers is the PM option of "Running to Failure" and that can be the outcome when knowledge is inadequate. It is hoped (if not expected) that RCM will uncover all the unnecessary tasks currently being carried out and magically reduce maintenance time substantially. This is a possibility, more likely with an inexperienced team, but it is just as likely there could be more *potential* PMs uncovered during the analysis.

Some companies, who operate blanket Reactive Maintenance systems, always allow the production equipment to run to failure. I don't think this is really the best type of maintenance system required to fulfill today's consumer demands. Most other companies already permit some parts to run to failure, provided they fail and have no immediate, negative effect. The decision on which parts are allowed to fail was probably never deliberate, but just evolved through working practices. The trick is not letting the wrong parts fail. A good illustration is corridor lighting. RCM would talk in functions. The function would be something like "To illuminate the corridor between A and B to a uniform intensity of X ± 100 candles per square meter." (The units might be old but the idea is correct.) This function can fail with different consequences depending on the number of lamps in the corridor. If there are 10 lamps, the failure of any one will affect only the uniformity of the illumination. For the corridor to fail to darkness, all 10 would need to fail at one time, so individual lamps can be permitted to run to failure. Now consider the same situation except now the corridor has only one light? Its failure would lead to total darkness. A team following the RCM process would evaluate the *function* of the light and the *consequences* of failure before deciding just to let it fail. If failure left the corridor in darkness and could present a safety hazard, RCM would recommend a "compulsory redesign." This would probably lead to a circuit modification to install an extra lamp, to ensure the corridor can never fail to darkness because of one lamp failure. The whole functional analysis would change if the lamps were in a greenhouse and the lighting was to provide plants with energy for photosynthesis. This time any one lamp failing might affect an area of plants. The *function* of the lamp is different.

Let's back up a bit and find out a bit more about functions and why RCM ignores the parts and components in a tool. Rather than say "...the tool has a thermocouple," RCM prefers to use a *function* similar to the following:

"To be able to measure the temperature of the center zone of the furnace within the range 400 to 1150°C with an accuracy of ±1°C."

Now, if this function fails, a *functional failure,* we would have a failure statement along the lines of

"Unable to measure the temperature of the center zone of the furnace."

Or

"Unable to measure the temperature of the center zone of the furnace with an accuracy of ±1°C."

Or

"Unable to measure the temperature of the center zone of the furnace above 500°C."

Notice that the single function above can fail in *at least* three ways. Failure of the function leads us to consider everything that could cause the failure. These are known as the *"failure modes."* The failure mode could be the thermocouple, but it could also be a cable, a connector, a thermocouple positioning problem, an amplifier, a failure in a temperature control board, or a temperature display problem. By looking at functions and not modules, we immediately start considering the whole tool and not just the thermocouples.

TABLE 9.2 Features Versus Functions of a Pen

Features	Features as functions
It feels very comfortable when writing.	To be able to write for a minimum period of 1 h with no discomfort.
It comes with a range of different ink colors.	To be capable of changing the color of the ink to red, green, blue, or black within a time of 10 ± 1 s.
It has a very nice appearance.	To have a styled appearance that will be acceptable by a minimum of 75% of the target audience. (The limit could also have a window: 72 ± 3%.)
It can write a thousand pages before needing to be refilled.	To be capable of writing a minimum of 1000 pages of lined A4 on a minimum 10 mL charge of ink.
It never leaks ink.	To contain the ink.
It is biodegradable.	To be capable of decomposing by 90 ± 1% of its volume in a maximum time of 300 years.
It is of low cost.	To cost less than $4.
It can be used to write with a thin line or a thick line.	To be capable of changing the writing tip size within the range 0.3 to 3 mm.
It writes with a nice smooth finish.	To be capable of completing an unbroken line over a minimum length of 500 in.
It can be changed from a ballpoint to a fiber tip.	To be capable of alternating the writing assembly between a ballpoint and a fiber tip in a maximum time of 15 s.

Did you notice how detailed the functions are? Why was the function not simply "To measure temperature"? Is there a difference between functions and features? Take a look at Table 9.2 which compares the features and functions of a pen.

Now, let's buy a pen.... Which one will I buy? When buying a pen, we probably buy it to fulfill only a couple of functions. In reality, it is likely we will not even make a list of the features we want, but we will know why we want one. Assume the pen is being bought as a gift; we might choose an expensive, smart-looking, high-quality pen with a nice box. However, if the pen is being bought to mark tiles for cutting, we might choose a cheap, nonpermanent felt tip or even decide that a pencil is more appropriate. So, the pen we buy will depend on why we want it: that is, the *function* we want it to fulfill.

Pen manufacturers must make the same choices as we do when designing pens, but they have to cover more options to make it an attractive purchase to a range of buyers.

The functions we want our pen to fulfill will have different levels of importance, with some being more critical than others. For example, the pen might look fabulous and come in a very slick box but if it is a terrible writer, it will be of no use. So there is one function that the pen must have, its *Primary Function*: the pen must be able to write. (Unless, of course, the pen has been bought for your partner to use with her/his checkbook....) The appearance of the pen and the nice box are still important but they are *Secondary Functions*. They could be classed as primary functions (it is possible to have more than one), but I cannot see any gain in it. RCM uses the same criteria when analyzing equipment. It looks for the Primary Function and then the Secondary Functions.

Notice the list of features in Table 9.1. Features are very similar to functions with one noticeable exception: features tend to be very general. Functions are intended to be as precise as possible, since they also enable the definition of failure. What is easier to confirm: "It has a nice appearance" or "To have a styled appearance that will be acceptable by a minimum of 75 percent of a target audience of 18- to 25-year-olds"? The latter is easier, because it is more specific. If we carry out a survey and only 60 percent think our pen has a nice appearance, then it fails. Even if 74.9 percent think it is nice, it still fails; 75 percent passes.

The Primary Function
The primary function is the main reason that the owner/user bought the item.

Secondary Functions
Secondary functions are all the other reasons for buying an item (Table 9.3).

TABLE 9.3 Secondary Functions

Safety	Appearance	Build quality
Safety interlocks	Cost of ownership	Efficiency
Environmental safety	The control system	Footprint size
Ergonomics	Reliability	Disposability

A secondary function is not an unimportant function. A car buyer will be unlikely to buy one he does not like the look of, even though it is not the primary function of a car. The looks are very important, even if the other features, the ones that make it run properly, the number of seats, miles per gallon, and quality of drive were all highly acceptable. Equally, the car having four doors, being an estate to allow for the dog, having ABS brakes, or the price could all be serious considerations, but they are all secondary functions.

The book by John Moubray refers to a handy pneumonic called the ESCAPES (Fig. 9.7). The ESCAPES are a list of functional groups that can be used as a memory prompter during the brainstorming session for identifying all of the functions of the asset.

When listing functions, the "ESCAPES" or another list of categories of your own choice, should be printed out and positioned on the wall, where it can be easily referenced by the team members. Remember to include the superfluous functions. These tend to be functions that are redundant or no longer used. They might be upgraded remote controllers, replaced by a central master control system; a circuit function that has been incorporated on to another board; a bank of pneumatic switches that have been replaced by fiber optics lines; or an internal pump that has been replaced by an external unit. If they are not removed, they still have the potential for failure and could have a negative effect on reliability.

Secondary Function(s)
"ESCAPES"

E = Environmental Integrity
S = Safety/Structure
C = Control/Containment/Comfort
A = Appearance
P = Protection/Protective Devices
E = Economy/Efficiency
S = Superfluous Functions

Figure 9.7 The ESCAPES as a memory prompter.

Functions should not be generalizations unless there is no other way to define them; accuracy is far better. This is a very important consideration when defining functions. By the very act of defining the function we are declaring what is acceptable and, by deduction, what is not. That is, we are also defining the conditions that constitute a failure of the function. Having an agreed standard about what constitutes a failure removes any dubiety about whether or not an asset needs to be repaired. This has frequently led to production disputes in the past. Because one defines the other, be certain the function is accurate and not hypothetical. Proper, agreed functions with acceptable values will avoid production disputes.

When listing the functions, you must state them as they are actually used on the tool—not as the user would like it to be or as listed in the asset's specification or on a quality list.

Consider the following function; it can have two failure modes: complete or partial. "To measure the temperature of the center zone of the furnace within the range 400 to 1150°C with an accuracy of ±1°C."

The *total failure* would be

"Unable to measure the temperature of the center zone of the furnace."

The *partial failure* could be

"Unable to measure the temperature of the center zone of the furnace with an accuracy of ±1°C."

The total failure cannot be ignored: it stops the tool running. With no temperature measurement, production cannot continue and the tool must be repaired. The partial failure, however, tends to be regarded more flexibly depending on the immediate circumstances. Provided the product is still capable of moving through the tool, it is often tempting to accept "a tolerable error" of, say, 1.5°C—just to relieve the pressure of production—even though 1.0°C is specified in the function. After all, it is only out by half a degree! That is hardly anything at all. The partial failure can have little or no effect on product throughput, but can have an effect on quality. If the half-degree change in temperature does not affect the quality (or *overall equipment efficiency*), then the wrong tolerance has been included in the function. Remember RCM is about how the tool is used *now*, not about the perception of the ideal.

If the current practice is to accept that an error of 1.5°C will not affect product and needs no intervention, then 1.5°C is the tolerance that must be defined in the function. The function must be valid.

If we change the function to ±1.5°C and it turns out that there is a quality issue discovered further down the line, then at least we have a possible cause. But if we have hidden our increased tolerance, we have obscured the cause of the fault and will be unable to avoid a repeat situation, and more scrap, in the future.

Defining functions is simply trying to find an easy way to document what something does as precisely as possible and in as short a way as possible, while avoiding waffle. I think everyone uses the following format. I have made changes when I could not get them to fit. The following list is a few examples of functions:

1. To heat the floor to a minimum of 20°C.
2. Not to heat the floor to a temperature greater than 20°C.
3. To heat the floor to a maximum of 20 + 0.5°C.
4. To cool the compartment to -20 ± 1°C.
5. Not to cool the compartment to less than $-20 - 1$°C.
6. To spin the disk to 1200 ± 3 rpm.
7. Not to spin the disk faster than 1210 rpm.
8. To spin the disk to a maximum of 1210 rpm.
9. To contain 1 pint ± 0.5 Fl Oz.
10. To contain a minimum of 1 pint.

There are several types of performance standards that can be used.

✓ Quantifying
These have values.
9.5 cm/s, 20 l/min, 400 sccm, or 7000 rpm.

✓ Qualitative
These are more descriptive but are difficult to be precise.
Looks acceptable, soft to the touch, tinted glass, warm coloring, pale blue.

✓ Absolute
To contain the gas, waterproof, nonstick.
This means that any leakage is a failed state.

✓ Variable
The quantifying value can change during use so the worst case must be used:
  The weight (downward force) of a person as a plane takes off, the height at which a plane flies, the stress on the fuselage of a rocket, the fuel consumption of a car with speed, the wear of a car tire with road conditions, the stress on a car's suspension with surface uniformity.

✓ Tolerance: Upper and Lower Limits
  300 ± 10 l/min, 500 ± 50 g, $1000 + 1, -0$°C, 25 ± 5 per hour.

If information is required to make the function more specific and there are no details available, make enquiries with the vendors, the process engineers, the operators, facilities, the quality department, purchasing, or the HR department.

There are a few useful points to consider when defining functions.

- A function can always be redefined at the team's discretion.
  If you discover the original is not good enough.
    Functions can become way too complex if too many conditions are in one function. Consider splitting it into two.
- Functions are defined from the *user's* perspective.
- The function must be written to reflect the tool as it is currently used.
  There is often the management standard and the currently accepted ones.
- For large area analysis, take photographs of the areas being analyzed and divide them into smaller areas to help with finding all of the functions.
- As a team, brainstorm the functions and standards.
- Record everything on flip charts or Post–it Notes to avoid missing or losing any functions.
    Post-it Notes are an excellent way to reorganize data when it becomes advantageous. Just move it.
- Start with the primary function of the tool.
  The primary function of the boatloader system might be
    "To be capable of moving silicon wafers up to a maximum load of $20 + 0.25$ lb, on a silicon carbide paddle, a distance of $195 + 0.5/-0$ cm at a rate of 9.5 ± 0.5 cm/min into and out of the furnace tube."
- Record the primary function on the RCM Information Worksheet. (Fig. 9.8.)
- Only identify functions to the level to which you would faultfind.
  If a power supply would be replaced rather than repaired then replacement would be the level of the analysis . There is no need to drill down to fans or internal circuit boards.
- Include all system functions.
  (Emergency stops, fuses, vacuum pumps, power supplies, pneumatics, etc.)
- Secondary functions are everything else that the tool/system does in addition to the primary function.
  These are features that you might want the machine to have, but the machine could still carry out its primary function without them.
- Consider the secondary functions by brainstorming.
  List all of the functions on the flip chart or Post-it Notes. This causes duplication, but when analyzing the data later, the duplication can be filtered out.

274 Chapter Nine

Information Worksheet On Excel

Unit or Item:		Unit or Item No.		Sheet:
Item or Component:		Item or Component No.		Date:

No:	Function	No:	Functional Failure *Loss of function*	No:	Failure Mode *Cause of failure*	Failure Effects *What happens when it fails*

Figure 9.8 The Information Worksheet.

- Look for hidden functions.
- Stay within the boundary defined by the operating context.
  However, the boundary can be redefined if the team agrees. An example might be to include a remote roof fan if it affects the stability of the tool being analyzed.

Features of a machine can be converted into functions for use in the analysis. Look for the following common features:

- To have an appearance that will impress customers.
- To have easy-to-operate controls.
- Covers are two-directional: to protect personnel from touching hazardous components and to protect components from external sources of damage. For example, electrical panels from water or short circuit.
- Areas specially designed to make them acid-resistant, fire-proof, resistant to chemicals, to heat, etc.
- Safety warnings.
  To warn personnel at the tool or remotely when the tool is in a failure condition or of any hazards present through, for example, notices or labels.
- The capability to run processes automatically—if desired.
- Automatic failure detection to tell you when the tool is operating out of its functional limits.
- To protect the environment.
  Toxic exhausts, abatement systems, filters, RF shielding and x-ray shielding, drains to catch liquid spills, sensors to detect leaking gases.
- Interlocks that give you warnings of failures and stop the system in emergencies.

Then there is the category discussed in Chap. 8 while looking at the decision diagram: the hidden functions. These are protection devices, similar to those in List 9.1, invisible to the operation of the equipment and operators during normal operation. An example would be the overtemperature sensor in an electric kettle. It can fail to operate or have already failed and the user will have no idea of the failure until the kettle boils dry and overheats. Ask the question

"If the function fails during normal operation of the unit, will the operator be made aware of the failure?"

A "No" answer means you have identified a hidden function.

List 9.1: Hidden Function Examples

Overpressure switches	Underpressure switches	Overrun switches
Overcurrent trips	Earth leakage detectors	emergency lighting
Low-flow switches	Fire alarms	Emergency off switches
Relief valves	Smoke detectors	Inflatable life boats
PPE	Nonreturn valves	Overtemperature switches
Parachutes	Breathing apparatus	Fire extinguishers
Backup pumps	Vibration switches	Shear pins
Toxic gas detection devices	Backup power supplies	

The interesting thing about hidden functions is that many engineers don't know they exist until they have failed. It is essential to talk to the vendor and confirm that they have all been identified. So, if the function is hidden, the only way to check it is to test it. This is known as *failure finding*. Functions requiring checking would include

- Manually checking an EMO (emergency off switch).

- Increasing the heater temperature to a level greater than the overtemperature limit or, if the sensor is adjustable, setting the limit lower to see if it trips.

- Checking the operation of the upper/lower level switch in a water tank.

- Checking the operation of a backup generator or pump.

Safety devices are designed to fail in such a way that there is no danger to the user or the environment. A device that fails in this way is called a fail-safe device. A fail-safe device in RCM is *not* the same as the normal, non-RCM definition. It is

"A device that fails in such a way that it will become obvious to the operators under normal circumstances."

Where possible, no protective devices should ever be used that can fail in such a way that there is no indication or warning of the loss of the function.

The hidden function is defined in a slightly different way. This is because it is only intended to operate in the event that something else fails, i.e., the device it is protecting.

"To be capable of protecting the ... in the event that the ... should fail to operate."

The function of an overpressure relief valve would be defined as

"*To be capable of* opening valve V9 *in the event that* the pressure within the system becomes equal to or greater than 120 psig."

To avoid extra analysis work, we try to avoid having the same function listed more than once. This does not mean simply to ignore all switches—quite the reverse, in fact. "Switches" can be found all over a tool but the *function* of each will be different, so they will not be deleted. Never list the function as a "switch" but as the purpose of the switch:

"To illuminate indicator L1 and sound alarm B3 on the control console when the cooling water level is less than 25 L."

Or

"To reset the digital position counter to 20.0 cm when the boatloader is traveling in the outward direction."

Or

"To reset the digital position counter to 20.0 cm when the boatloader is traveling in the inward direction."

Every area will have protective covers, but their functions and consequences will also differ. The covers will be there to protect the user and the equipment from electrical hazards, mechanical hazards, chemical hazards, or other dangers. The function of the cover could be

➢ "To prevent the operator from touching Contactor CB-3 in the remote power distribution panel."
➢ "To prevent access to the rotating fan F3 while it is operating."

A duplicate function is one whose failure will prevent the operation of another function. For example, all three functions below will cause the same total failure: "Does not provide any water flow."

1. "To provide a minimum water flow rate of 100 L/min."
2. "To be capable of removing electrical power to pump P1 in the water cooling circuit in the event that the feed current increases above 15 amps."
3. "To supply a current of up to 15 ± 1 amps DC at a maximum voltage of 14V DC to pump P1 in the water-cooling system."

It would be necessary to include only the first function. The second and third functions will both cause the first function to stop. (Beware of their impact on other functions too.)

When identified, do not delete the duplicate failures—just in case they are not duplicates. Score them out, move the Post-it Note to a "parking bay" or, if using a PC, gray the text. They can be reviewed at the end of the analysis to confirm the decision was correct. Figure 9.9 is a flowchart that demonstrates the method and questions to ask when testing for duplicate functions.

The RCM worksheet is virtually an FMEA (Failure Modes and Effects Analysis) worksheet. I have found that the teams could complete the sheet accurately until they reached the failure effects, where they

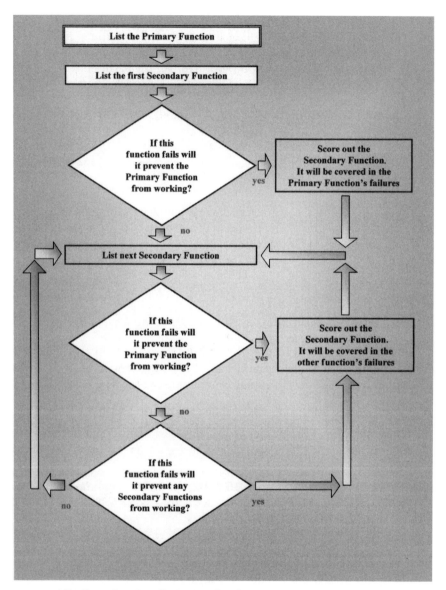

Figure 9.9 Duplicate function elimination flowchart.

started to miss details. Sometimes it was plain forgetfulness but other times, it was deliberate: ignoring the *unnecessary information*. As a memory jogger and a means of simplification, I added four columns to an alternative worksheet (Fig. 9.10): "Secondary or Consequential Damage," "Safety Warning Signs or Switches," "Safety Hazard Created," and "Human Error Input."

Master Inf W-sheet unlocked

	Labor Rate (Average)		Scrap Cost for Main Unit Reviewed		Consequential Scrap Cost per Machine		Lost Production Main Unit Reviewed (Hourly Rate)		Lost Production (Consequential) (Hourly Rate)
	Unit or Item:	SVG Furnace system	Unit or Item No.	LD6.3	Sheet:	See Footer	Cost and Consequences of Failure		
	Item or Component:	Boatloader	Item or Component No.		Date:	25-May-01			

No:	Function	No:	Functional Failure *Loss of function*	No:	Failure Mode *Cause of failure*	Failure Effects *What happens when it fails*	Secondary or Consequential Damage	Safety Warning Signs or Switches	Safety Hazard Created	Human Error Input
1		A		1						

Figure 9.10 Modified Information Worksheet.

Even this did not draw all the details from the team. They still ignored specific costs like downtime hours, labor-hours cost, cost of lost production, parts cost, waiting time for parts, requalifying times, and so on. So I created a new master spreadsheet to track the cash values. See Fig. 9.11 for the extra columns.

The columns in Fig. 9.11 make up a functional spreadsheet that calculates the costs for both sides: the maintenance cost and the cost of the consequences.

They include

- Downtime hours including waiting
- Total repair cost
- Cost of scrap
- Cost of lost production
- Maximum waiting time before starting repair
- Diagnosis time
- Repair time
- Parts delivery time
- Parts cost
- Test run time
- Repair labor-hours
- Number of assets affected
- Waiting time and cost for lost assets

These columns are the ones I thought suited the purpose I had. If you are intending to carry out your own analysis, you have the freedom to make any changes that might make your own calculations more accurate. Not only that: if you want to try out a few variations that might help even more, then you can do so.

Identifying Functions and Labeling

The functions are analyzed and recorded on the Information Worksheet. The worksheet identifies each function and links the relevant parts. Figure 9.12 is a flow diagram of the steps we have already taken and those up to the use of the Information Worksheet. However, before filling in the sheet, we need a method for numbering. For standardization, this is the same method I was taught and is used in *RCM II* by John Moubray.

Figure 9.11 The extra "numbers" columns on the Information Worksheet.

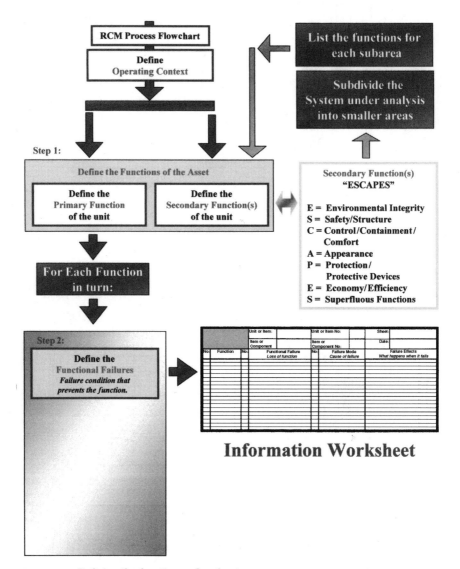

Figure 9.12 Defining the functions—flowchart.

- The functions are normally numbered 1, 2, 3, 4, 5, 6, 7, ... *This assumes that there is only one primary function (Function "1").*
- The functional failures for the functions are numbered A, B, C, D, E, F, ...
- The failure modes for the functional failures are numbered 1, 2, 3, 4, 5, 6, 7, ...

≫ The failure effect refers to the failure mode to the left of it and has no number—other than that of the failure mode.

So we have an absolute ID number for each failure mode linking it to the function that it refers to.

1A1, 1A2, 1A3, 1A4, ..., 1B1, 1B2, 1B3, 1B4, ..., 1C1, 1C2, 1C3, 1C4,...

"1" is always the primary function. If there are more than one primary functions, then you have the option of numbering them as you would a secondary function (that is 2 or 3)—since, in reality, they will be treated exactly the same way. It can be recorded that these are primary functions in the Function column if desired. Limit the number of primary functions to no more than three, my preference is one.

Equally (but I do not favor this one) you could call it "1" and have bullets to show it is connected to the primary function. Then the IDs might be "1a" - A – 1, "1b" - A – 1, "1c" - A – 1. I suspect that this might make the failure effect link a bit tenuous.

There are more than 100 functions in Fig. 9.13. To simplify finding them, take photographs of the system and divide the tool into smaller functional areas and then analyze the smaller areas. Then brainstorm each area.

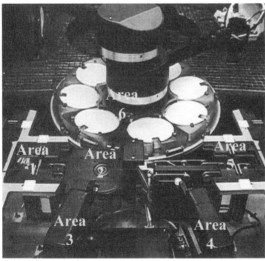

1. Wafer Loading
2. Wafer Orientation
3. Wafer to Disk
4. Wafer from Disk
5. Wafer Unload
6. Disk

Figure 9.13 Simplifying finding functions by subdividing the areas.

Use extra photographs and drawings as needed to make obscured areas more visible to the team and make the components more easily identifiable.

Functional Failures to Failure Effects

Functional failures are functions that do not fulfill their description as defined. They are usually the opposite of the functions. However, we need to look a bit deeper, particularly at the partial failures. For example, consider the function

"To be capable of illuminating indicator L1 and sounding alarm B3 on the control console in the event that the cooling water level falls to less than 25 litres."

This function has failed if the water level drops to any value less than 25 L, even 24.99 L would be a failure. This is the reason that functions try to be as accurate as possible: to remove any argument about whether a failure has occurred. If, in normal operation, 24 L would be acceptable for running the equipment, then the function must be changed to the following:

"To be capable of illuminating indicator L1 and sounding alarm B3 on the control console in the event that the cooling water level falls to less than 24 L."

It is important that the failure limits are agreed in conjunction with management and then are adhered to. There will always be the desire to run "just a little under (or over) the limit" making it essential that the value chosen for the performance standard is valid. If ignored, it risks having an effect on the process. If they are agreed now, when there are no production pressures, the decision will be based on facts and logic. There should be no need for further argument later.

Just as functions are precisely defined, so are functional failures. The two types of failure we will meet are "total" and "partial" failures. The total failure is usually easy to define; it tends to be the exact opposite of the function. To illustrate this, consider the function

"To coat the surface area of the product to a depth of 3.0 ± 0.5 mm with liquid chocolate."

A total failure could be

"Unable to coat the surface area of the product with liquid chocolate."

The partial failures are often a bit harder to define. The same function could have a partial failure of

"Unable to coat the surface area of the product to a depth of 3.0 ± 0.5 mm with liquid chocolate."

Or

"Unable to coat all of the surface area of the product to a depth of 3.0 ± 0.5 mm with liquid chocolate."

Both are partial failures and both are different. The chocolate could have lumps in it, have holes in it, be smooth to look at but be too thick or too thin, or it could even have ripples across the surface. Not only that, I have only suggested two failure modes, what if the chocolate was the wrong flavor, did not have enough sugar, had too much sugar, was too dark to be milk chocolate, or even have small lumps that are within the defined limits. The answer lies in the functions. The failures are defined on the basis of them. If they are not well enough defined, we need to recognize their limitations and make corrections. Missing a detail is not a catastrophe, we just need to make the correction and check how it affects our analysis. Always remember the rule: Plan-Do-Check-Act.

The test that I learned for confirming a functional failure is to ask "Do these words describe an observable failed state, associated with the function statement?"

	Actions
9-6	☺ Confirm that the operating context has been completed.
9-7	☺ Create the Information Worksheet. There are three options to choose from: Figs. 9.8, 9.10, and 9.11.
9-8	☺ Organize a meeting room with flip charts, Post-it Notes, pens, and other display media as required (TV, overhead or LCD projector, PC, etc.).
9-9	☺ Print out the ESCAPES or an alternative list of memory prompts on a large sheet of paper and attach to the wall of the meeting room.
9-10	☺ Photograph the areas under analysis and print on a large sheet of paper. Attach to the wall of the meeting room.
9-11	☺ If complex, divide the area photographs into manageable functional areas. (Fig. 9.13)
9-12	☺ Define the primary function by brainstorming.
9-13	☺ Enter the primary function on the Information Worksheet.
9-14	☺ List the secondary functions by brainstorming.
9-15	☺ Identify all hidden functions, including those that are written into software.
9-16	☺ Score out any functions that are obvious duplicates. Do not erase them as they can be referred to in future.
9-17	☺ List the functional failures on the Information Worksheet.

Failure modes

When we looked at functional failures we defined them as all the ways that could prevent the function from working as described. For each functional failure there can be a multitude of causes of the failure as

No:	Function	No:	Functional Failure	No:	Failure Mode
1	To measure the temperature of the center zone of the furnace within the range 400 to 1150°C with an accuracy of ±1°C.	A	Unable to measure the temperature of the center zone of the furnace.	1	No thermocouple in position.
				2	Thermocouple TC2 not connected to Terminal Block TB4.
				3	Cable break between thermocouple TC2 and display module DM1
				4	Furnace is off.
				5	Furnace is at a temperature less than 300°C.
				6	The Temperature Control Board p/n 200035 in the Control Console has failed.
				7	Temperature Display module DM1 in the Control Console has a fault.
				8	TC AMP 2 (Thermocouple Amplifier) power supply PS1 failed.
				9	Control Console power supply PS3 failed.
		B	Unable to measure the temperature of the center zone of the furnace with an accuracy of ±1°C.	1	Failure of the analogue to digital coverter IC-7 on the Temperature Control Board p/n 200035.
				2	Failure of the thermocouple amplifier TC AMP 2.
				3	Poor connection to thermocouple TC2.
				4	Thermocouple TC2 incorrectly positioned at the furnace tube.
				5	Thermocouple Amplifier TC AMP 2 power supply PS1 is unstable.
		C	Unable to measure the temperature of the center zone of the furnace above 500°C	1	Thermocouple Amplifier TC AMP 2 power supply PS1 has failed.
					Failure of the analogue to digital coverter IC-7 on the Temperature Control Board p/n 200035.
					Failure of the thermocouple amplifier TC AMP 2.
					Thermocouple TC2 incorrectly positioned at the furnace.
					Temperature Control Board p/n 200035 in the control Console has failed.

Figure 9.14 Example of functions, functional failures, and failure modes. (Data is for illustration only.)

shown in Fig. 9.14. We must identify all of the functions and the functional failures before we start looking for the *failure modes*. Analyzing failure modes is a team task. The best way to find them is by brainstorming.

The team does not have to start gathering information from scratch. Data will have been recorded before that can be easily accessed and can also be used to trigger new ideas.

➢ Which failures have happened before?
  ⇒ Fault history reports
  ⇒ Personal experience
  ⇒ Vendor engineers
  ⇒ Other users
  ⇒ The manufacturer

> Which parts do we already maintain?
> ⇒ Check with other users, they might maintain different parts.
> Are there any parts that are likely to fail or we know fail often?
> Each component in the tool will have a probability of failing. Some similar parts might have failed in other tools. By studying the area photographs and brainstorming for ideas or by visiting the machine and having a look around, it will be possible to get inspiration that will help generate more ideas. Consider
> ⇒ Power supplies
> ⇒ Circuit boards
> ⇒ Cable runs, connectors, pneumatic systems, components with o-rings and seals, moving parts, etc.
> ⇒ Some failures are very likely to fail, but some will have a very low likelihood and could be neglected—but not all!

Only analyze the failure modes to the level at which you maintain or faultfind your equipment. For example, board level and not component level; the pump and not the pump gasket; the computer and not the circuit board. Going too deep will extend the analysis time. However, RCM is flexible. It is up to the team to choose to look deeper where they believe it will help. Just remember to modify the operating context.

Apart from the usual causes of failure—mechanical, electrical, electronic, design, and deterioration—failure modes can often be due to human error (anthropomorphic). In most organizations, a very high percentage of issues are traced to people problems.

> Incorrect fitting of components.
> This can be attributed to poor alignment, using the wrong part, fitting the wrong way round, making incomplete connections, using the wrong tools, using excessive force, etc.
> Fitting faulty or poor quality components.
> Continuing to use components that have been damaged during fitting.
> Once it has been recognized that a part has been damaged and it is probable that it will fail, it would be folly to proceed with the assembly. In this situation, use a new part and find out how it will be possible to avoid another part from becoming damaged.
> Complicated assemblies—parts that are difficult to set up.
> If the assembly cannot be changed, then it boils down to a detailed procedure and a lot of practice. Try a technique like 100% Proficiency to repeat the task until it can be carried out correctly every time.
> Misleading procedures that do not explain exactly what to do and can lead to mistakes.

➤ Bad design: indicator lights and alarms that are difficult to see or hear because of their position or can simply be ignored by the operator.

With RCM being based on probabilities, there is one situation that must be taken seriously despite a low probability of occurrence. This is where the consequences are potentially very serious. Some failures are very unlikely to happen and can be ignored, but not all. Occasionally there will be one that can have unacceptable consequences. The example everyone uses to illustrate this point is aircraft flying zones. The chance of a plane crashing into a nuclear power station is very low, but the consequences if one was to crash would be immense. So aircraft are simply not permitted to fly anywhere near them. A new one has arisen since 9/11: planes flying into buildings. Other consequences could include chemical or gas leaks, oil spills, fires, contamination of food, causing accidents or injuries to personnel, loss of a very important customer account, etc.

Severe consequences do not have to be immense in magnitude, just important enough to the manufacturer that they cannot be allowed to happen. If something could affect the company's best customer, it might be regarded as unacceptable. It is up to the user to determine the probability of the risk versus the consequences. Whatever is decided, it must be certain that the logic is sound. The team does not have to make all the decisions on their own. They must remember they are part of a larger organization and can seek advice from anyone who might be in a position to help.

Failure Modes Summary

1. When listing failure modes, always work as a team and record every mode on a flip chart or a Post-it.
2. Number each failure mode so that it can be identified. It should be linked to the parent function and the functional failure.
3. If unhappy with the choice of a particular mode, don't delete it, just score it out and come back to it later if necessary.
4. Consider whether to list all the failure modes first, then to list the effects of each failure mode as it is discussed later, or whether to use a combination of both.
   Sometimes the team will hit a flow of ideas that might be lost if a strict regime has to be followed. Do what is best for the team, you can always revisit them.
5. Failure effects must be linked to the correct failure mode.

Failure effects

A chain of events follow every failure, if they didn't, the movie industry would have made some really short films. Each *event* will have its own equivalent $cost. Imagine we were intending to present a bill to the company for every failure effect. We should include all the technical details of the cause of the failure and the costs that the failure incurs—particularly labor and production losses. To get an idea of how much can be involved in the *consequences* of a failure, let's consider a hypothetical power supply failure in a temperature-controlled, acid-cleaning bath. This has not been written in the absolute detailed format that would be required for an Information Worksheet, it is written simply for illustration. Let's pick a simple reason for the power supply failure (the failure mode): loose output connections. We have all seen them in a car battery. The output current will arc between the supply contacts and the cable. This arcing eventually causes overheating which can melt the insulation. It can also cause smoke. In our example, we will assume it also burned a small hole in an adjacent polythene drain pipe.

Immediately following the initial failure, the system's automatic procedures were executed. These are controlled by the hard-wired circuitry and software in its control system. The operator watched the stirrer in the bath juddering before stopping. The console display dimmed for a second before the red warning lamp lit and the audible alarm sounded. This was followed by a muffled bang from below the machine and a burning smell. There appeared to be slight traces of smoke escaping at the rear of the tool. There were also possible acid fumes.

After a short delay, the external alarms sounded and initiated the fire and toxic gas warning lamps. The area was immediately evacuated. The emergency response team (ERT) checked out the area and isolated the tool power. A pool of liquid, assumed to be acid was forming on the floor below the bath. The ERT called the equipment technicians and engineers to investigate, plug the leak, and drain the acid from the bath. The tool and the other equipment in contact with the acid puddle were all rinsed, decontaminated.

Next, the maintenance group analyzed the fault, traced it to the power supply cables, and began to repair the damage. Three men worked on the tool for 7 h. There was an additional 4-h delay waiting for a new power supply, cable loom, and replacement pipe section to be fitted. We now have 11 h of production downtime for the tool and $3400 for parts. While waiting for the new supply, a meeting was held between the vendor, the equipment technician involved in the fix, the tool engineer, his manager, and the safety department manager to review the need for modifications required to prevent the same failure from happening again. A new PM inspection step and layout change was planned. After making the modifications and restoring the tool, a couple of days offline were needed by the senior engineer to create new procedures and document the new layout.

There was secondary damage to the floor and to the cables on adjacent tools, but the product was salvageable. It took one man 5 h to check whether the other tools were safe and repair any damage. The cost was $150. Six production tools had to stop running with two operators waiting to return to work—cost $220 labor and $850 in lost production.

The labor involved in the main incident and the repair will need to be established. It has to include the meetings and the redesign. The costs are very high, but apart from the parts no "real money" has been paid out. Before RCM, the failure cost would probably have been put at $3400 for parts. No consideration would have been given to any manpower, product lost on other tools, or the cost to remake the replacement product.

What was the total financial cost of the loss? What could we have billed the company? There will be a cash cost for the work by the purchasing team, the safety group for decontaminating the floor, and the vendor for replacing the floor tiles. There might even have been an overtime cost to catch up on lost production if an urgent delivery was required or possibly even a lost customer if the product could not be delivered on time.

While the costs in the example are not typical—although I have been to faults of the same magnitude—virtually every fault has a higher cost than is normally considered. The "normal" is only for the cost of the part and vendors, but it is always more: maintenance labor-hours, preproduction costs (time), quality check costs, operator's waiting times, parts collection time, and lost production on other tools. Some smaller costs might be ignored, but evaluate them to be certain. The cost of the failure in the example totaled more than $7000. Compare this cost to the preventive cost for the cables to be inspected during a PM: around $30. Is it a fair trade-off?

The above cost areas are begging to be used as headings for a spreadsheet that would display text and calculate costs.

When estimating the potential failure costs for RCM analyses, always assume that the failure occurred during normal operation of the tool and that no preventive measures were taken to limit the damage. If a failure can occur in a number of different ways that have differing degrees of consequences, it is not necessary to consider all of the different modes, but only the one that incurs the maximum costs. Consider brake failure in a car. The car can be parked, moving slowly, driving along a quiet road, driving in a busy town, or driving at high speed on a motorway. The most serious failure is the only one that has to be considered, as any improvements would need to be based on that option.

A major part of RCM consists of comparing the cost of checking a tool against the total cost of a failure with all of the associated consequential costs. To quote John Moubray, before committing to carrying out a maintenance task or a check, it must be "...technically feasible and worth doing." It is reasonably obvious to say that most people do not have any idea what the real cost of a failure could be; what is less obvious is that most people probably do not know how much each current scheduled maintenance task is costing them.

To evaluate the cost of a *failure effect* we need to ask a few questions. Actual data for some of the failure modes will not be available, so a team consensus of hypothetical data will need to be assembled using the technical expertise of the RCM team. Extra information must be sought from other sources if required.

- How do we know that a fault has actually occurred?
  What messages or alarms displayed at the tool? Are there any more alarms? Does anything stop working that is visible to the operator; are there any noises, smells, or backup systems that come into operation? Does the product show any defects?
- What do we need to do to repair the failure?
  What are the faultfinding, repair, and testing times as a labor-hours cost? What is the cost of the parts?
- Is it possible that the failure could cause secondary damage to other components or other tools?
- Could the failure present a safety risk to personnel either immediately or later?
  Could the failure mode affect the safety of the operator, other persons in the area, or any person likely to enter the area?
- Could the failure damage the environment in any way—internal or external—by releasing materials into the atmosphere?
  A material might be inert on its own, but what happens to it when it is ignited and burns or it comes into contact with other materials? Look for sources of danger: chemicals, gas, biological, electromagnetic radiation, or any other method.
- Could the failure mode affect production on this tool or any other tools?
  Requalification times and test piece costs, operator waiting time, and the time spent setting up for production.

	Actions
9-18	☺ List the failure modes and failure effects; do so as part of a group and note every mode on a flip chart.
9-19	☺ Decide whether to list all the modes first, whether to list the effects of each failure mode as it is discussed, or whether to use a combination of both.
9-20	☺ Number each failure mode so that it is easily traceable to the function and the functional failure.
9-21	☺ Number each of the failure effects so that they are easily traceable to the failure modes.
9-22	☺ Transfer the data to the Information Worksheet.

Figures 9.15 to 9.17 are guides as to what a standard Information Worksheet would look like. Notice in figure 9-17 that functions 12-A-1 and 13-A-1 are grayed out. These are duplicate functions that will be covered in previous failures, but have been retained in case subsequent analysis is required. "Graying" them out or scoring them out is preferred to deletion as deletion is a wee bit too permanent.

Where Did RCM Come From?

So where did RCM come from? Who would be so unhappy with a system of maintenance that they would go out and develop a completely new one? Well, it could have been Taiichi Ohno, but he went for the Toyota Production System. So, who else could it be? It is almost a trick question. I kind of wonder if the original faultfinding engineers working on the equipment did not faultfind to root causes. Maybe there was no need in their industry. It could just have been that money was no object and so expectations were set too low. Had they repaired to a deeper standard, would they not have realized—eventually—that the breakdown was due to the part being incorrectly assembled, wrong for the purpose, or maintained before it needed to be? I normally pose the question, if the guys who fixed the tools were due to get on a plane and fly home, could they be certain that the equipment will not break down again soon after they leave? Personally, I am glad RCM was developed. I believe it is a very useful technique. One major advantage of RCM is the concept of analyzing consequences and having to take actions—like create backup systems—to eliminate the effect of serious failures.

So let's go back a step. Who would be so unhappy with a system of maintenance that they would go out and develop a new one? The answer is the American Aircraft industry.

Boatloader Info Worksheet Rev 6

No:	Function	No:	Functional Failure Loss of function	No:	Failure Mode Cause of failure	Failure Effects
1	To be capable of moving a load of silicon wafers and baffles not exceeding 20 lb, on a silicon carbide paddle, a minimum distance of 195 cm at a rate of 9.5 ± 0.5 cm/min in and of the furnace.	A	Unable to drive the paddle into the furnace.	1	Leadscrew Motor failure.	Boatloader does not move in. Alarms on the Tube Computer after 30 s. The alarm warning lamp flashes as does "Event Zero." The process time remaining also appears. "B" (Boat Stall). Motor requires replacement. Before repair can begin the three other tubes cannot be running and must be returned to standby. Time lost before the repair can start — Tubes stay down during repair—production lost. If operator misses the alarm, wafers will be destroyed. Equally, after wafers driven out, operator can unload them by mistake—assuming process complete and feed them through the system underprocessed. Costs: One day lost production $3000, Worst case scrap $10,000, 9 labor-hours cost $270 - Plus motor cost.
1		A		2	Drive belt broken	Boatloader does not move in. Alarms on the Tube Computer after 30 s. The alarm warning lamp flashes as does "Event Zero." The process time remaining also appears. "B" (Boat Stall). Belt requires replacement. Before repair can begin the three other tubes cannot be running and must be returned to standby. Time lost before the repair can start — Tubes stay down during repair—production lost. If operator misses the alarm, wafers will be destroyed. Equally, after wafers driven out, operator can unload them by mistake—assuming process complete and feed them through the system underprocessed. Costs: 30 h lost production $3600, Worst case scrap $10,000, 9 labor-hours cost $270 - Plus belt cost.
1		A		3	Dirty leadscrew	Boatloader does not move in. Alarms on the Tube Computer after 30 s. The alarm warning lamp flashes as does "Event Zero." The process time remaining also appears. "B" (Boat Stall). Leadscrew needs removed to clean. Before repair can begin the three other tubes must be IN the furnace and not running product. Time lost before the repair can start is 2 h minimum. Production lost only during removal and repair time. If operator misses the alarm, wafers will be destroyed. Equally, after wafers driven out, operator can unload them by mistake—assuming process complete and feed them through the system underprocessed. Costs: 9 h lost production $2000, Worst case scrap $10,000, 7 labor-hours cost $210.
1		A		4	Rohlix bearing failed	Boatloader does not move in. Alarms on the Tube Computer after 30 s. The alarm warning lamp flashes as does "Event Zero." The process time remaining also appears. "B" (Boat Stall). Leadscrew needs removed to fix. Before repair can begin the three other tubes cannot be running and must be returned to standby. Time lost before the repair can start. Production lost only during removal and repair time. If operator misses the alarm, wafers will be destroyed. Equally, after wafers driven out, operator can unload them by mistake—assuming process complete and feed them through the system underprocessed. Costs: 26 h lost production $3600, Worst case scrap $10,000, 36 labor-hours cost $1080 - Plus bearing cost.

Figure 9.15 Information Worksheet Example 1.

In America, people were reluctant to fly because of the perceived unreliability of aircraft. Today, we expect to turn up at the airport, wait a few hours, get on the plane, and fly off with no problems—other than scheduling issues or industrial action. This was not the case in the 1960s and 1970s. The industry was looking for better availability, better reliability, and fewer failures, so they did what virtually everyone else would probably have done. A review of their systems and setup was initiated and they carried out more PMs. If it is failing, it is overdue for

Chapter Nine

Boatloader Info Worksheet Rev 6

No:	Function	No:	Functional Failure Loss of function	No:	Failure Mode Cause of failure	Failure Effects
1		A		5	Failure of Tube Computer - 24V PSU	Boatloader does not move in. Alarms on the Tube Computer after 30 seconds. The alarm warning lamp flashes as does "Event Zero." The process time remaining also appears. "B" (Boat Stall). Leadscrew needs removed to fix. Before repair can begin the three other tubes must be in standy — risk of electrical damage to other computers. Time lost before the repair can start. Production lost during removal and repair time. If operator misses the alarm, wafers will be destroyed. Equally, after wafers driven out, operator can unload them by mistake — assuming process complete and feed them through the system underprocessed. Costs: 50 h lost production $10,000, Worst case scrap $10,000, 12 labor-hours cost $360 - Plus parts cost.
1		A		6	Zero Speed programmed in error	Loader will not move in. "Error 60" alarm after 20 min: Incorrect program or missing parameter. Costs: 2 h production lost = $400, $10,000 scrapped product, 2 labor-hours = $60; $60 parts.
1		A		7	Failure of Tube Computer - Loss of 110VAC power	UPS Power Supply detects mains failure. UPS alarms, triggers Tube Computer alarms. UPS can run for 20 min Process gases are switched off. The other tubes are not affected but must be in "Standby" before repair can start. There is a danger of shorting the other Tube Computers. During the repair time, the tubes cannot run production. Costs: 64 h production lost cost $12,800, $15.909 scrapped product, 16 labor-hours cost $480.
1		A		8	Failure of Tube Computer - Vendor Repair	PC dead—No alarms. Only discovered by operator who finds blank display. The operator should notice that there are no displays. Spare not held, so if repair on site not possible, it must go back to vendor for repair. Downtime potentially very high unless part substituted from other tool. Intensive fix required: testing and a Tube Profile following this repair. During the repair time, the tubes cannot run production. Costs hard to estimate but worst case would be: Costs: 600 h production lost cost $120,000, $10,000 scrapped product, 20 labor-hours cost $600 - Plus repair cost.
1		B	Not capable of driving boatloader out of furnace.	1	Open circuit failure of Over Travel switch LS1	No drive. Overtravel alarm at console. Boatloader Alarm lamp "B" flashes. Boatloader must be driven out manually. Possible product damage depending on extra processing time. Must assume worst case. Costs: 25 h production lost cost $5,000, $10,000 scrap, 6 labor-hours cost $180.
1		C	Boatloader speed greater than 10 cm/min.	1	Boatloader Drive Board out of calibration	Operator will not know a problem exits until product tests surface or customer complains of faulty chips. Furnace requires speed calibration and profiling. The highr speed of the paddle might cause thermal damage to wafers. Costs: 60 h production lost cost $1200, $10,00 scrap. 25 labor-hours cost $750.

Figure 9.16 Boatloader Information Worksheet Example 2.

Boatloader Info Worksheet Rev 6

No:	Function	No:	Functional Failure Loss of function	No:	Failure Mode Cause of failure	Failure Effects
11	To prevent process gases from entering the Load Station at a level greater than the TLV.	A	Does not prevent gases from entering the Load Station above the TLV level.	1	Extract too low	POCL, Phosphoric acid, and HCl can escape through the tube opening and coat the surfaces on the load station. All tubes in standby. All four extracts are interconnected so adjusting one can affect the others. At least 1 test run required per tube following adjustment. Modus Alarm if extract low enough. Potential for chemical contact causing burns and tool corrosion. Costs: 100 h production lost = $20,000, $10,0009 scrap, 30 labor-hours cost $900.
12	To confirm that the tube endplate is closed at the Zero Position.	A	Does not show that the tube endplate is closed at the Zero Position.	1	Posi-indicators jammed.	The posi-indicators are spring-mounted & extend when the endplate closes. Gives improved closure alignment and visual indication if compression amount. 1 cm movement – It is only a visual indication, there is no safety interlock. No impact on production. Same as 1C3.
13	To close the opening between the Scavenger and the Load Station when the Paddle is at the Zero Position.	A	Does not close the opening between the Scavenger and the Load Station when the Paddle is at the Zero Position.	1	Tube Door screws loose	Process gas flow and product will probably be affected. Heat lost proportional to gap size – also possible effect on product. Modus Alarm will sound if extract pressure falls enough Not noticed till final test or when customer's chips fail. Same as 7A1.
14	To be capable of electrically isolating the Load Station from the supply in the event of an emergency situation.	A	Not capable of isolating the Load Station from the electricity supply in the event of an emergency situation.	1	EMO failure	There is no dedicated EMO on the tool under analysis. Isolation is via the main breaker. The introduction of an EMO could only make the tool safer to personnel in the event of a situation arising. Estimated Cost $500: All four tubes would need to be off during the installation.
15	To indicate the Tube Throat extract pressure.	A	Does not indicate the Tube Throat extract pressure.	1	Modus Gauge failure	"Enable" and "Boat Out" when pressed together should stop the boatloader when driving IN and reverse the direction. Need to switch OFF Tube Computer Costs: 48 h production lost cost $9600, $15.909 scrapped product, 27 labor-hours cost $810.
16	To be capable of automatically aborting the process in the event of the extract pressure being less than 0.5 mm H_2O.	A	Not capable of aborting the process automatically in the event of the extract pressure being less than 0.5 mm H_2O.	1	Extract measurement— Modus Gauge signal not used.	This is an old tool with limited switching inputs. The system would need modification to enable the capability. The manual "Stop" and "Stop Stop" buttons, pressed simultaneously is likely to stop the boatloader drive and return it to the 195-cm position. Need to confim. Potential cost if extract fails and not detected $10,000 scrap, if one tube only affected.
17	To be capable of manually stopping the boatloader in the event of an error.	A	Not capable of manually stopping the boatloader in the event of an error.	1	Operator Panel failure	No alarms—display should be blank—which operator might notice. Panel needs replacement. Remaining tubes must be in "Standby" to avoid shorting danger. During the repair time, the tubes cannot run production. Costs: 36 h production lost cost $7,200, $10,000 scrap, 12 labor-hours cost $360.

Figure 9.17 Boatloader Information Worksheet Example 3.

a PM ... right? The sad fact was that rather than reducing the rate of breakdowns, the extra maintenance caused even more. The engineers had identified the components that failed, but not the *reasons* for the failures. (Does this part not just remind you of similar issues in TPM?) So, a new study was set up to do just that: to find out why things failed.

This study looked in minute detail at every failure and tried to establish a root cause for each. What they found came as a real surprise to

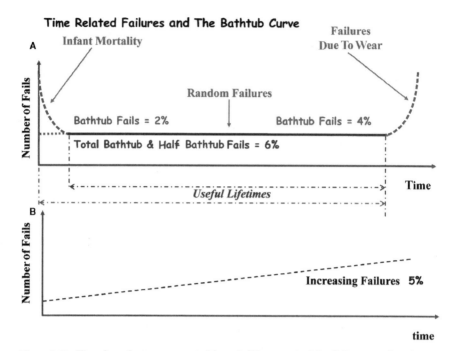

Figure 9.18 Time-based causes occurred in only 11 percent of the failures analyzed.

them. They discovered that time-based maintenance procedures would have helped in only 11 percent of the failures. See Fig. 9.18. It is important to bear in mind that this survey was carried out on aircraft components and so the failure distribution is likely to be different than for other industries. Nonetheless, the one overwhelming discovery made is familiar to field service engineers all over the world; the fault is often manmade.

In Fig. 9.18A, the bathtub curve is shown in two ways. First, the full bathtub, which comprises of three sections:

1. Components that suffer a high failure rate when they are first installed. This high rate is caused by infant mortality.
   This is represented by the falling dashed curve that lasts until the steady state is reached.
2. The steady state in point 1 is a period of random failures.
   It is represented by the horizontal straight line that runs until the rate starts to rise.
3. The rising curve starts when the parts reach a point where the wear of the component takes over. At this point, the failure rate starts to increase rapidly.
   This rise is represented by the rising dashed curve.

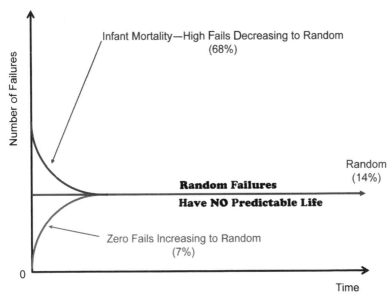

Figure 9.19 Failures that are not time-related occurred in 89 percent of the failures.

The shape of the main graph described in the three points above is that of a bathtub curve. There is also a second bathtub, or to be more precise, a half bathtub. The first part of the graph, the infant mortality part, is replaced by an extension to the steady-state horizontal line. This is represented by a horizontal dotted line in Fig. 9.18A. After the steady-state region, the half bathtub still follows the curved rapid rise at the end.

Figure 9.18B shows a steady, increasing rate of failures with no distinct failure point. Components that follow this pattern have maintenance periods based on the acceptable risk rate, the cost of checking, and the cost of losses caused by the failure. If you would like more details on how this data came about, I would recommend *RCM II* by John Moubray.

Non-Time-Based Failures

Figure 9.19 incorporates three separate graph patterns, all of which have a common feature: apart from the initial period they can all virtually run indefinitely. Perhaps indefinitely is too strong a word for it, but parts that follow these patterns usually fail at random. Their demise cannot be predicted although sometimes the onset of failure can be detected.

The chart comprises of three traces.

1. Infant mortality with a high failure rate falling to random.
   The curve with the decreasing slope that starts with the infant mortality has the highest failure rate of all. It accounts for 68 percent of all of the failures and yet it follows the same random failure pattern after the initial fails.
2. Random.
   The random fails account for 14 percent.
3. Zero fails increasing to random.
   This pattern accounts for only 7 percent of the fails.

If we could eliminate the infant mortality failures, we could dramatically reduce the overall failures. So what is infant mortality, why does it cause so many faults and what can we do to limit its impact on performance?

Infant mortality

Production people never want to let tools go down for maintenance. Either they instinctively know or their experience tells them that they will not get the machine back for ages or they will get it back but it will run badly or they will get it back but it will run for a while and will fail soon afterwards. What the production people don't realize is they have a gut feeling grasp of infant mortality.

Infant mortality is the name given to the failure of new parts that have just been installed and have an initially high probability of failing. The failures can be due to badly machined parts, parts with very wide tolerance, poor installation of the part, bad electrical contacts or wiring, flaws in the original materials, damaged bearings in the part, and so on. Electrical components and heating elements are prone to infant mortality and so the manufacturers have introduced a "burn in" period designed to capture and eliminate these early failures before the parts are released to the customers.

The big picture is that if the components that failed because of infant mortality had been left alone and had not been maintained, there could have been many fewer failures. To avoid failures due to infant mortality, we need to establish the quality and reliability testing of the incoming parts and upskill all of the engineers. If we increase their technical skill, we should be able to eliminate all of the damage caused by poor engineering standards.

Chapter

10

Time- and Condition-Based Maintenance

A new piece of equipment will arrive with a few manuals and a list of recommended maintenance procedures. The tasks in Fig. 10.1 are a small part of a set based on a maintenance schedule recommended by a vendor. They are carried out at specific, predetermined times and so are cleverly called "time-based" maintenance procedures. My original version of this tabular format included all 52 weeks, about 100 tasks, and required a signature to verify that the work had been completed.

The theory behind a *Preventive Maintenance* (PM) schedule is simple; if each task is carried out when the manufacturer recommends it, then, provided the work is carried out correctly, the equipment will not break down. The manufacturer's recommendations are likely to become the basis of a company's equipment maintenance, just as did the one in Fig. 10.1. The problem with the vendor-supplied list is that it is intended to be a baseline for all similar tools sold. Even with the best intentions, it will be based on the worst case and is likely to err on the side of some parts being overmaintained. This has obvious implications, none the least being increased downtime, increased labor usage, increased cost in the four of parts that have been replaced too soon and new parts that have been installed to replace them, and of course, lost production. This might be a slight exaggeration, but it certainly seemed true enough: to reduce the impact of the PMs, many sites quickly doubled the recommended times and started using them. Companies, like the one whose schedule is used above, will work with customers to adapt schedules.

How the tool is used (the use conditions) also has a major effect on the time intervals between the required tasks. Compare the example of two identical cars, one is a taxi driven only in the city; the other is driven

DESCRIPTION / WEEK NUMBER	1	2	4	6	8	10	12	26	52
COUNTER BALANCE PRESSURE (30 psi)									
ARM PRESSURE (60 psi)									
AT4 AIR PRESSURE (25 psi)									
AT4 INPUT AIR PRESSURE (23 psi)									
FACILITIES AIR PRESSURE CHECK (90–100 psi)									
FACILITIES NITROGEN CHECK (15–30 psi)									
DI SYSTEM WATER LEVEL									
DI SYSTEM FILTER PRESSURE (<45 psi)									
CHECK/CHANGE SOURCE ROUGHING PUMP OIL									
CHECK/CHANGE OIL LEVEL OF RP2, RP3, AND RP4									
REGEN CRYOS									
NOTE CONSOLE HOURS METERS									
SERVICE DIFFUSION PUMP, CLEAN CHRISTMAS TREE	■		■		■		■	■	
SERVICE HEAT SHIELD	■		■		■		■	■	
PM SOURCE LINER/SOURCE HOUSING	■		■		■		■	■	
REMOVE EXTRACTION ELECTRODE AND PM	■		■		■		■	■	
REPLACE SOURCE									
PM FLAG FARADAY									
CHECK E.S. BIAS (–300 V AND –130 V)					■		■	■	
CHECK EXTRACTION BIAS VOLTS (–1.8 KV)					■		■	■	
CHECK EXCHANGE									
PM DISK DOOR HUB AND TIGHTEN				■			■	■	
CHECK SOURCE DEFINING APERTURE				■			■	■	
CHECK DISK GROUNDING SYSTEM (<15 Ωs)				■			■	■	
LUBRICATE CHAMBER DRIVE AND BALL SCREW ASSEMBLY				■					
PM ELECTRON SHOWER									
REPLACE E.S. FILAMENT, GRID, AND INSULATOR BLOCK	■		■		■		■		
PM SLIDING SEALS	■		■		■		■		
PM V3									
CHECK HIGH VOLTAGE CALIBRATION AND LEAKAGE									
CHECK EXHAUST PIPES	■							■	

Figure 10.1 Time-based maintenance example.

only on the motorway. Assuming they are both driven the same distance, which car will produce the greatest wear on parts? The town car would. We would expect to find more wear on the tyres, suspension, brakes, and even the engine—unless, of course, the motorway was the M8 or M25! Driving in town requires more braking, more starting, more stopping, and more turning, and motorways have a better class of potholes. So the town cars would need to be inspected more often if we wanted to detect the wear in time to prevent a failure. Does this mean we have to inspect both cars at the same frequency? No. In the case of the motorway car, when we carry out the first time-based check, we would consider the amount of wear found. If, as expected, the wear was light, we would increase the time interval to the next inspection, to reflect the amount of deterioration found. Provided the "use conditions" remain the same, with a bit of experience and tweaking, we should be able to tune the checks to optimize the inspections. The overall goal of any maintenance is to maximize the time between inspections, while ensuring that the tool, safety, and the quality of the product are not compromised. This is known as *Predictive Maintenance* (PdM).

Consider the example of an oil pump, a piece of equipment that runs nonstop and is guaranteed to wear out eventually. For the sake of argument, assume the vendor's time-based maintenance recommends annual checking. What would we expect if we change the function from pumping clean oil to pumping dirty oil? The added abrasive action of the dirt will force deterioration and the pump will require a shorter time between inspections. If it was pumping unprocessed oil, extracted from an oilfield, which is abrasive enough to grind the inside walls of pipes, the time between inspections might need to be decreased even more. How do we find out the correct time interval? By experience and educated deductive reasoning.

Designing a practical maintenance schedule affects production too. One of the goals should be to optimize the routine maintenance downtime periods (scheduled downtime) so that they minimize disruption to production. Look for periods when the tool is in standby. Where possible, avoid having a maintenance period spread across two shifts: it detracts from overall job responsibility, loses extra time at the beginning and end of the shift and, if a task is complicated, it makes it more difficult for the person who has to reassemble the module without having the reference point of taking it apart.

My preference is to target more frequent, smaller PMs, but everything depends on circumstances. Sadly, there are always exceptions, for example, where an overriding practical advantage exists. If a tool is used only at the beginning of a production run, consider carrying out the maintenance in larger chunks, once it has completed its part of the production runs and would normally be sitting idle. Another situation where a larger PM would be considered is when one takes advantage of a lengthy preparation time or time required to gain access to the part being maintained. If you are going to dig up a road to lay a cable, it makes sense to check the pipes at the same time. The street where I live was resurfaced. It was a beautiful job and lasted a whole 2 months. After the 2 months, if it was even as long as that, a 4-ft wide trench was dug along the length of it (and for a few miles in either direction). Enough said!

When a tool is constantly in use, it is best to agree to a series of short interventions with production. Try coupling the PM with a changeover. Until maintenance becomes completely reliable, short PMs limit the impact of poor maintenance and bad luck. When you start *Total Productive Maintenance* (TPM), you will quickly discover that one of its aims is to develop procedures that ensure tasks are carried out correctly—the first time and every time. Even then, in the event of any errors, like a vacuum leak or a damaged cable run, the smaller the area to be checked, the easier—and the faster it will be to locate the problem. It is easier to find something we have lost when we have only been to few places. The real trick is to set up the maintenance in such a way that the opportunity for mistakes is minimized. One of the best ways to limit mistakes is the use of preassembled, tested, turnaround parts. Another advantage of short PMs is that they create natural windows for operators to have meetings, attend training, and to carry out 5S or other continuous improvement duties.

Introduction to On-Condition Maintenance

Before moving to the next section, there is another type of maintenance to be considered. To return to the example of the car, do we really need to inspect the brakes at all? Car designers now have sensors that detect brake wear, so it is now possible to drive until *the car* tells you it needs to have the brake pads exchanged. The system is not maintenance-free, but it is inspection-free.

Using on-condition maintenance can dramatically cut the amount of work and, consequently, the equipment downtime required. At its best, condition monitoring is a unit that continuously checks itself and needs no maintenance intervention or downtime until work is required. With the introduction of better electronics and the microcomputer, data can be continuously recorded and made available for retrospective analysis. Improved sensor technology with modern electronics can simplify modifying systems to use monitoring. Consider the other advances in cars; more and more have electronic ignition systems and sensors fitted throughout the engine. There are sensors in the wheels for tire pressure, others for fluid level measurement, and even sensors for detecting skids and wheel locking. Consequently, some cars are able to tell you when they need to be serviced. This concept has spread to equipment and, with improving technology, it is becoming more versatile.

If we were to continuously monitor the condition of parts in equipment, some would be found to perform perfectly over a long period, exhibiting only a tiny rate of deterioration. Then, at some point, the *rate* of wear will noticeably change; it increases and becomes detectable. As the tool continues to run, the deterioration rate increases even more rapidly and performance continues to drop off until total failure is reached. In Fig. 10.2, the detection point for a deteriorating condition is shown as "P," and "F" represents total failure. The graph is cunningly known as the P-F curve.

The response of the P-F curve for equipment is analogous to the body's reaction to drinking alcohol. Everything is fine for the first few drinks. Performance is almost steady with a slow, imperceptible decrease. Eventually, the point "P" is reached, at which point a change in performance becomes noticeable. The first symptom is usually in the form of increasing speech volume and generally talking rubbish, both of which continue deteriorating rapidly to the point "F," where there is normally

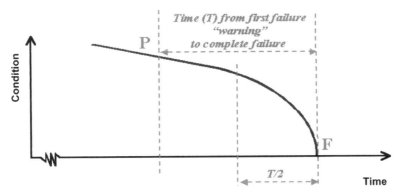

Figure 10.2 The P-F curve used for on-condition maintenance.

a total lack of function. The point "P" could be identified using sensors: acoustic for the noise level and speech recognition systems linked to grammar checkers for the content. On detecting the oncoming failure, appropriate action can be taken to prevent or at least limit the impact.

For equipment, the point "P" can be signaled by vibration in moving parts, heat generation due to friction, noise, dust, or leaking fluids from seals. (It still sounds like people, doesn't it?) The time T in Fig. 10.2 is the time interval between being able to detect a failure and the unit failing completely. The time T can vary from virtually instantaneous to months. If the time period is known, repeatable, and is long enough to be helpful in an equipment situation, then the unit can be manually inspected within the period T at normally around 50 percent of the interval. Depending on the result, either an estimate of the remaining lifetime can be made and another test scheduled or the unit can be repaired or replaced. The beauty of it all is that the checks are minimal until the warning is given and the fix can be planned.

If we can identify the correct sensors needed for the specific part and get some experience in using them, we can then take appropriate action when the point "P" is reached and avoid an unplanned failure. There are vast selections of sensors available: vibration sensors, particle counters, strain gauges, acoustic sensors, leak rate measurements, ultrasonic, x-rays, vacuum monitors, flow meters, or motor currents. There are many others, some require sampling oil for purity or to measure the rate of gas release or some that use chemicals to look for cracks in casings. I don't know all of the tests, but these days it is relatively easy to find out what is available. Try your local university or the Web; someone out there will be desperate to sell you a solution.

Vibration monitoring is very popular, although the analysis can be very complex. If used properly, the data can pinpoint deterioration to specific gears, shafts, or bearings. The trick is recognizing if the noise pattern, or change in pattern, indicates a real failure looming or just a change. If a high degree of resolution is required, then it is necessary to train people thoroughly in its use or hire a contractor. Sometimes, a simple noise level jump or frequency shift will be enough to tell you that it is time to start making regular checks or plan to replace the part.

Some semiconductor furnaces have heaters that resemble giant electric fire elements. *(One is used as the example for RCM.)* From school-level physics and broken filaments in light bulbs, we know that the resistance of a wire (element) increases with use. This means that to maintain a constant temperature, the electric current has to decrease. Experience suggests that once the current falls below a certain value for a specific heater, there will only be a limited number of weeks life left before failure. So the current can be monitored and set as a flag to warn the user when it is time to schedule an element change. After replacement and setup, no measurements are required until the next flag is made. Take care though, the temptation is to try and get as close to the "F" point as possible before replacement, but to do this greatly increases the risk of failure in use. My recommendation would be to estimate the potential savings against risk of lost product and don't be too greedy. Then have the courage of your convictions and stand by your decision.

Automation is not essential; cost is a major consideration. The condition of a tool can even be checked manually for vibration, noise, temperature, accuracy, backlash, and so on. Replacement is based on the results. One advantage of automated condition monitoring is that it only requires a low level of human interaction. It can be very cost-effective, but don't assume it will always be a cost saving. Before installing any monitoring, it is necessary to evaluate *if* it will do what we need it to do. Next, we need to evaluate if the task is possible. The vendor might be very positive that the modification will work. If he is, make sure he is liable if it does not work. Next, we have to establish how much it will cost to fit, to train the users, and how the costs will compare with the old-fashioned manual checks over a realistic time period. There has to be a positive gain of some kind to justify the cost. Avoiding any issues that can affect customers is a positive gain.

Friction between Maintenance and Production

There is often a common dispute between the users and the fixers. Production groups are reluctant to hand a tool over for maintenance because "It never comes back on time . . . " and/or "It always breaks down when we get it back. . . ." The fears of the production group are often well founded. In some sites, tools go down for maintenance with no expectations when they will be returned. There are several reasons for this:

1. *No one has thought about how long it should take.*
   This should be easy to overcome by specifying how long each "task" will take.

   In one site, it took a number of years to establish task times. Not because no one knew what they were, but because it was felt it would upset the technicians to be told how long they should take. Often PMs took "a shift" to complete and were carried out at weekends—usually on overtime. Even after removing a 2-h task from a weekend "shift" PM, the new time became—a shift. Interesting…

   The PM times should initially be agreed through discussion between the technical staff—engineers and technicians. It should be understood at the outset that this is only a starting point and that the times will be regularly monitored and revised to take account of any unanticipated issues. However, as these issues are identified, resolved, and the methods improved, it would be expected that the times could also decrease accordingly. (Plan-Do-Check-Act.)

   The total PM time will not just be the sum of all the tasks. Depending on the content of the PM, the total downtime will need to include

   ➢ Breaks
   Assuming the maintenance work does not continue across break periods—even though it should.
   Bottleneck tools, in particular, should NEVER be left idle across breaks.
   ➢ Time for minor repairs discovered during the PM.
   ➢ Extra time to replace a part that has only been allocated an inspection time, should it be found to need replacement.

2. *Turnaround parts are not used.*

   Turnaround parts are one of the most valuable tools available to the asset manager. Be warned, I will mention turnaround parts as often as I can. If I wasn't already married, I would marry one …

   Parts that need to be removed, dismantled, cleaned, reassembled using new component parts, and then reinstalled should, wherever possible, be replaced by complete working units.

   In a Dutch plant, one standard PM took a quarter of the time it took for an equivalent PM in a UK plant that used the same equipment. Actually, that is not quite true. The time the machine was *not available for production* was reduced to a quarter of the time it was

No Turnaround Parts		Using Turnaround Parts	
Task	Time, h	Task	Time, h
Removal of unit	0.5	Removal of unit	0.5
Dismantle	0.5		
Clean	10		
Reassemble	0.5		
Reinstall unit	0.5	Reinstall unit	0.5
Total Downtime Time	12	*Total Downtime Time*	1
Off-line Maintenance Time	0	*Off-line Maintenance Time*	11

Figure 10.3 Downtime saved by replacing an "electron shower" module compared to removing, servicing, and reinstalling.

unavailable in the UK plant. The reason was the use of turnaround spares, which can save literally hours of downtime—I have seen tools being down for days more than necessary as a result of the time spent in dismantling, cleaning, and reassembling units.

The example in Fig. 10.3 was a particularly dirty module and probably should have been serviced sooner. Manual cleaning also increased the downtime. A "wet" bead blaster would have (and eventually did) significantly shortened the cleaning time.

"Who would do the work on the module that has been removed?"
Apart from not having considered them, this is normally the first argument for not using turnaround parts. When I said that the Dutch PM was not reduced to a quarter of the time, the reason was that someone still had to carry out the refurbishment of the spares. The belief is that the used part will just be put aside on removal and sit there waiting for someone to voluntarily accept responsibility for cleaning. Remarkably, no one will! There is a simple answer, though: manage the situation and control it properly.

As a rough guide, once the tool is back in production, the parts should be cleaned by the person who would have cleaned them initially, since he would have been doing it anyway had turnaround parts not been used.

However, because of the opportunity for cost savings on a grand scale, consideration should be given to who does the cleaning from a cost and efficiency perspective. It is not necessary to use highly trained technicians for cleaning and rebuilding. Depending on the complexity of the tasks, semiskilled personnel or trainees, who have been specifically trained on the items, can be used for the cleaning stages, and hence increase their experience, reduce repair costs, and free up skilled technicians for more suitable issues.

Consider areas where parts might need replacing and cleaning every few days or where there are several tools. There could be justification for having a dedicated cleaning operation with its own people.

There can be advantages in setting up a joint cleaning facility. If you decide to share an area between two or more maintenance groups, ensure you have a plan to cover the following responsibilities:

- ✓ replacing used parts
- ✓ cleaning tools and work surfaces
- ✓ documenting faults and repairs
- ✓ organizing maintenance for equipment

What about the extra cost of turnaround parts? There will be an initial outlay required. Calculate the cost of the lost production for the extra cleaning time of the modules over a time period and compare it to the cost of the replacement units. I suspect that the dollar costs will balance out pretty quickly and lead to a gain not long after—to say nothing of the improvement in equipment reliability.

Hourly production losses can vary significantly. Some products have a much higher value than others in terms of production rate per hour, particularly items like microprocessors. Let's analyze the example of the electron shower in Fig. 10.3. Twelve hours of cleaning equated to approximately $36,000 of lost production—I picked a value of $3000 an hour based on a sign stuck to a tool—one which always sticks in my memory. However, to avoid disputing the large sum, I will cut the cost to a tenth of that: down to $300 per hour. (Whereas this seems like a lot of money in anyone's language, I have seen lines that lose in the order of $10,000 an hour.)

At the 10 percent loss rate, we still save $3600 of lost production each time the task is carried out. (By the way, in the year 2000, the hourly rate for a technician alone, the guy cleaning the unit, was estimated at $20 to $30 an hour and so the cleaning cost alone was around $360.) The turnaround electron shower module in the example costs about $20,000 and was cleaned *at least* seven times a year. This would make the annual saving—per tool—about $25,000. This is a real saving of $5000 in the first year. If we had two tools, the real saving becomes $30,000. Factor in that multiple tools use the same part and it can be seen that huge savings can be made annually.

Now, imagine the savings at the real rate and not just at the 10 percent level...

To make these comparisons valid, it is necessary to know as accurately as possible the hourly cost of lost production on each tool. Often, it is only a wild guess.

Knowing the cost not only lets you evaluate the benefits of turnaround parts but also helps in decisions on carrying higher value spare parts. While recommending spares kits in the past, I have costed some of the spares in "hours of downtime." If the risk of failure is high and the delivery time from the vendor is long, then a straight comparison can be made.

3. *Turnaround parts are faulty.*
   The turnaround parts must be properly tested before reinstalling.
   An analysis of the failure must be carried out and once the causes have been found, appropriate actions taken. Problems like poor electrical connections and bad seals can be eliminated with a bit of effort and off-line testing. I have seen faulty parts being used several times as replacements because of the lack of a basic system. This is a completely unnecessary cause of extra downtime.

   The avoidance system is simple and please note, this is not standard practice in many sites:
   - When a suspected faulty part is removed from the tool, a label must be attached. This should include
     ⇒ The date of removal.
     ⇒ A note of the suspected fault with symptoms.
     ⇒ The name of the person who removed it.
   - When the part has been serviced, more information is needed. This should include
     ⇒ The date the service was completed.
     ⇒ Confirmation that a fault was found and repaired.
     ⇒ Details of the actual fault that was found.
     ⇒ Confirmation that the unit has been tested and is fully functional
     ⇒ The name of the person who repaired, serviced and tested it.
   - Turnaround parts should be given their own serial number or some form of ID to enable their history to be tracked.
   - There should be a master log for tracking all repair information. This is invaluable for identifying recurring faults and possibly identifying process or procedural issues.

4. *The PM is started before checking if parts are available.*
   Make sure that all of the parts required to carry out the PM are available before the tool is put down for maintenance.
   This one really gets me. Who would ever consider putting a tool down for a PM when there are no parts available? Well it really happens—often! The sequence is the tool is shut down for a PM and

then the technician goes to stores to collect spare parts, leaving no one working on the tool. Frequently, "to save time," the trip to stores is just before or after a break—even when they have to wait a while before the break. To maximize the time saving, they go for the break earlier or return later! If there are no parts in stores, we have wasted the shutdown time, the waiting time, the time for the trip to and from stores, the run-up time for the tool, and the requalification time. There are varying degrees of this problem. Sometimes the technician gets only the parts for the area he is currently working on and not all of the units scheduled for maintenance in this PM.

The worst offenders of all actually remove the part to be serviced, dismantle it, and then go to stores for the parts. If the parts are out of stock—and this is not an unlikely scenario—we now have a seriously dead tool. The worst example of this I have seen was with the same electron shower mentioned above. (I am beginning to feel I have a fixation for this part.) It was removed and dismantled. The technician then went to stores to get the replacement parts and discovered that the main part was out of stock. Thanks to Murphy's Law, it was not possible to reuse the old part so the tool was down for more than 2 days until the vendor was able to deliver the part. We have now lost nearly 60 h of production time, with all of the associated costs.

How do we overcome the problems?

My favorite solution is a spare parts trolley. This is a trolley filled with boxes of spares, each box being a kit containing all of the spares for one of the units / modules to be maintained. There should be a kit for each of the units that need to be routinely maintained in the tool.

When a given module is serviced, any part that needs replacing can be used. This has the added advantage that, in the event a part not scheduled for PM is found to be faulty, depending on the importance of the problem (that is, the damage it can cause to the machine or to the product), it can either be replaced from the trolley with minimum downtime or scheduled for repair at a later date.

Any parts used during the PM are recorded on a stock sheet supplied with the trolley. The stores personnel replenish all the parts kits when the tool trolley is returned.

5. *The technicians working on the tool find problems they did not expect.* This was briefly mentioned in point 4 above in the discussion of tool trolleys. While carrying out maintenance work, it frequently arises that some other issue is identified. Sometimes it is a "must do" task like a broken wire, a safety issue, or an item that is likely to fail imminently with a potential loss of product. This is a "no brainer" choice. However, if the issue is unlikely to have an immediate effect on the product or quality, it can be rescheduled for future repairs.

There should be a predefined window of understanding with production that the planned time for the PM enables short unplanned work to be carried out.

However, the production supervisor should always be kept informed of any potential delays so that, if necessary, he can plan alternate routes for product or request extra manpower to expedite the repair.

6. *The technicians who carried out the job introduce a problem that prevents the tool passing a post-maintenance test run.*

   This can be caused in several ways:

   ➢ By accident: for example, damaging a wire or a connector while working in a tight space.
   ➢ Inexperience or lack of skill of the technician.
   ➢ No documented procedure for the tasks.

   If it is possible to damage other components during repairs, and the fault is known to have happened more than once, then any documented procedure must highlight the possibility so it can be checked as part of the job. If possible and cost-effective, the tool could be modified to eliminate the possibility of the damage. Consider a connector that is regularly being damaged, perhaps it could be moved, protected by a cover, or changed to a less vulnerable type of connector. *Training and proper illustrated procedures will overcome inexperience.* A method of writing procedures that dramatically reduces the possibility of errors is discussed in Chap. 5. Technical staff love them, but the PC engineers hate them because they are usually very large files.

7. *The technicians doing the job introduce a problem that does not appear until after the tool has been returned.*

   This has the same causes and cures as point 6, but causes much longer downtimes, as the tool has to be shut down again, repaired, and then requalified.

   Sometimes replacing a part can cause failures due to the "infant mortality" of new parts. Infant mortality often occurs in the following:

   ➢ Electrical systems or parts that have not been adequately tested or "burned in."
   ➢ Parts that have been badly installed, using too much force, which has damaged moving parts or bearings.
   ➢ Substandard parts.
   ➢ As a result of carelessness or distractions while working. This includes being called to another problem.

This can include missing steps, not tightening bolts properly, forgetting to fit lock nuts or washers, or forgetting to support cable assemblies.

The use of pretested turnaround parts, the correct tools, proper training, and use of procedures will reduce all of these issues.

It is also beneficial to monitor the type of problems that are introduced and try to identify the causes. Once found, it should be possible to change the procedures or modify the tool to prevent further issues. And remember, if you change a procedure, all of the technicians need to be retrained.

What if we were starting from scratch?

This is an interesting problem. If it was a new facility and it had a computerized maintenance management/production system in place, I would ensure that it was used properly from day 1. If it was an established facility but just had no organized system, the task would be much harder. However, we would have to start somewhere. Where would be the best place? The options are to start with the equipment or with the people.

The first thing to do, if there was no computerized system, would be to set up a fault logging system with the usual data being recorded. We would need to collect data on how the machines operate, particularly where the failure areas are. However, it will take time to gather the data. The maintenance data must be transferred from the sheets to a database or an Excel workbook. It would be necessary initially to carry out the data transfer personally to get a feel for the way the sheets are being completed. This also provides the opportunity to ensure the engineers are providing the correct data and not omitting details.

The next task must be to become familiar with the equipment and to establish if there are any maintenance schedules in operation. If there are none, they will have to be established. The best way to do this is through the vendors: find out from them their recommended maintenance schedules. The tasks should be split into three levels:

1. Basic skills
2. Intermediate skills
3. Advanced skills

The engineers should be involved in organizing a plan to start implementing a preliminary maintenance schedule, based on the vendor's input. We would need to get instructions on how to carry out the PMs from the vendors until we are in a position to improve the ones we need to.

If we grade the maintenance tasks, we can also compare the skills we need to maintain the tools with the currently available skills of the maintenance group. It should not be too hard to create a maintenance task list with the guidance of the engineers and vendors. There are always simple tasks that we use to train apprentices. They usually include the machine setup, operation, and running product and tests. The intermediate checks are for technicians and engineers. The third level tends to be the highest skilled tasks and these are reserved for the most experienced engineers.

Next we would need to establish in more detail the skill level of each of the maintenance engineers. This could be roughly carried out using a simple questionnaire based on the manufacturers' maintenance schedules. For each task, we could apply the competency test explained in Chap. 5, which looks for five levels of skill. The levels are

1. Knows nothing.
2. Knows theory.
3. Can do with supervision.
4. Can do unsupervised.
5. Knows well enough to teach.

When we compare the PM tasks with the skill levels, we should be able to identify what training we need to concentrate on and who would be the best person to carry it out.

As the maintenance data starts to grow we should be able to establish which tools are the worst performers. This can be combined with the skills data and the problems can be prioritized for resolution. Naturally any improvement will require more in-depth training to be carried out. Some of the training might have to be supplied by vendors, but if it is, the standard and content will have to be targeted at the areas that need it most. If it is to be internal training, we would need to create standard procedures to be used in the training.

It would be worth setting up a group to be involved in developing a plan for the organization of the engineers. I would like to dedicate engineers to specific tools or groups of tools, depending on their experience. I prefer that to the situation where anyone can answer any breakdown, whether they have the experience or not. It might also be prudent, manpower permitting, to have the engineers work in pairs when possible to speed up learning.

Although I would already be applying selected TPM methodology from the beginning, as soon as it was practical, I would start to lay the groundwork for starting TPM in the maintenance department and would plan to deploy Zero Fails teams. In this case, I would have to give the teams responsibility for establishing the integration of the maintenance schedules that the vendor has recommended. If it was possible, I would try and tie in the schedules on a larger scale, the big picture, to spread the load amongst all of the equipment.

It would take coordination with production and maintenance at a couple of levels to get the system running properly, but most of all it will take determination and time.

In Summary

1. Use the vendor-supplied maintenance schedule as a baseline.
2. Consider the processes likely to be run on the tool and use this information to discuss with the vendor any changes that could be made to the schedule.
3. Make sure that your team is properly trained.

   This includes operators and process engineers. A vendor course lets the candidate see the job carried out or do it once. This is not even close to making them experts.
4. Check the documented procedures in the vendor manual used to carry out the maintenance tasks.

   Ensure that they are correct and have enough detail to prevent any errors. If they are not good enough, ask the vendor to supply better instructions or, alternatively, use one of the most experienced technicians to rewrite them, using the vendor for advice.
5. Every time a maintenance check is carried out, evaluate the condition of the part in relation to tool use.

   Decide if the interval between services needs to be optimized. This can be carried out as a team analysis.
6. Do not let the tool deteriorate to a state where it has a series of known faults that are bypassed or accepted as normal.

   Apart from any safety implications, there is normally a price to pay for every fault.
7. Use Total Preventive Maintenance (TPM) or Reliability Centered Maintenance (RCM) to develop the PM schedules.

   My recommendation would be TPM to restore and embed the basic condition, followed by RCM to optimize the type of maintenance tasks used.

   The RCM methodology and the decision diagram are explained in Chap. 8.

Chapter

11

Fault Analysis: A Few Ways to Help Find Root Causes

In this chapter we will discuss various techniques for solving problems. In Chap. 1, we discussed a few production and equipment examples while explaining the basic condition and use conditions. I don't want to become too theoretical, but we have two observations to make. First of all, what is a problem? How do we recognize we have one? The second is how do we solve it?

If we work to standards, we know that this television, monitor, machine, bicycle, camera, or laptop computer has a list of functions it has to perform and the standard, or specification, should tell us what they are. Suppose our monitor is a TFT screen (like the screens used on mid-to-high range laptop PCs, one that can be viewed from wide angles). If we display a blank, black screen and see bright stars, do we have a problem? If you don't like the bright dots, you have a problem: but does the machine? The short answer is yes, but only if there are more dots on the screen than the company has said there could be. The solution, or best compromise, would be to choose a monitor that has a higher quality control or only buy one you have visually inspected.

So it seems a problem can have two main criteria: the user perceives there is a problem which, whether it is valid or not, is the basis for the complaint or the need to resolve it; secondly, the equipment, or whatever, is not operating to its specification. The first case is what Six Sigma calls "the voice of the customer." This is what the customer perceives is a problem. Often the customer and the manufacturer have different ideas of what a problem is. The company has to target solutions for all of their problems, but I would suspect they would concentrate or prioritize on safety, legislative, quality, and customer issues, with the customer issues probably being put very high on the list. If the safety issue affected the product, I suspect it would have a higher priority than one that did not, because of its potential impact on the customer. Companies with satisfied customers, irrespective of the legal right or wrongs, tend to have customers that will return again and again. Companies that do not resolve problems tend not to see the same customer loyalty. Of course these are only my views.

I once had a problem with a car dealer, who maintained my car and left the oil cap off. The dealer had also made a few mistakes previously, but the people in the dealership were very nice. I had planned to leave on a trip to the Scottish Highlands after I picked up the car. When I set off on my journey, I could smell oil. I suspected it was due to a minor spill and sloppy maintenance. After traveling only 50 miles the smell of oil was not getting any better, in fact it was becoming too much to endure. I had a look inside the engine compartment and found the oil cap was missing and oil was everywhere. On discovering the error, I had to drive the 50 miles back to the garage, get the oil refilled, and have all traces of the (now) dripping oil removed from my engine compartment and bonnet. I had to abandon my trip. Guess who does not use that garage any more? The voice of the customer has spoken again.

Having had problems getting domestic equipment issues resolved, now, when I plan to buy a high-cost item from a company, like a computer, VCR, or camera, I do not call the sales department, which will always be well supported; I call the customer support line. If I have to hang on for an unrealistic time, I assume that either they have a whole lot of problems or they just don't care about the customer. Either way, I do not buy from this company.

Problem-solving techniques are also applied to major issues as a way of identifying all the minor components that contribute to the overall problem. In this case, teams of cross-functional employees are used to analyze the causes. Total Productive Maintenance (TPM), Reliability Centred Maintenance (RCM), Six Sigma, Single Minute Exchange of Die (SMED), 5S, Brainstorming, and Kaizen all use teams for problem solving. Basically they all use the same methods, except Brainstorming which is a method in itself.

The reason that problems become a serious issue is that, more often than not, they are not resolved the first time around. In some cases this is due to the wrong choice of solution being selected out of a range of possible options. In other cases it is due to the fault symptom being tackled and not the cause of the fault. In the latter case we are guaranteed to have a recurring problem. For some the issue might become embedded and accepted as inevitable; for others it will be tolerated until someone gets fed up living with it or takes the time to calculate how much the problem is actually costing him in dollars. When the cost is established, the problem can say goodbye to its job security.

Root-cause analysis is the Holy Grail for problem solving. It is not always easy to identify it outright because in some cases there can be more than one cause that accounts for the symptoms. There are a number of ways to help find the root cause, including Fault Tree Analysis, Fishbone (Cause and Effect) Diagrams, CEDAC, Brainstorming, Six Sigma, and Why-Why Analysis—also known as the "5 Why's." Because of its simplicity, the 5 Why's is often the first problem-solving tool a team is taught. It is the one I would recommend all teams use initially. The 5 Why's is not dependent on technical skill and all of the team members will find it easy to use. This single feature ensures that everyone stays involved in the problem-solving process. Later, as the teams progress through their projects and they need a more complex method, they will learn the other techniques covered in this chapter.

The 5 Why's

To illustrate how it works, consider the following problem: a networked computer has just failed.

1. **Why** has the computer failed?
   Answer: The fuse has blown in the chassis.
   This is often where the faultfinding ends. The engineer will fit a fuse and say, "Let's see if it happens again..."
   What would have happened if we had asked "why" again?
2. **Why** has the fuse blown?
   Answer: The hard drive is overheating and taking too much current.
   Wow. The fuse might have blown for a reason. This is often not a consideration.
   What would have happened if we had asked "why" again?
3. **Why** is the hard drive overheating?
   Answer: It is generating a lot of heat.
   This is where the hard drive would be replaced. If it is overheating that has to be a bad thing.
   What would have happened if we had asked "why" again?
4. **Why** is it generating a lot of heat?
   Answer: It is a high-speed drive and generates more heat.
   Interesting, but if the supplier knew it did that would he not have taken preventive measures?
   What would have happened if we had asked "why" again?

5. ***Why*** is it not being cooled?

   Answer: The fan is not cooling it.
   We have reached the fifth "Why" but have not reached a solution. It does seem as if we are making progress, though.
   What would have happened if we had asked "why" again?

6. ***Why*** is the fan not cooling it?
   It has been mounted so that the airflow is in the opposite direction to the desired flow. In place of cold air, it is being "cooled" by hot air, which is being pulled from the circuit boards and power supply.
   The computer was upgraded recently and the fan was changed to a more powerful one that was recommended by the hard drive supplier. However, the instructions for the airflow direction were overlooked.

The "5 Why's" will force the team to look deeper than the "obvious" initial symptoms. The number "5" in the "5 Whys" is not cast in stone: The answer might not take all five "Why's" or, in some cases, like the one in the example, it could take more. Let's look at the example again and see what happens if we ask "Why" more often. It might be interesting to find out more details.

7	**"Why** is the fan blowing in the wrong direction?" [We need to ignore the extra information written in *italic* in point 6 above to see how the conclusion would have been reached by continuing to ask why.] *The computer was upgraded recently and the fan was changed.*
8	**"Why** was the fan changed?" *To improve the airflow for the high-power hard drive. The manufacturer of the hard drive recognized the need for the fan.*
9	**"Why** did it not improve the airflow?" *It was mounted in reverse.*
10	**"Why** was it mounted in reverse?" *The technician did not follow the instructions that were supplied by the hard drive supplier.*
11	**"Why** did he not follow the instructions?" *We have now found the real root cause. It is a case of not following the instructions.*

There is no need to proceed; we have discovered that had the IS department or PC vendor followed their own procedures, we would not have had a problem. In TPM terminology it was a case of *operating standards not followed*. This is a significant cause of equipment failure. Another clue that we had reached at the end of the trail was that the answer pointed directly to a specific person.

By going through the "5 Why's" rather than the hit-or-miss method used in real life, a significant range of savings would have been made.

1. A saving was made on the time the PC was off-line.
   There was a large amount of time off-line. This stopped the user from accessing his current e-mails, files, and even his network drive.
2. Time was saved on repair labor.
   The engineer made several visits. The fuse blew twice before he became suspicious that the fuses were actually performing as they were intended to. Then he changed the drive, which would have taken out the fuse again, if he had not noticed the details about fitting the fan.
3. Waiting time was saved for the PC operator.
4. A new hard drive would not have been needed.
5. The data from the old drive did not need to be transferred to the new one.
6. The big saving...
   If the fuse or the hard drive had been replaced, the PC would just have failed again! Except this time another cause might have been looked for—or possibly not!

As part of a TPM procedure, the "5 Why's" analysis should always be documented so that the reasoning can be checked in the event that the fault returns (Fig. 11.1). One way of checking whether the logic works is to read the "Why's" in reverse order. They should still sound logical.

In the PC example above a Kepner Tragoe fault analysis, based on what has changed in the system, would have worked just as well—provided that the analysis included all of the items that were changed and did not stop when someone spotted the hard drive upgrade. Kepner Tragoe has not been included in this volume. It is a nice problem-solving tool but it is not necessary at the point. However, since I have whetted your appetite, I will simply say that it works by looking for changes. What is different between the situation as it stands now, where we have a problem, and the situation before the problem was first seen. Kepner Tragoe tends to look for changes in procedures, methods, processes, parts, work carried out that might have had an impact, engineer or operator skill levels, and so on. It looks for anything that, if carried

Step	Why?	Answer	Comment	Final Action
1				
2				
3				
4				
5				

Tool ID:
Date:
Problem Description:
Team:

Figure 11.1 Example of a "Why-Why" Analysis Sheet.

out differently, could create the issue (and work being carried out incorrectly is definitely different).

The "5 Why's" will probably solve most of the problems the teams will encounter, but here are a couple of very simple examples of more powerful solutions.

Fishbone Diagrams

The fishbone diagram, which is also called the cause and effect diagram, looks like—well, fish bones ... a fish skeleton. The main spine points to the original problem, which is written on the "head" and the major bones that join to the main spine are the first levels of failure options. Each bone can be fed by smaller bones and so on, working down to deeper levels of analysis (see Fig. 11.2). When developing a fishbone diagram, expect to use a lot of paper.

When an analysis starts to move down too many levels for a single fishbone diagram to handle, simply take the offending "first level" spine that has all the substages and make it the main spine in a second "close-up" diagram. The description in the "head" will also have to be changed to reflect the label that identified the "first level" spine. For example in Fig. 11.2, the new "head" might become Methods. Zooming in on spines can be used as often as required as long as they are suitably labeled and it is obvious where each spine originates. Ensure that the *parent* and *daughter* relationship can be easily identified at all times.

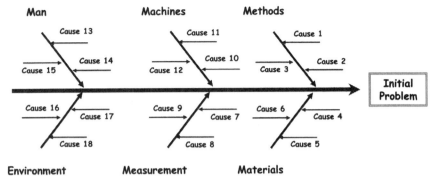

Figure 11.2 The general fishbone diagram.

The headings above are general, but you should recognize them. They cover most reasons for problems (including the 4M's ... Ah but, there are 5M's ... which is not a problem. Measurement is not one of the 4M's.) Measurement has been included to allow for quality and tooling issues. Environment has also been included to account for use conditions. Measurement would normally be included with methods and environment would be linked to machines. They are an ideal source of ideas when brainstorming for causes. In reality, it does not matter how the spines are labeled. The best way to label the spines is the way that suits the team and how they see the problems. Besides, worst case, the spine can be repositioned and redrawn if desired.

Let's analyze a real problem like not getting up on time or as we like to call it, "sleeping in." The first step is to look for reasonable excuses—sorry, primary causes. The alarm didn't work being the most common cause, as used by the hard of thinking. However, the more enterprising might blame the alarm being set incorrectly. The more honest might hint the alarm might not have been set at all. The really honest one will admit the alarm probably worked but it didn't wake him: he is fully aware that large amounts of alcohol can do that to you.

Notice the smaller spines. These are the possible embellishments that can be used to add realism.

It is worth taking time to restate the importance of the group when creating problem-solving charts. The chart in Fig. 11.3 was created by a one-man team (me) to illustrate a point, but although I thought it was reasonably comprehensive, when I spoke to a team of people on a Kaizen course, they all had different perspectives. They included causes I had not considered:

☹ Children
   Not sleeping, ill, sleeping in the parent's bed, wanting drinks, ...

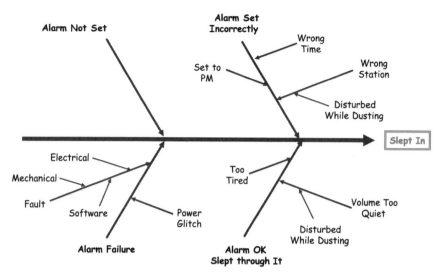

Figure 11.3 The alarm fishbone.

☹ Shift working

☹ Illness

Having a bad night's sleep, unable to sleep, having to go to the hospital in an emergency, ...

☹ Being young and "fancy free"

Not wanting to sleep, having visitors, ...

If having an input from teams on an example as simple as "sleeping in" opens up so many new areas to be considered; just imagine how valuable a team will be when faultfinding for real in a complex problem situation.

There is one other type of fishbone diagram that can be used with positive benefits: the CEDAC chart (*C*ause and *E*ffect *D*iagram with *A*dditional *C*ards). Basically it is still a fishbone chart with spines, but rather than add text to the drawing, Post-it Notes are used. I have previously explained the benefits of moving data to make a fishbone easier to follow. Similarly, being able to move entire spines of data with ease makes zooming in so much easier. They are a perfect marriage in which one enables the other to be as effective as it can be.

One point to remember is to follow up on all suggestions made during Brainstorming. If one member suggests "changes," look for any ways that changes could have been made. Write "Change" on a Post-it and give it a spine. Changes can either be known to the team or unknown to the team. What I am getting at is consider the impact other departments can have on the equipment without the team's knowledge. Is it possible that the facilities department has reduced extraction pressure or cooling flows: could that affect stability? Could the purchasing department have changed a supplier and hence one or more spare parts, without informing anyone who might be affected by the decision? Is it possible that a part has been exchanged by an engineer and not logged. Look for everything.

Personally, I only use the fishbone to get discussion going. My preference is the fault tree diagram.

Fault Tree Diagrams

The fault tree diagram is a flowchart like a family tree, except that it starts with the failure and is then drilled down into the various causes (see Fig. 11.4). Each level that the tree moves downward increases the precision of the option. This is my favorite method of faultfinding. I ask, "If I wanted to cause this problem, how would I do it?" and start thinking about all the possible ways I could do it.

It is also possible to attach probabilities to each option. Naturally, these will be guesses, although sometimes the guess will be reasonably accurate. In the case of the alarm clock example, it would be safe to assume that clocks fail very infrequently and that lots of people forget to set the alarm. I had one clock for more than 20 years, so I would put the failure rate at about 1 percent. On the other hand, I would guess that about 85 percent of people forget to set their alarms at some time or the other.

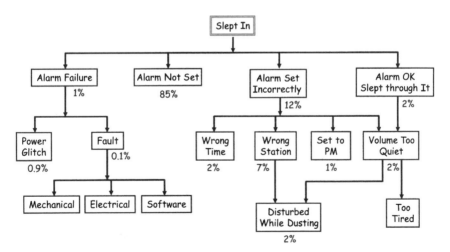

Figure 11.4 The fault tree diagram.

We have already discussed that the causes in the alarm fault tree are only some of the possible causes; there are others that have been missed (for example, not at home, no alarm clock, or the dog ate the alarm clock). The purpose of the fault tree is to identify the options and then select a best guess option to check first. It is not necessary to put percentages on each of the options. It can be very difficult to make accurate guesses on some subjects. The only thing the team has to decide is which failure option to try first. This comes down to experience, the exact fault symptoms, the actions the machine carried out just prior to failure, any unusual indications, the skill of the engineers, the simplicity of the particular checks, and a degree of probability. The team need to review the options and allocate priorities for the options to be checked.

OCAPs: Out-of-Control Action Plans

In just the same way as we discussed above, the team needs to set priorities for the options to check. Consider the example of sleeping late as analyzed in the above methods. Assuming that sleeping in was not a deliberate choice, then some other factor was the cause. What could be the options?

1. The alarm failed.
2. The alarm was not set.
3. The alarm was set incorrectly.
4. The volume of the alarm was too quiet.
5. Everything worked, but you slept right through it.

If each of the above steps was a machine fault, we could easily waste time by checking them in the wrong order. Unfortunately, not everyone will follow the correct order if checking on their own. This is even more likely when it is a complex machine issue or a process issue.

An even worse point to consider is that the first engineering shift will check them out in the order: A, B, C, and D. Then they will pass the problem to the next shift who will check them out in the order A, B, D, and E. Then the third shift will check A, B, ... Can you see a pattern emerging? Each shift appears to have no faith in the work of the previous one, or they did not bother to read the pass-on notes. In addition, even where they do check the handover logs, they never seem to check back any further than one report. I once discovered a tool was down for 7 days. I was a tad concerned that I had missed it. In my own defense, I was working on other problems and the fault reports didn't flag any issues as the machine was down for a number of small repairs and not a large one. It appears the same part was replaced 15 times during the week and *no one* suspected the cause of the problem was not that part. Not one person. Even when I asked the guys to change a different part, they changed the good one three more times! As you can see, my word is law . . . . Normally, I insist on a changeover meeting as opposed to pass-on notes. They are far more effective, particularly when there is a serious issue.

So, when we have a fault that has more than one possible cause, we need a system to guide the engineering team through the faultfinding options. We need to develop a standard method of fault diagnosis for resolving problems in specific areas. We need to use an OCAP—an Out-of-Control Action Plan. This will ensure that the teams follow the correct actions, in the correct order, every time. An OCAP considers all the possibilities and defines the agreed sequence of the checks; the most likely cause being the first one to be checked. The intention is to find the fault in the fastest possible time and with the minimum number of checks. The OCAP also details how to carry out each check. While reading the following steps, don't get bogged down with the details, the example is only intended to illustrate the logic of the thinking.

Our OCAP might look something like this:

1. The alarm was not set.
    1.1. Check if it is in the Set position.
        Details in the manual on p. 14.
2. The alarm was set incorrectly: wrong time or station.
    2.1. Confirm the alarm set time.
        Details in the manual on p. 223.
    2.2. To check the station, switch the alarm to "Radio On."
        Check p. 10 of the manual for details.
    2.3. Check if the station selected would have been "on air" at the time set.
        Use the Radio Times or Television & Radio guide for information.
3. The volume of the alarm was too quiet.
    3.1. Switch the alarm to "Radio On."
        Check p. 10 of the manual for details.
    3.2. Listen to the volume and evaluate.
        If it is quiet, then it might be the cause. However, if it is loud, the volume may have been increased in the interim period.
4. Everything worked, but you slept right through it.
    4.1. Review points 1 to 3 above.

4.2. Is the alarm currently working?
If no problem showed in 4.1 and 4.2, consider 4.3.
4.3. The problem was manmade.
4.3.1. Boozed the night before.
4.3.2. Worked late and needed sleep.
4.3.3. Was just really tired and did not wake.
5. The alarm failed.
    5.1. Is the alarm working now?
        5.1.1. If "yes"
            Has anyone repaired it?
        5.1.2. If "no"
            The reason has been found.
    5.2. Power failure or glitch.
        5.2.1. Are the digits flashing?
            5.2.1.1. If "yes"
                The reason has been found.
            5.2.1.2. If "no"
                No solution can be found.

It is unlikely that the steps of any OCAP will be in the correct sequence at the first attempt. It will take experience and time to establish what needs to be corrected. It could be that the symptoms of the problem are not completely understood. The order of the OCAP tasks could be wrong. If the solution is always found at either Step 3 or Step 4, then these are the two steps that should be checked first. Equally, the failure options can always be refined if they are found to be wrong or any are missing. Finally, the instructions for checking can be improved as the system develops and the teams provide feedback.

Chapter

12

Team Objectives and Activity Boards

In this chapter we will discuss the purpose of *activity boards* and how they will be used to promote all of the improvements of the teams. The boards are also the way the team display their goals and targets, so it will also be necessary to look at how the targets are set. Since we will have to evaluate the current state of the equipment to get a baseline from which the team targets can be evaluated, a method for the calculation is suggested. We will also take a quick look at the sort of savings possible from an RCM (Reliability Centered Maintenance) analysis. There will also be a section on lean manufacturing. To the best of my knowledge, lean methodology is found in every efficiency technique in this book. It is like the glue that binds them all together.

Lean is the parent, 5S and SMED are two of its children. It uses a whole range of methods to seek out and eliminate waste. By definition, waste absorbs resources and gives nothing of value back to the company. We will also briefly discuss the different manifestations of waste and then how value and its two variations, value-added and non-value-added, impact the way a company is run. The former is the one companies should be seeking to add to their systems and non-value-added tasks have to be identified and removed. One other component of lean will be discussed as it, too, is frequently referred to in all of the continuous improvement techniques: the customer. The entire purpose of business (apart from making money) is to supply a product to the customer. It should be reasonably obvious that the customer should be considered to some degree in virtually every decision made. In fact, the customer should be the number one consideration.

Activity Boards

Total Productive Maintenance (TPM), Autonomous Maintenance (AM), and 5S teams all display their progress using notice boards known as *activity boards*. Each team designs and maintains their own board and uses it to highlight their successes, improvements, and learning. The board is the team's advertising poster, a visual display with "before and after" photographs, graphs, Pareto charts, tables, and training data. The location of the board should be a place where everyone can see it: at the tool, near to the tool, or on a canteen wall. Other employees should be encouraged to go and take a look at the boards. The purpose of the teams is not to create activity boards; it is to carry out whichever task they have been asked to perform: TPM, AM, 5S, and even SMED (Single Minute Exchange of Die) which is not normally associated with boards. The boards are only for promoting the teams' progress.

There are a number of things the board has to promote. These include all of the team information and what their current project is. Since the board is an advertising tool, the project should be made to sound as interesting as possible, like a headline from a tabloid newspaper: "Eliminate all problems from the Orange Squasherizer and cut annual running costs by £500,000 every year." The progress of the team should be tracked on the board, with any milestones and targets clearly identified. Where "before" and "after" images are displayed, the most impressive photographs should be given the greatest prominence. The board should also give consideration to the fact that it will be viewed by people who are not company employees, customers for example. Consequently, the board should always have a positive tone.

There will be a lot of basic team information needed:

☺ Team name.

☺ A list of the team members and which departments they work in. Photos are good, but there are those who dislike being photographed.

☺ Targets and objectives.

☺ Project timelines.

Including the main milestones.

☺ Meeting information.

☺ "Before" and "after" photos of the initial clean and clean cycles.

☺ F-tag logs; Graphs and who is responsible to complete the tasks. A planned completion date would be good.

☺ PM maps.

☺ Defect maps.

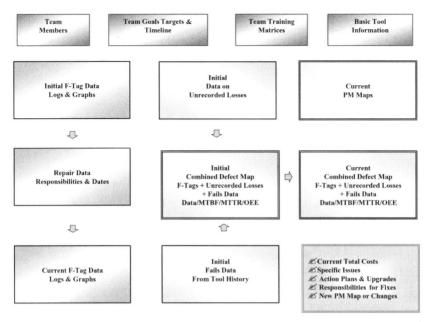

Figure 12.1 A sample layout of an activity board.

☺ Quality information if possible.
Customers love to see how operators are working to improve the quality of their product.

A board layout will be needed. It can be along the lines of the example in Fig. 12.1, which is a very simple layout that includes the major stages in the application of zero fails. The best format for the activity board is the one that the teams choose for themselves. Whatever is decided, it should become the format for all boards. Having said that, some teams like to stamp their own personality on it. The only hard and fast rule is that it must be easy to understand, up-to-date, and good to look at.

	Actions
12-1	☺ Decide on the location of the activity board(s).
12-2	☺ Decide on a standard format.
12-3	☺ Start populating the boards.
12-4	☺ Ensure they are regularly updated.

Team Goals

The management will set the target for the improvements. It will be a challenging target. If it is too simple there will be no challenge. It will be necessary to target at least a 50 percent improvement. The goals will include

- Collecting the information needed to establish the current performance for the tool being analyzed. This has to include basic information:
    - Current failure rate
    - Current Mean Time Between Fails (MTBF)
    - Current Mean Time To Repair (MTTR)
    - Current Overall Equipment Efficiency (OEE)
- A target for reducing the failure rate and a time to achieve it.
    - Be realistic; don't set up the teams for failure.
    - Consider the complexity of the issues, how hard they could be to fix, and the time allocated.
- Improvements in equipment performance that will be measured by MTBF, MTTR, and OEE.
    - The MTBF will decrease automatically.
      This will happen as long as the root cause for each fault is found and resolved. It is amazing how often people will still settle for the quick fix. If they even remotely believed the fault would not return, it would be possible to see their point. I could even understand one or two quick fixes if they were only trying to make time until the real fix is carried out. Even worse, they complain incessantly about the "crap tool that always fails...."
    - Training and "turnaround" spares dramatically improve MTTR.
- OEE will also improve as the uptime improves and the failures decrease.

  If the teams are following TPM, they will also be improving productivity and quality issues. These are issues that previously sat on a back burner waiting for time to be allocated. They were also likely to be "partial failures" that tend to be speed reductions and areas where performance has been running just under the vendor's specification. Unless the equipment came to an abrupt halt, these issues were rarely considered.

When a team finds a solution to a problem, it must be applied to all of the other tools of the same type. There might have to be checks to make sure there are no variations in the equipment that will affect the solution. If there are, then modifications to the solution will need to be made. We will want to evaluate the improvements seen by each individual tool and have an average for the complete toolset. Each team has to make improvements to achieve overall success, but never forget that an outstanding performance by a single team, or even a team member, should also be recognized.

Monitoring Progress

It might be necessary to audit the progress of the teams to prevent stagnation and keep the momentum up. This can be carried out by the pilot team or the TPM supervisory group. A project review format can be used. If there are other teams working on similar tools, the teams must all be trained in any solutions and they should be involved in the creation of the new standards.

A manager or a section head could carry out a system of interim audits to confirm that the teams are still moving forward with no obvious outstanding issues or roadblocks. There should be a standard evaluation form to ensure uniformity of assessments across all of the teams.

What do we monitor?

- ☺ Team attendance at meetings.
  Poor attendance at the team meetings will affect the motivation and progress of the team. If it is allowed to continue with no comment, personnel who do not attend or even managers who will not allow members to attend will not be recognized. An official attendance record avoids conflict and will illuminate underlying issues with the adherence of TPM by the area managers.
- ☺ Team attitude to TPM.
  If we do not appreciate there are any problems, we will be unable to correct them.
- ☺ Team training logs.
- ☺ The activity boards.
- ☺ Are the teams' milestones being met and are the improvements planned for the period been achieved?

Initially, rate of fails, MTBF, and OEE.

- ☺ Are there any roadblocks or issues?
- ☺ How do the teams compare across any given toolset?
  This is about team standards. If there are variations, is there a reason?

How do we calculate the failure rate and the target improvement?

We need to be able to evaluate the improvements made on a tool. When we look at Fig. 12.2 it will be hard to avoid complaining about how complicated it is. It was originally designed to set a target for a reduction in fails, which is the simplest measure we can apply in TPM, since the target is zero fails.

332 Chapter Twelve

Step	Data Only for Tools in Toolsets that have Zero Fails Teams		
1	Date of Start of Data Collection:	T1	
2	Date of Start of Initial Clean:	T2	
3	Data Collection Time Period:	T	$T2 - T1$
4	Number of Fails for All Toolset ZF Teams during Time T:	F	
5	Total Amount of Product for Toolset ZF Teams during Time T (Moves)	M	
6	Fails per Move—Baseline Fails Ratio in Time T:	Fm	F/M
7	Target Reduction = 50% of Fails/Moves Ratio:	Rt	$Fm/50$
8	Target Fails per Move:	Rn	$Fm - Rt$
9	Number of Fails in 1 month (or 1 week) Since Start of Initial Clean:	F1	
10	Number of Product Moves in 1 month (or 1 week) Since Start of Initial Clean:	M1	
11	Fails per Move:	Fm1	$F1/M1$
12	Improvement in Failure Rate Compared to Target:		$Fm1 - Fm$
13	% Improvement on Baseline:	Fb	$(Fm - Fm1)/Fm \times 100$

Note: Fails per move might be a decimal number and should be scaled up to make a whole number. That is multiply by 100 or 1000.

Figure 12.2 Example of an initial improvement sheet.

1. How may fails did we have before TPM started?

    Use the data collected over the 3-month analysis period, including the minor stops. We have to stop when the initial clean starts, since this is the point from where we start resolving the issues.

    If the minor stops are not included the initial fails, data will be too low.

2. Establish the total number of product moves for the tool over the same period.

    We need to be able to see an improvement in productivity.

3. Standardize the data; the easiest way to do this is to calculate the ratio of "fails per move."

    This will be a fraction. If it is not a fraction then possibly Semtex could be used for the fixes.

    It is always easier to view ratios as a whole number. Fortunately, this is easy to do. To convert the value to a whole number, scale up the fails rate to something like "fails per hundred moves" or "fails per thousand moves." In Six Sigma we have a term called "defects per million opportunities."

4. Set a reduction target (40 to 60 percent) and calculate it as "fails per moves" so it can be added to the improvement graphs.

    This is a good range for a reduction. Fifty percent is a challenge and a good target and yet it also leaves room for more.

5. Now collect the improvement data, starting from the initial clean.

    Use a 1-week or a 1-month period and find the total number of failures. The data can also be plotted as a rolling average to smooth out the lumps and bumps.

6. Find the total number of completed product (moves) for the same time period.

    This is also going to be of value when the company starts to consider OEE.

7. Calculate the improved ratio of fails per moves.

8. Compare to the improvement target.

    Is the team making good progress? A reduction in failure rate is the target. This will follow if the initial failures are fixed to a root cause.

Graph all of the data; a visual display is better than a table or a list of numbers. Seeing a reduction at this point is a good morale booster for the team to say nothing of a real time improvement in OEE and productivity in particular. Display the data clearly on the activity board.

Authority for Working in Specific Machine Areas

Within practical limitations, equipment is divided into functional areas in much the same way people are. People have heads, arms, legs, and a main body. They can be further subdivided into hands, feet, upper and lower body, and so on. With the exception of the sinuses, the different parts (areas) of people have been developed (or evolved) to be in the most suitable locations to carry out their functions. So it should be the same with equipment: each bit should be where it has to be in order that it can be used to properly fulfill its function. Even tools evolve. With each subsequent model or version there will be changes designed to improve some of the shortcomings of the previous tool. Very often these improvements are fed back through discussions with the customers, who know what they need for the next generation of product. (They are another example of the voice of the customer.)

The *responsibility map* defines the skill level required to work in each area, based on the complexity of the tasks and the potential hazards in each area.

Safety. Safety. Safety.

The areas that operators are permitted to work in have to be assessed according to the specific skills and aptitudes of the operators in the same way that a technician trainee is slowly exposed to more complex tasks. Operators and indeed unskilled technicians, where necessary, must be approved to work in specific areas. Their skill level must be such so as to ensure there is no possibility of them being involved in an accident while working. Equally, some areas might require an operator and a technician working as a pair with the technician supervising the operator. This is the favored method as it continually boosts the operator's skill level. In some situations it might not be permissible to allow any operators at all.

Figure 12.3 and Table 12.1 combine to illustrate how a tool might be divided into "owners" who would have responsibility for their areas. If there are five team members, try to allocate the responsibilities evenly. For variation, the areas can be rotated later, but never forget, the selection and qualification criteria must still apply.

The initial selection should be made on the basis of the skills required to work within the area. Although it is possible to teach an operator how to carry out particular tasks and follow instructions, they would never be permitted to work within the gas box (Area 4). If any accidental damage was to be caused to the internal components, the consequences could be very severe, so only persons with the *technical training, skill, and the ability to respond to unforeseen circumstances* should be permitted access.

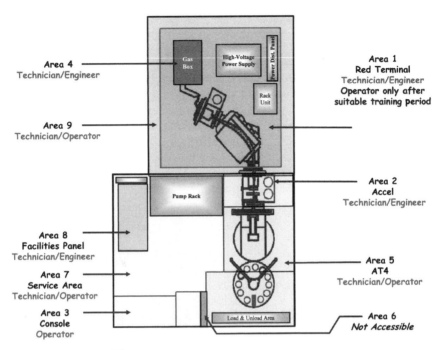

Figure 12.3 Responsibility map.

When setting up teams, where possible, technician and operator teams should be used. This way both parties gain knowledge and experience from the other.

	Actions
12-5	☺ Understand the goals for the factory and for the tool.
12-6	☺ Decide on how the teams will be audited for progress.
12-7	☺ Decide which team members are responsible for the various areas of the tool.
12-8	☺ Create an area ownership map.
12-9	☺ Create an Area Responsibility and Certification Table. (Table 12.1)

What Do the Results of a Real RCM Analysis Look Like?

The boatloader example we used in Chap. 9 was initiated to prevent the losses caused by the product being incorrectly processed because

TABLE 12.1 Example of an Area Responsibility and Certification Table

Area	Functions	Access levels	Certified owner	Certified date
1	High-voltage supply for Gas Box. Contains Gas Box Power supplies for Gas Box, Source, and Terminal components. Terminal vacuum system Analyzer magnet and power supply Terminal electronics and Power Distribution Panel.	Engineer Technician Operator/ Technician Team	Technician #1 Technician #2 Operator #2	01-Feb-04 01-Jan-04 03-Mar-04
2	Vacuum cryopums and compressors. Power supplies. Accel Electrode. Flag Faraday and Electron Shower. Pneumatics for chamber door.	Engineer Technician	Technician #1 Technician #2	01-Feb-04 01-Jan-04
3	Control Console Load and Unload electronics and pneumatics	Engineer Technician Operator	Technician #1 Operator #1 Operator #2	01-Feb-04 01-Jan-04 01-Jan-04
4	Gas Bottles Source Electronics Toxic Exhaust	Engineer Technician Operator	Technician #1 Technician #2	01-Feb-04 01-Jan-04
5	Wafer handling Pneumatics and hydraulics for handling and chamber door. Power distribution	Engineer Technician Operator/ Technician Team	Technician #2 Operator #2	01-Jan-04 01-Jan-04
7	Rear of Console Vacuum roughing pumps Power supplies Pneumatics	Engineer Technician Operator/ Technician Team	Technician #1 Operator #2	01-Feb 04 01-Jan-04
8	Isolation transformer for terminal power Terminal high voltage power supply Input facilities (water, compressed air, Nitrogen, electricity)	Engineer Technician Operator/ Technician Team	Technician #1 Technician #2	01-Jan-04 01-Jan-04
9	Area between outer doors and terminal used for enclosure and high-voltage arcing protection.	Engineer Technician Operator/ Technician Team	Technician #2 Operator #2	01-Jan-04 01-Jan-04

of the load and unload system stopping in the wrong position. It could stop in a range of different ways with different failure consequences. For example,

1. It could fail to start moving into the tube.
2. It could stop while driving into the tube, but before entering the furnace.
3. It could fail with wafers exposed to heat, but not fully in the furnace.
4. It could fail with all the wafers in the furnace.
5. It could fail on the way out of the furnace with wafers still exposed to heat, but not fully in the furnace.
6. It could fail while driving out of the furnace with all of the wafers clear of the heat.

From the RCM point of view, the worst possible failure is the one we would analyze—point 4. We would possibly list the others, but score them out. They would have a cost proportional to point 4, but usually lower. Points 1, 2, and 6 might have no product damage but they would have investigation, repair, and lost production costs, including operator waiting time. If we analyzed them all, we would end up with an array of prices, but in every case, all the actions would be the same.

Summary of the boatloader analysis

Figure 12.4 is a general view of a boatloader. The furnace illustrated is a TMX Series furnace. The one shown is an old model, but they have such a high reliability and reservoir of knowledge in the field that they are not only still in use, but are in constant demand by manufacturers. It is interesting to note this fact because many companies believe the only way to solve maintenance issues is to purchase new, modern equipment. This is the way to go if the decision is made based on process requirements and the need for current technology. However, we have shown in our explanation of TPM and RCM that many (if not most) of the problems with equipment are related to the way the tools are used and maintained and are not due to the equipment itself. I understand Tetreon Technologies have recently bought over the company supporting TMX equipment worldwide.

The RCM analysis of the boatloader took approximately 3 months to complete and an average of 16 h of meetings each week. The attendance figures were not as good as would have been wanted, with process support and operators finding time to attend the meetings the most difficult but, nevertheless, we still managed to get a satisfactory set of data. The shift members, both technicians and operators, also

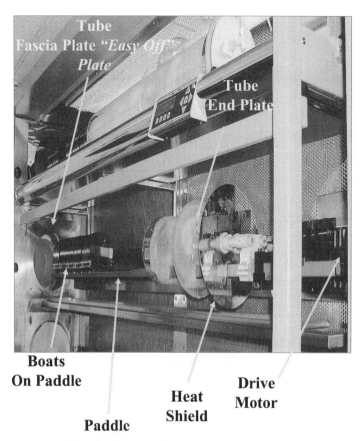

Figure 12.4 A boatloader assembly.

used any spare time they had for making checks between meetings. If the time constraints had been thought to be too demanding, they could have been changed at the behest of the management team. It would simply have made the analysis take a longer time to reach completion. In the review of the results, the costs and the number of fails have been concealed as a courtesy to the companies involved, but to enable you to have faith in the results I can safely say that they are all lower than the real values. The relative cost of fails to maintenance costs, however, is accurate.

After completing the Decision Worksheets, we had a large list of tasks to review and plan for implementation. Even though task transfer is a TPM task and not an RCM task, one of the objectives was to find suitable tasks to be considered for passing to operators as autonomous maintenance. Just over 10 percent of the tasks were found to be suitable for consideration for transfer.

A further 10 percent of the checks were calibrations that would improve performance. These calibrations were not previously carried out although they were specified in the manuals. The remainder involved safety and redesign. The safety recommendations must be carried out, even though they were not particularly serious. The redesign recommendations need to be prioritized, scheduled, and controlled. After confirming that the redesigns will solve the problems, each one would need to be replicated on the other tools of the same type.

It is always advisable to be wary of adding new parts to a tool, even in the form of redesigned components or new safety devices. New parts introduce more functions that can now fail and that need to be analyzed for reliability. Make sure the new components are suitable, evaluate their reliability and failure patterns, and devise an appropriate maintenance or failure-finding routine.

It is likely that more than one RCM analysis will be running at any given time. In some companies there might also be TPM, 5S, and SMED teams to be considered, so the limiting factors to the number of tasks that can be completed are manpower and cost control. Of the furnaces being analyzed, most of them were older models, many of which have been retrofitted and upgraded. Basically these tools have very good reliability; the problem is that the cost of a single failure can be high because of the amount of product that can be affected at one time. There is also a high downtime cost in fixing the tool.

It is worth remembering that the original reason for selecting the boatloader for an RCM analysis was the unreliability and cost of the loading/unloading "boat stalls." These were caused by processes that ran particular gases that can escape from the front door of the furnace. They create a sticky residue and affect the drive assembly, particularly the leadscrew. If the boat stops inside the furnace and no major damage has been done, all of the product can be lost at an estimated cost of around $15,000. The analysis identified contaminated leadscrews as a primary cause of the failures. This cause was already well known to engineers and technicians. However, the RCM analysis quantified the cost of the damage and identified the best way to prevent the fails. The introduction of new cleaning and lubrication cycles on one tool over a year amounts to about 5 percent of the cost of one single failure.

The paddles that carry the wafers are also subject to a few failure modes that could cost a substantial amount. These costs could be much higher than the cost of the stalls. Two of the consequences of the paddle breaking are damage to the original tool and, in the worst case, damage to the other three tools on the same stack. Normally the damage is caused by gravity, so the damage rarely affects tubes above the one with the problem. The cost of daily visual checks for a year, carried out by the operators, is less than 2 percent of the cost of one single failure.

The majority of the failure modes uncovered by the analysis could each be avoided for less than 5 percent of the cost of their consequences.

The high number of redesigns sounds a bit scary until you look deeper. One of them, a completely new procedure for one of the main tasks, would resolve almost 20 of the failure modes. This issue became obvious very quickly: within a couple of weeks. We issued an interim report with recommendations that enabled this issue to be tackled immediately, even though we did not have all the necessary data. We did have enough, however, to know something had to done as soon as possible. This is a point worth remembering; it is not necessary to complete an analysis before identifying issues and getting benefits. One other redesign, involving monitoring motor current, would take care of a few more failure modes. Between them, the two redesigns solved almost half of the issues.

The safety recommendations were all minor and were prioritized for resolution. A few of the redesigns had also been identified previously, but had never been effectively evaluated until RCM provided the cost data.

The actual analysis will have identified some shortcomings in the systems and should maximize the tool uptime and the quality of the product. The number of improvements eventually completed will be the responsibility of the management. The other gains will include a few unnecessary PMs that will be eliminated and plans can be made to enable failure prediction. Where routine maintenance is still required, we will have identified the best option between scheduled maintenance and scheduled discard.

The RCM analysis also picked up on a few procedures that needed improvement. The final advantage is that we identified a number of failure modes that we did not know existed and could have been very costly.

One other benefit is that the people who carried out the analysis will never look at failures the same ever again. Why did it happen? Is it a part we maintain? Do we do the correct maintenance? Is it maintained at the correct frequency? Was it our procedure? Are the guys trained to do it? Have we changed anything? Has it happened before? Can we predict this failure in the future?

The quick fix will most likely be gone, except in the case of one-time emergencies. The foundation has been laid for continuous improvement at the equipment level.

Lean Manufacturing

Lean manufacturing was developed by Toyota in Japan. It was originally known by its other name the Toyota Production System (TPS).

The system was so successful that it was studied in the United States by the Massachusetts Institute of Technology. They were very surprised to discover that although it was a very Japanese system, it was actually possible that the system could be used by anyone and the same successes could be achieved. One of the most interesting things about the TPS (it is peppered with interesting things) is that it was developed from American methods. The major differences were that the American system believed that only technical people and engineers could make equipment improvements. The Japanese, on the other hand, believed that everyone in the company had an impact on machine performance, particularly operators and production people.

Although Japan had two major car plants in the 1920s (one of them was Ford and the other was General Motors), the Japanese wanted their own motor industry. A member of the Toyoda family (Kiichiro) visited both the UK and the US car plants as far back as 1929. But the Toyota Motor Company (Toyoda changed the name) was not formed until 1937. It was not a large-scale production organization at that time. After the war, it was time to improve their productivity, so a second member of the Toyoda family (Eiji) made a visit to the United States. Toyota was manufacturing cars in very low numbers, while Ford was making around 200 times more. The Japanese cars were not being made well, with a high percentage of jobs having to be repeated. This was seen as waste by Kiichiro and had to be overcome.

Then along came Taiichi Ohno. To say he had a powerful character would be like saying Mother Teresa was just a nice person. Taiichi was reputed to be a very forceful character. However, when it came to business he was highly innovative. He was tenacious, efficient, set high standards for himself and others, and, my favorite of his qualities, he questioned everything. Nothing was sacred. He was the driver for SMED and 5S although it was Shigeo Shingo who carried out the development work. Taiichi hated having to pay for cars to be made that had no customers lined up to buy them. This seemed to be reasonably logical. Except, he knew that the current logic for making cars in bulk was well established and accepted universally by everyone—except him. He did not agree. The reason for the overproduction was attributed to the long times it took to change from one car setup to another. He studied the operations and concluded that if he could reduce changeover times, he could make cars at the rate required by the customers and he could make them when the customers wanted them.

Apart from improving the service to the customers, this SMED technique saved all the labor costs, the parts costs, the material costs, the storage costs, the facilities costs, the quality control and repair costs, and just about every other cost involved in producing all the cars that were destined to sit in the car parks, like sunbathers on a crowded

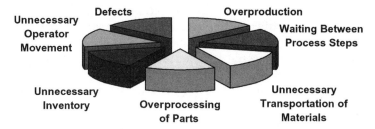

Figure 12.5 The 7 wastes as defined by Taiichi Ohno.

Spanish beach. In short, he saved lots and lots of money. By watching how things were done, correctly and otherwise, he identified a whole series of areas where losses were being made. These can be seen in Fig. 12.5.

By targeting the 7 wastes for reduction or elimination, any company stands to save a vast amount of money. Many companies never really consider waste or its associated costs. Unless a check has to be written, it is often assumed that no money has been spent. Lean promotes a framework to overcome this shortcoming. There are variations of the 7 wastes or at least different interpretations, but in essence, they usually look for the same things. This is a good time to take a more detailed look at the wastes.

Where to start? I guess I will opt for the easiest one first, which is most likely "defects."

Defects

What is a defect? It is a task, a stage in a process, or a final product that has not been completed to the proper standard. If we have a 10-step process, we have 10 potential situations where we have a chance to create faulty product. In reality we have more: if we have hidden steps like moving the parts around or storing them, we are adding to the potential for damaging the product. There are so many options; I can only mention the obvious causes for defects:

- Poor handling techniques at every stage of production.
  This includes unpacking the raw materials, any automated transport, quality inspection methods, loading and unloading, and moving from process to process.
- Faulty equipment.
  If equipment is not maintained it is likely to have issues that will directly affect the product. This could be wrong dimensions due to the machine calibration, uneven cutting, incorrectly printed labels

due to faulty print heads, damaged caps caused by faulty moulds, bashed tins, bubbles in spray painting, dust on semiconductors, badly soldered electronic contacts, broken stitches, damaged threads on screws, and so on. A badly maintained tool can also damage entire batches of product if it fails during a production run.

☹ Badly setup equipment.

This tends to cause alignment problems, incorrect proportions of ingredients, wrong temperatures, wrong colors, wrong stitch types, labels in the wrong places and even wrong labels, over- or underprocessed product, using blunt or damaged tools, incorrectly positioned weights on wheel balancing machines, and so on.

☹ Not having any standard manufacturing procedures or not following the ones we do have.

Apart from poorly maintained equipment this is pretty high on the list of causes.

☹ Poor hygiene control.

This can cause problems in companies, pharmaceutical, and food processing and preparation companies.

☹ Incorrect materials being used.

Defects cost money. The problem with defects is that they have to be remade or reworked. If they have to be scrapped, we effectively have to pay for production twice: that means the labor and the cost of the materials, including all of the overheads. If we only have to repair it (rework) then we only have extra labor-hour costs, repair costs, and delivery delays. Don't forget, there is also the potential for defects to sneak through the detection net and find their way to customers, where they can reduce the reliability of their product or, if the customer is the consumer, they can receive faulty or unreliable goods.

Overproduction

Overproduction would not be a problem if companies were supplied all of their materials free until the goods they make have been sold to a customer. If there was no cost for storing the goods: no rent, no heating cost, no lighting costs, or cleaning costs; if the stored goods could not rust or be damaged while sitting, waiting for sale and if the company's employees all agreed to postpone their wages until the goods were sold; if the electricity and all of the other production materials did not have to be paid for until after the goods were sold—then overproduction would not be a problem. Overproduction is a waste of money because the company has to pay first and then wait. Depending on how long they have to wait, they might even have to pay for modifications, upgrades, or drop the price to get a sale. All of which costs money and labor.

Waiting

Waiting between process steps can be literally that, but it can also be interpreted as not being able to continue along the production line. The causes for the holdups are listed below:

- There might be no operator available.
  This could be a labor shortage or bad organization.
- The equipment could be down for scheduled or (worse) unscheduled maintenance.
  Breakdowns are a plague to production. Enter TPM to reduce failures.
- There might be operators waiting for materials to be delivered.
  Check materials before planning production runs. There must be a way to ensure the parts are available and there are enough of them to complete the runs.
- The operator cannot run until pre-production quality test results are approved.
  This is often due to legislation, but it can also be due to historical needs within a company or a lack of confidence in the equipment. Check why the test is being carried out. Is it necessary? Do we have to wait for the result before starting? How likely is it to fail? What is the risk versus the delay cost? Is the equipment unreliable? Can it be improved? If it can, make the improvements. Enter TPM.

Transporting

Transporting materials when it is not needed is forbidden. This comes down to line layout. Many companies operate with production villages, rooms of equipment that make only one part and they put them on shelves or move them in bulk to new shelves in a different area. They sit in their new areas, waiting to be processed in their next village, after which they will eventually leave and will be taken to another new storage rack to await their next process. Lean does not like the production village format unless the product is being manufactured in thousands or millions of units. Even then, I suspect it might prefer to find a way to use parallel cells. The transportation issue is worse when product has to move between buildings. Moving between towns is next on the undesired list. The worst transportation offender is when goods have to move between countries.

Apply this to stores. Imagine a worst-case scenario and all the most commonly used parts are stored furthest away. Every day, the storeman will need to run a marathon. It seems a bit silly, does it not? Try using a spaghetti map to track the movements over a few days and see if it can be improved. Position the goods such that they minimize movement. It is not as simple as it seems as some components will be small, some will have to be stored in special areas, or some large items might be stored on pallets. It will not be the end of the world if we do it wrong, it can always be corrected. There will also be other departments that might benefit. Have a look around and check them out. Think about offices, where are invoices stored? Is the photocopier suitably positioned or is it well out of the way? How easy is it to find a document? Use 5S to get the unnecessary items out of the area and use the free space to reorganize and minimize movement.

One company I visited showed me a row of tables that were sitting against a wall. "This is the first production stage," explained my guide. He then went to stages two, three, and four, where he pointed out a 12-in. gap on either side of the next table. "This is where the product is shipped to Taiwan for...." Moving between countries is normally based on the cost of labor. However, there are other considerations that might make the decision less desirable. There is a time and a cash cost in actually transporting the goods. This pushes up the product's lead time. There will be a new factory to build and equipment to install. It will be commissioned using high-cost vendors, with air travel costs, all living in local hotels, meal expenses, and hiring cars. The overseas company normally has to import managers and labor from the parent country and pay for expenses and relocation. There will also be a cultural difference, a learning curve that includes the need to learn the language by the nonlocal employees. Technical training will also be necessary and vendor support can be costly. Finally, I suspect that the low wage rates will have a short half-life. Will the time be long enough to recoup all the setup and operational costs? What would be the product *profit* if the cost of not moving abroad was compared with the cost of staying at home or building the factory where the goods are going to be sold?

Manufacturing in cells is a goal of lean. A cell minimizes the distance the product has to move. Ideally, the product should finish one process and jump to the next with no manual assistance. Another advantage of a cell is that it can be run using reduced manpower. Once the equipment runs reliably and the operator has been trained, he can support multiple machines. Take a look at Fig. 12.6. It shows two systems: one contains four villages, which for the sake of illustration contain four tools each. The second section contains four cells, using the tools from the villages. The process tools should be side by side, or more accurately head to tail, with no storage racks. The ideal layout for a cell is a "U," although

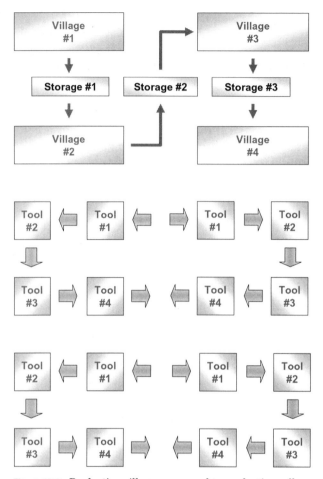

Figure 12.6 Production villages compared to production cells.

an "L" is also favored. The idea is for product to enter, run through its allotted processes, and leave where it came in.

In Fig. 12.6, we have four cells. If the product demand suddenly drops to 50 percent, we can simply shut two production "lines" down. Equally, if we were running only two lines, we can choose to expand to three or four as the customer demands. Where the lines are capable of making different product ranges, we can chop and change lines in tandem with the demand. In periods of low demand, when a line is shut down, it saves on all of the facilities. Having multiple lines enables time to be used to carry out maintenance. It is amazing how many people argue they are too busy to carry out maintenance: "This tool runs 22 h a day...." In fact, the average OEE is often as good as 50 percent. They seem surprised when I counter with the fact that airlines, hospitals, ambulances, and trains can still do their maintenance. If there are four lines, we are also in a position to schedule slots to carry out the routine maintenance, or it can be coupled with a changeover. There is always the option to switch production to one of the parallel lines.

Overprocessing

Resolving overprocessing of parts involves looking for unnecessary steps. The customer is looking for a particular product; it is of no interest to him if there is a label inside the product that has the manufacturer's name on it. To the customer, the label adds no value even though it costs the manufacturer money and time. This is a processing step that is not required. There are other processing steps that need to be considered. Storing is a non-value-added task; it makes absolutely no difference to the customer whether the goods are stored or not, unless perhaps it is to allow vegetables or fruit to ripen. We have already discussed the negative side of transporting product.

Inspection is a bit different. There will always be a need for inspections, but the question is, "How many inspections?" Some inspections are not legislated and are carried out only to confirm that the equipment is still operating to specification. If it is possible to increase the equipment reliability, it might be possible to reduce or eliminate some of these process steps. Another possibility is that the operator might be able to become his own inspector. It is not unreasonable to expect that an operator should make the product as it is specified and so should not have to check his own work. This is possible provided the equipment is accurate enough, the operator is suitably skilled and competent, and the raw materials supplied to the operator are correct. The next operator in the production line might also be in a position to pick up any errors. There could also be occasional independent sampling. One other potential source for eliminating production steps is historical checks that have been added as a result of a previous problem. If a root cause was found, is it still necessary to carry out that check?

Transforming the product is the one that actually does something that the customer needs. It has value. This includes tasks like connecting components, adding switches and displays, putting the contents into the box, cooking the food, shaping the component, drilling the holes, designing the modules, stitching the cloth, sewing on the buttons, calibrating the modules, ordering the components, packing the goods, shipping the goods, and so on. These are the things that define the product as seen by the customer. Even so, can there be any unnecessary steps? Remember that the processes are not limited to production: what about purchasing or stores. How many people have to approve a purchase?

When a customer contacts the company, how many people does he have to speak to find out what he needs to know? How many people does it take to make a quotation? How many desks does the quote have to sit on? Checking out these steps is known as *value flow analysis* or *value stream flow analysis*. It is a cross between a flowchart and a spaghetti diagram, which is designed to find unnecessary or duplicated process steps. Does Fred have to pass this document to Bert: can he make the decision by himself? Even better, could Jane have made the decision before it landed on Fred's desk in the first place? Question everything and think SMED applied to a process.

Unnecessary inventory

Next we have to look at unnecessary inventory. This is basically stuff that your company owns or owes money for. It is either

1. Sitting in your "goods in" stores, waiting for you to do something with it.
2. Sitting in storage racks scattered throughout the production area, waiting for you to do something with it or something else to it.
3. Sitting in a storage warehouse as finished goods, waiting for you to do something with it.

Can we see a trend here? If we need the goods we should order them, but not enough to last a year! The Japanese like to operate using a system called *Just In Time(JIT)*, where only the amount needed is delivered. As opposed to having a 10-ton truckload arriving every 10 days, we get 10 daily deliveries of 1 ton. I guess I am sitting on the fence where Just In Time is concerned. I would need to be convinced that my suppliers could deliver the parts reliably. Many of today's successful companies use JIT. I understand the logic, but I would need to see the figures. I have heard of situations where one major company, say Company A, gets JIT deliveries from, say, Company B. However, Company B has rented local storage, where it stores the goods in bulk waiting to be taken to Company A in bite-sized chunks. I cannot imagine the system not working. And yet, when I think of the delays in transport, space shortages in Goods In, the risk of running out of raw materials, or product being scrapped for whatever reason and needing replacement, I get cold feet. Deep down, I am probably one of the *Just In Case* people, although I would try and work as close as possible to the exact levels. It is possible to apply JIT methods selectively. Perhaps initially they can be applied to some of the less critical materials and consumables, and then eventually, as confidence in the suppliers increases or new suppliers are found, can gradually include others. Being a tad pessimistic, I think I would also need a backup plan for emergencies.

In addition to cost, there are other problems with ordering too many parts. If the part becomes obsolete it will have to be written off. If it has been made obsolete and still lives in the stores, we run the risk of using the wrong components in current product. We also have to have storage areas that are much larger than we would really need. Refer to the chapter on 5S. The parts that have been assembled either partially or fully run the same risk of not being sold and simply running up losses as scrap. These parts also run the risk of becoming obsolete.

The solution is to make only what we can sell, with a bit of smoothing out. Buy only what we need ... and use techniques like SMED and 5S to ensure the product changes are as smooth and as fast as possible.

Unnecessary operator movement

The last of the 7 wastes is unnecessary operator movement. This one is a bit confusing. It can be confused with unnecessary transportation. But much of this is covered in 5S. We want the operator to be able to carry out his/her job without unnecessary effort. The use of shadow boards, positioning parts where they should be for the operator, and using open storage are all positives. However, this waste goes even deeper. It is looking at the level of ergonomics for the elimination of

- Unnecessary twisting
- Bending
- Crouching
- Stretching
- Changing hands
- Having to use two hands
- Having to decide which hand to use to pick up an item
- Having difficulty actually picking an item up and getting it positioned for use

Not only do these movements waste time, but they also tire out the operators. We need to actively limit the amount of movement that operators have to do to get the tools and parts they need on a day-to-day basis.

This movement control is not limited to operators. It should also be applied to stores and the layout of the parts. Consideration has to be given to manual handling techniques and where parts are stored. Think about the way your kitchen is organized. The things we need most are at the front, the less frequently used are at the rear.

Value

Value is what the customer sees in the product. Value added is good and non-value added is bad. Lean manufacturing seeks to identify all non-value-added tasks and eliminate them. It sounds easy to do, but is not always that simple. The 7 wastes listed above would be non-value–added. They just have to be recognized when looking at a huge list of tasks. Figure 12.7 is a force diagram showing the balancing of revenues with running costs. Notice, though, that the costs comprise of two vectors, one of them being waste, or to put it in a better way, unnecessary costs. When the starting point is what the customer wants, we have to make that consideration ourselves. Of course, we could try and find out from the customer, or the prospective customer. This comes down to some kind of product marketing. Often a customer only has a rough idea of what he wants but knows what he doesn't want. It is often easier to help the customer when it comes to the next generation of product. He can be involved in the design. If we can make his job easier or his product more attractive then we have added value. Working to this level is virtually the same as having a partnership with the customer: the supplier and the customer working together and setting

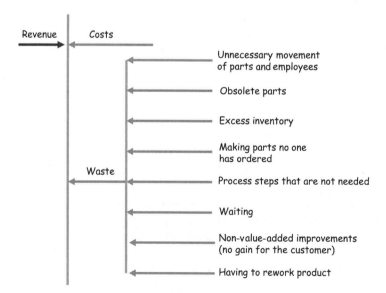

Figure 12.7 Balancing costs.

the specification for the product. This can be effective in both directions of the supply chain.

One of the more complex issues of value concerns the complexity of design. As a customer, does he really care how many moving parts are used to make the wobbly bit or does he only care if it does or does not wobble. This is even more obvious if all of the parts are concealed and the customer never sees the masterful design. Too many parts in an assembly can push up the cost and introduce a whole range of parts that are potential sources of failures. All the customer wants is for it to wobble when he wants it to and not break down. What it boils down to is that the difference between the expensive wobbly bit and the cost-effective one is non-value-added. This does not only refer to the part but also to the time taken to plan and design it, to make the parts, to change all the tooling, train all the assembly operators, modify the casing, and so on. They are all extra, unnecessary costs—that is, non-value-added.

My nephew has a mobile phone. It can take pictures, it has an alarm clock, a diary, an organizer, an address book, games, can play real music, send and receive photographs and e-mails, access the web, send text messages, and, I nearly forgot, it can be used to talk to people. How many of these functions will he actually use? If he does not use them, they are non-value-added. They used resources to develop and cost the company time and money. When the functions the phone offers are considered, we need to ask, are they a good or a bad idea? Was it a waste of money for the company to incorporate them? Just because they are on the microchip, should they be used? In this case they are not functional "value" but fashion value. This is what the customers want. Fashion is also a key illustration of an area where the customer's idea of a problem is different from the functional perspective normally seen by the company.

Take software as another example. Most people use less than 50 percent of their capability. Should the extra features be included or would they be used if they were explained better? Why buy a program costing oodles of money and not use it all? The unused programs are all non-value-added. I guess the hard part for a software company would be deciding what parts of a program are used most often. They would have to decide what to cut and yet not make it too limited. In short, they would have to ask the customers what they want and make it as they want. This would be value-added. What we have to do is look at all the tasks the company carries out and evaluate which ones are necessary and add value. If we decide one is not needed, we must consider what the consequences of eliminating the task will be and how easy it would be to correct if we made a mistake.

Lean also applies to management. Unnecessary managers are non-value-added. It is not unusual for a company to have too many managers all operating in their own specialisms. The difficulties often arise when their responsibilities overlap. They prefer to exist in their own fields and be their own bosses. The standard management format is to have a production department, an engineering department, a purchasing department, and so on, all working in isolation and being contacted when needed. Lean prefers to have product groupings, like cross-functional teams, all working together. One of the benefits of cross-functional teams is that all of the members are capable of working outside their designated labels: engineers have ideas on production and both production and product engineers have an understanding of equipment. As a product-based management group, barriers between groups are dissolved and avoid the "them and us" blame scenarios. They all work together and are not limited to the areas of the process they can work on. It sounds a bit touchy-feely, but it works.

Figure 12.8 is an example of value stream flow analysis. If the customer has to put his car in for a service, he must arrange to start late and leave his work early on the day his car is booked in or, as used in this example, he can take the day off. He has to drive to the garage and wait at the service desk to book his car in. He then has to phone for a taxi and wait for it to arrive before he can get home. He has to wait until the garage calls him before he can arrange his taxi to return to the garage. Once he has ordered the taxi he has to wait again. When he arrives at the garage he has to wait to pay for the service and again to collect the car, before he can drive home. In Figure 12.8, all of the waiting steps are non-value-added. Getting the taxi home and back again could be non-value-added depending on whether he was just going to wait at home. The final column on the right side has only two steps: pick up car and return car. Both are value-added to the customer and he does not waste hours of his day.

Value flows should be applied to all areas of the company. It is necessary to check

- ☺ The procedure used for purchasing parts
- ☺ How the parts were chosen
- ☺ How the supplier was chosen
- ☺ How many parts will be ordered
- ☺ How often they will be ordered
- ☺ How the orders are raised
- ☺ How many telephone calls and meetings are needed to finalize the order

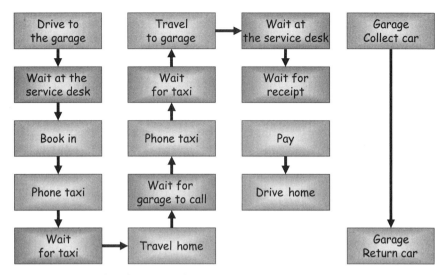

Figure 12.8 A simple value stream flow.

☺ How the parts are billed and who gets the invoice
☺ Who has to authorize and approve the quality of the parts
☺ What happens if the person is off work
☺ Who, if anyone, checks the parts are suitable

I don't want generate a never-ending list, but what I am trying to highlight is what has to be done, what can be avoided, what can be changed, and so on. In short, what adds value and what is simply a waste of resources.

Value stream flow analysis is carried out on every system in the factory. Naturally they will be prioritized. When we choose the one we are going to analyze we start with a high-level flowchart showing the major steps from start to end. Then we investigate the stages in between the major steps. We are looking for duplication, doubling back, waiting, unnecessary approval routes—we are looking for waste. Think about the SMED technique of elements and microelements in case there are steps that are not documented or are automatically carried out without any specific reasons.

Equipment

Buying a new piece of equipment is a high-cost activity that is fraught with pitfalls. Some companies buy only new equipment, some buy only secondhand, and some buy only the cheapest one they can get. One thing I have seen often is a trend toward people buying the machines they know about. If they are the same as the ones in their current company, this is a good thing, but if it is the one they used in their old position and the current employer does not use it, then it might not be such a good idea. If the company already has one tool for carrying out a particular process, then their engineers and maintenance facility will be geared up in maintaining it. (Notice I have not assumed any competency in skill level, as I don't want to complicate the issue.) If the company needs to buy a second machine, it is tempting for the engineer to buy the one he prefers, which will often be the one from his old job as he has experience of it and the one his current employer has does not seem as good. What we often end up with is a company with a whole range of different tools, made by different manufacturers, requiring different skills and different everything else.

Did you notice the word "different" was used four times in the previous sentence? In the example used above this translates to money: lots of it. The bad news is that the movement is in the wrong direction—out of the company. Unless the tool must be different, to provide a particular capability, then it has a series of hidden costs. The first cost being spare parts. You will already have a complete set of spares and consumables for the original tool sitting in stores. The new tool will need to have its own set of spares, but no one will have any experience of the best parts to purchase, how likely any part will be to fail, and what the rate of use of consumables will be. If they had stuck with their original tool type, they would only have had to increase their maximum and minimum stock levels and they might have been able to negotiate a cost reduction for the extra parts.

The next issue will be the training needed for the engineers and the technicians. A vendor course would be essential for a new tool but, as I have explained in other parts of the book, this is not an instant cure to the skill problem. There will be a prolonged time during which the new tool challenges, and defeats, the skills of the maintenance group, particularly in the area of breakdowns. It will also create difficulties for the operators and the process group. This could easily be as much as 1 year, during which the machine will be cursed as being unreliable. It might be necessary to rely heavily on the vendor, which also has an associated cost. If the original tool type had been purchased, this could all have been avoided. The only training would have been in the areas of differences between this tool and the old one. Depending on the age difference or model, this might not be very much.

The processes used by the company might not be instantaneously transferable from tool to tool. In fact, when equipment is transferred between companies the same processes can take a long time to get running again. The move is rarely simple. With a new machine, if there is no in-house experience, it will take time to learn the idiosyncrasies of the tool and what has to be done to eliminate their effects. There are likely to be far fewer issues if the original kit type was purchased.

When buying a new tool, there will be all of the obvious specifications to consider, but I hope I have highlighted a few of the less obvious money pits. It might be that the old tool cannot do what is needed and there is no option but to buy from an alternative supplier. In this case, think of the big picture. What are your plans for the future? Will the one you want to buy be able to make the next product? Does the tool do everything you need? Does it do too much?

There is a feature in lean manufacturing which identifies "right-sized" tools. This does not refer to fitting the tool into the room, although it might surprise you how many people buy tools and have to cut holes in the roof or remove walls to get them in. Right sizing refers to the amount of product a tool can make, the number of functions it can carry out, and the degree of automation. If you sell a few hundred parts in a week, do you want to buy a tool that can produce a thousand in a day? Would it be a better option to buy two tools that make four or five hundred a day? Is there a tool that makes the number you need? Big, all-singing, all-dancing tools that can carry out all the main processes needed tend to be complex and might make a greater demand on the maintenance group, as discussed above in the case of buying a different manufacturer's tool. The hard decision is to consider what are the alternatives. Would it be better to have cells of simple tools, with skilled operators and engineers, where the cells can be expanded or collapsed as required to accommodate the needs of the customers? The decision on which setup to use has to be one made by the company. Just make sure all the evidence is available. Before buying the complex tool, ask the equipment manufacturer for a list of users of their equipment (including any unhappy ones), ask for a guarantee of uptime or OEE and a cost of ownership, and, finally, ask them directly if they have had any reliability issues with the tool and what they did to resolve the problems.

Pull

Lean manufacturing is based on an ideology. We have already covered a few of them: achieving the highest quality of product, by eliminating *muda*; providing value to the customer; the stream of value and how it should flow smoothly in one direction—forward; and finally *pull*. Pull is the force the market (your customers) makes on your company for the products or services that you provide. It does not matter what the product is: it can be double glazing or window cleaner, cars or windscreen wash, pull is the rate of demand for your product and the goal is for you to be able to respond to the "pull" at the same rate. Some companies can respond to the pull simply because they have large buffers of stock, sitting in warehouses. This buffer, as explained previously, is expensive to run, but it has the ability to hide inefficiencies like poor maintenance and long changeovers. But we now know better. Supplying the customer demand is the most efficient way to run a business.

The customer calls and asks for the goods and the shipping department is asked for a shipment to be sent to the customer on the 5th. He asks the production department for the parts he needs. Each stage of the line asks for the parts he needs from the previous production stage. Ideally, this chain stretches all the way back to the supplier of the raw materials, but there will be practical compromises that have had to be made, like having a minimum stock level in-house. Lean does not mean having no inventory, but having a controlled, rationally considered amount that will cope with any variation in the production methods. The difference between lean and not lean is that the company is aware of the costs and has considered all of the options before deciding on the minimum levels. However, in its purest form, pull makes only what the customer needs. If there is a minimum stock of completed inventory (for example, to enable immediate supply of a part), pull replaces the parts sold and restores this level. If there is no need, there is no production. Which brings us neatly back to right sizing and all of the continuous improvement techniques in the book.

One other point I have not discussed is the potential for significant reductions in employee numbers or the difficulties that will be encountered when applying lean to a company. Womack and Jones explain in detail in their books all of the difficulties that will be encountered and how important a lean expert is when applying the technique. I am not an expert by any means. My experience is improving daily and I have been on the receiving end of the application of lean methodology. However, there are areas that I am uncomfortable about, the main one being the need to "get the employees behind you." This is often achieved by pretending the company is in extreme difficulty, by creating a powerful competitor that you have beat to survive, or by having a customer who threatens to take away their business. There can be any combination of these features and some variation and enhancement in the detail. The key criterion is that the employees have to buy in to it. After all, it is the employees who will have to bear the stress of working harder and see their friends and colleagues losing their jobs.

I think I could be ruthless in the right circumstances, but I would prefer to try and grow the business through increasing quality and reducing prices, both of which are possible through the application of lean. If I understand them properly, Womack and Jones recommend the company should have one major jobs cut at the onset and then no more due to the efficiency improvements. I believe these are honorable guidelines, but I suspect many companies would fail to sustain this model and will eventually lose all of the confidence of their employees. The employees will only leave the company if other jobs are available, and in the meantime they will continue to become progressively unhappy.

These difficulties are common in most of the continuous improvement techniques. The drive must come from the top and there must be a determination to succeed. Every attempt will be made to win over the unconvinced or reluctant management, but if they become too disruptive it might be time for them to move on. Hopefully, there will be mechanisms where the managers can resolve any issues or fears they might have, but if success is going to be assured, in the end it will become necessary to remove any roadblocks.

It is not necessary for the company to jump head first into lean to make significant improvements. The application of the other methodologies in this book, many of which were developed from the Toyota Production System (which is also known as lean manufacturing), and the application of its best concepts, for example, finding waste, reducing changeover times, reducing work in process (WIP), improving layouts for efficiency and error reduction, and, of course, improving the maintenance and the production structure of the organization, can also be made. All of these methods plus the introduction or improvement of standards coupled with the introduction of better training methodology will make general skills improvements for operators and engineers, which will all target the goal of improving OEE and increasing profitability.

Chapter 13

Six Sigma: A High-Level Appreciation

Not too long after setting up my business, I spoke to an HR manager from a microwave components company. She had the responsibility for training. Even before we started talking, she mentioned, with obvious disdain, that she "had tried all the fads." It was a bit off-putting, but strangely enough, it was not the first time I had come across an HR person with the responsibility for training. The odd thing is that she appeared to have very little understanding about productivity improvement methods. She kind of surprised me and I am not exactly sure why. I guess it is my fault; I should be able to jump in there and take command, filling the gaps in her knowledge using my skill, wit, and repartee. Anyway, I digress.... While explaining the basic starting point for continuous improvement, 5S, I saw from her facial expression and the knowing glance between the Engineering Manager and her that I had discovered the main reason for me being there.

As it transpired, they had a disorganized work area, and some poor soul inadvertently used the wrong components in an assembly. The storage racks were largely unlabeled and had some very old components on them. There was also no specific organization. On one of the shelves, there was a box of parts that had become obsolete within the last couple of years. The assembly operator was either unable to find the correct part or picked up the obsolete batch of components from the shelf by mistake. In either event, he started fitting them into the product.

A few minutes of chatting about 5S had identified the exact cause of the problem and, had 5S been applied previously, it would have prevented the situation from arising. So it appears, then, that the easiest fad of them all would have saved the company an undisclosed number of thousands of pounds and avoided some upset customers and, possibly, a load of lost future sales. And it would have improved the working conditions of the employees.

The incident made me realize that there are at least three types of people who do *not* use improvement techniques: those who don't recognize that there might be a need for them, those who simply don't understand them and shy away, and the disillusioned. I was really caught off guard by her attitude: I guess I just expected her to know better. But then I suppose that if it is hard enough for some engineers to accept the ideas, it must be really difficult for nontechnical people, even if they are in charge of training. This brings me nicely to my investigation of Six Sigma. I had heard a bit about it and how it had saved millions of dollars for a range of companies, so I wanted to find out more about it. I wanted to find out if it was fad or if it was real?

My first step was to read a couple of books on Six Sigma. What caught my attention as I was reading was that Six Sigma did not seem very much different from just about every other kind of faultfinding of which I have experience. So I read some more. This time I pondered over the examples. It seemed to me that some of the problems were solved by lean management's value flow analysis technique.

My next step was to speak to a few people who had used Six Sigma. They agreed that it worked in pretty much the same way as the way I would attack an equipment process issue. How would I do that, I hear you ask? I do it in the same way that I have explained for virtually all the problem solving in this book, with three differences.

The first variation is the simple addition of a bit of mathematical analysis. I plot some data and graph it. "If I change this value, what changes at the other end?" A basic application of cause and effect. It makes sense when we say that turning up the volume on a music system increases the output sound level at the speakers. It does not, however, give any guidance as to how much to turn it up or what the expected output volume should be. One of the most fundamental techniques recommended to make a problem easier to understand and to clarify the issues is to write it down and express it in words with as much detail as possible. To do this effectively the problem has to be quantified and converted to an input in numbers: "Set the input to level 5...." Then the output has to be measured in some way to get another value, also in numbers. If you refer back to technical standards and to the difference between features and functions, they both explain why verbal descriptions are inadequate. The input instruction could include the position of the Volume knob or, for high-tech tools, the digital display reading. The output measurement has to be more complex: we could measure the voltage and current to the speakers which would give a power value, but the value could be wrong. The reason is that the voltage and current do not allow for the inefficiency of the speakers and so might not represent what the customer (the user) hears. If the problem was in the speakers, we would miss it. Alternatively, we could measure the speaker output power in decibels, which is right at the output end of the system.

The second and third differences are that I do not use Six Sigma terminology and that I don't understand the logic behind the calculation of the Sigma values. I would like to understand the table in Fig. 13.4, but to use Six Sigma, I don't need to.

Graphs and Their Use in Six Sigma

Six Sigma differs from most other problem-solving processes in that a mathematical link is needed to identify and to confirm the cause of and the solution to a problem. It assumes that normal faultfinding techniques are not based on proof. (This is an interesting observation. I had better go back and remove all the references to Plan-Do-Check-Act and root-cause analysis.) The lack of proof might be the case when a diagnosis is made by an inexperienced engineer or there are several causes that could create the symptoms seen. In Six Sigma, the fix must be verified graphically or statistically by collecting data that proves *causal* links between what is believed to be the cause of the problem and the manifestation of the fault. Woof, what a sentence! The word causal is important; finding a link does not always mean the cause has been found. Daffodils come out earlier in the year if the weather is warm, but having more daffodils does not warm up the planet. (I bet someone will argue with that!) Another example might be the link between being good at something or some task and how it boosts self-esteem. However, positive reinforcement techniques that boost self-esteem do not make the person better at the task. As the Six Sigma process reaches its final stages, it applies the same maths to prove that the problem has been resolved.

To collect the data needed for an analysis, it is often necessary to specially design tests and experiments. First we make a hypothesis and then we find a way to prove it is correct. Six Sigma looks at the *big picture* to find places where the data should be collected, including the documentation and the systems. Two of the most common graphs used are histograms like the binomial distribution and simple $x - y$ graphs.

Figure 13.1 is an example of the *binomial distribution*. This is frequently used by statisticians and engineers to evaluate the probability of an event happening. It is also known as the *normal distribution* or the *bell curve*. When you hear statistics about the average wage, the average height, average nose size, average family size, who is leading in an election poll, or the average lifetime of a light bulb, the

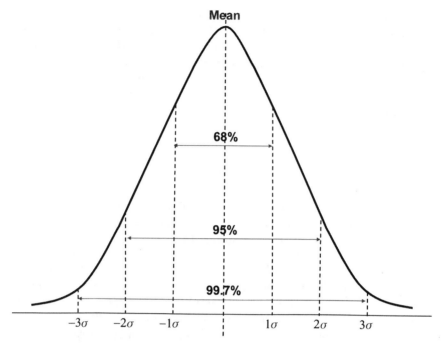

Figure 13.1 The binomial distribution.

binomial distribution is the shape of graph expected. The accuracy of any predictions is limited by the data sample size.

If we wanted to analyze the distribution of salaries in a country, we would plot the salary range on the x-axis, in a series of intervals. We use intervals to break down the range of data and control the number of data points. The y-axis would be the number of people earning a salary within each salary interval. As mentioned above, Fig. 13.1 is the pattern we would expect to see when the graph has been plotted: a normal distribution. The value of the *standard deviation* is reflected in the sharpness of the peak of the graph. If it is sharp like a mountain peak, we have a small standard deviation: a blunt, rolling hill would be a large standard deviation.

Average and standard deviation

Statistics can often lead to misunderstanding. To demonstrate my point, consider the average wage within a small company called Profit & Co. It is reported as $40,000. So you might expect if you worked in that company, you would be earning around that level. This is not the case.

The average can be drastically affected by the upper or lower points. In our example, we have two, exceptionally well-paid directors, each

TABLE 13.1 The Data and Histogram of the Wage Distribution in Profit & Co.

Salary in $000	Number at salary	Total $000
10	47	470
15	22	330
20	8	160
25		0
30		0
35	1	35
40		0
45		0
50	1	50
100		0
1100	2	2200
	81	3245
	Average =	40.1
	Standard Deviation =	169.800423

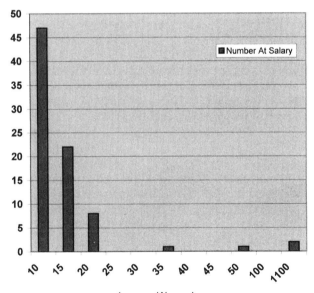

Number of Employees at Each Salary

Average Wage.xls

earning more than 1 million dollars. This has the distorting effect of pulling the average wage upwards. Forty thousand dollars would still be the average wage, but it is not the most common wage range in the company. The data is shown in Table 13.1.

Without even looking at the graph attached to Table 13.1, to see how many earn each salary, it is clear from the table that all the data is offset to the low end. The data has been deliberately chosen to illustrate the importance of the standard deviation. Mind you, would it be possible to find such a variation in life? What about top professional footballers when compared to the other members of their own teams, pro-golfers, sportsmen, movie stars or directors, designers, entertainers, or surgeons? There will always be industries and organizations where there are a few headliners and most of the others will be your everyday workers: sewing machine operators, dancers, support actors, nurses, technicians, or groundskeepers....

Let's revisit Profit & Co. To find the most likely wage, the one that we could realistically expect to earn, it is necessary to find the *mode* not the mean. To get this value we must list all of the salaries in a row, in increasing order:

- $10,000 would be written 47 times.
- $15,000 would be written 22 times.
- $20,000 would be written 8 times.
- $35,000 would be written once.
- $50,000 would be written once.
- $1,100,000 would be written twice.

There are 81 different wages, so the mode is the mid-value: the 40.5th number in the list: $10,000. Not really what we were hoping for! The problem is that the data has a huge standard deviation, nearly $170,000, which is more than four times the mean.

Standard deviation and z scores

Let's consider a courier service's promise to deliver a parcel from Glasgow to London in a certain time. Again, the binomial distribution is the shape of the graph you would expect to see for all parcels from Glasgow to London, provided a large enough sample of their delivery times was analyzed. The time intervals would be represented on the x-axis, and the y-axis would be the number of items arriving within each time window.

One characteristic of the binomial distribution is that its similarity (that is the shape of the bell curve) enables data points to be converted to "standard scores" (z scores); these are like ratios that enable comparisons of data. A z score of 0 is equal to the mean or average. If the z score is equal to 1, it means that the data point is 1 standard deviation (SD) above the mean. A z score equal to -1 places the data point 1 SD below the mean. Now, I guess you want to know what this means in practical terms.

Consider two students from different college classes sitting for their respective History examinations. Their exams were created by their own class teachers. If both students achieve a score of 90 percent, does this mean they are at the same level? At first glance, it certainly looks like it. But let's not be too hasty; we have already been a bit misled by the average wage at Profit & Co. What if one exam was much easier than the other? Would it make any difference? How can we be sure? If the two results are converted to standard scores, which are proportional to the average score *and* the standard deviation of each class, the higher z score would be the best result.

The horizontal divisions shown in Fig. 13.1 are increments of the standard deviation. Just like the z scores, standard deviations allow different delivery patterns to be compared. By comparing the standard deviations, we can compare the courier's times to New York, Munich, or Paris. A small standard deviation is better than a large one as it confirms less variation. Plus or minus three sigma is a fairly typical limit and should equate to approximately 97 percent of data points—or deliveries. So, if the company promised delivery in less than a certain time, using ± 3 SD as limits, they would expect about 1.5 percent of deliveries to take longer. This is the area to the right of $+3\sigma$ in Fig. 13.1: 100 percent less 97 percent, and the answer (3 percent) is then divided by 2 to get the right-hand side only.

The x–y graph

Not every graph follows the binomial distribution. Equally, some graphs are more complex than others. When the input is directly linked to the output, like the wear of the tread on a tire with distance traveled or the length of your finger nails with time, the data would give a straight line graph, like the sloping section of Fig. 13.2. Different graphs will each have their own gradients (slope), or they might be linear over a certain length only, or the line might start at a different value on the y-axis. The data can be plotted on graph paper or by using a spreadsheet program like Excel. Graphs do not all end up as straight lines; they can also be curves of all shapes and sizes.

Imagine that Fig. 13.2 represents the power output curve from a new battery design. The output power is shown on the y-axis and the concentration of electrolyte mixed with the gel solution is shown on the x-axis. There is only a slope up to a concentration of 60 percent, after which point there is no increase in output power despite increasing concentration. If we had a quality problem with the battery, which had the fault symptom of a variation of output power, we could be looking at a concentration problem. It would be necessary to check all the processes that contribute to measuring and controlling the concentration.

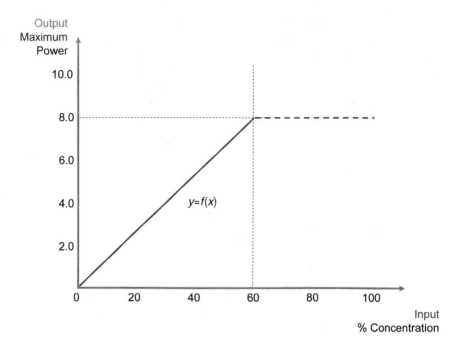

Figure 13.2 A simple $x-y$ graph.

The full analysis should lead to a root-cause solution. Alternatively, as a stopgap solution, we could increase the gel concentration and move to a safe point on the "flat" part of the graph that is less sensitive to variations. The final solution would depend on the costs involved in running at the higher levels.

In mathematics, graphs normally have two axes (as seen in Fig. 13.2) but they can also have a third axis, like a 3D line drawing. The third, the z-axis, would be perpendicular to the paper, but is drawn at about 45° on sketches. The point where the axes meet is called the origin.

Graphed data is rarely a nice straight line. In fact, it often looks like a wall when the dart board has been removed. In Fig. 13.3, the points plotted are not so obvious. It is possible to see there is a connection, but to get an accurate link a maths program like Excel or SAS JMP would need to look for a "best straight line" and give an estimate of the degree of accuracy. The graphs can be much more complicated than this. This is why there is a need to be trained on the software to be used and how to understand what information you need to make sense of it and to be able to understand what the data is telling you. I found SAS JMP to be relatively easy to use although the explanations on how to use the summary data could have been simpler to understand.

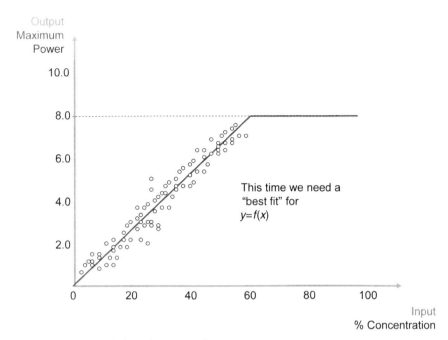

Figure 13.3 A slightly less obvious graph.

The Main Terms of Six Sigma

The main terms of Six Sigma are as follows:

The customer

The voice of the customer. The voice of the customer (VOC) refers to all the ways (the mechanisms) open to the customer (singular and plural) that allow him to have his opinions heard by the supplier. This does not simply mean that the company has a complaints department, but that the company has a method for canvassing the customer to identify what he feels about the company's service and performance. His concerns could be related to the product, the service he receives, the ordering system, delivery times, his input to the future development of new products, and so on.

Critical to quality. These are the factors from the point of view of the customer that account for quality. They are the standards he will accept. Interestingly, customers can have a different perspective of what is a failure than the supplier or manufacturer might have. There can be failures that are not seen as an issue by customers and also serious problems to a customer that a manufacturer cannot see. Fashion is a key example. It is one area where having the item while it is in fashion supersedes quality considerations. I guess this explains the market for counterfeit goods.

The teams and the leaders

Green Belt, Black Belt, Master Black Belt. These are three levels of involvement by the team members.

The Green Belt is a part-timer. He can be an employee working on his own job in addition to his part in the team. He *can be* trained to a lower standard than the Black Belt, but probably has not been.

The Black Belt is a full-time team member with no other work commitments. This allows him to concentrate on his Six Sigma duties. His responsibilities will be similar to a TPM or an RCM team leader or facilitator.

The Master Black Belt is a manager. He will have a higher level of training than the Black Belt particularly in the use of statistics. He is expected to guide all of the teams, advise them on problems, and keep them motivated.

The Champion or Sponsor

In all of the processes we have discussed so far, we have promoted the idea of a manager with the authority to clear roadblocks. This is part of the duty of a sponsor. He works with the other managers to select the projects for the teams and establish what the company expects as a solution. Six Sigma also needs to have training, training time, and project time for teams and it must rely on the rest of the company's manpower for support. This is the function of the Champion.

Six Sigma Controller

This position is the equivalent of a continuous improvements manager except with responsibility for the Six Sigma program. He confirms the projects are in line with the goals of the company.

The Rules and Expectations

The Six Sigma Charter

The Charter is to Six Sigma what the Operating Context is to RCM. It is the team agreement with the company. It formally allocates roles and responsibilities like a project plan. I am not sure why it feels the need to do this.

The Charter identifies the extent of problem and what would be an acceptable solution in terms of costs and benefits to the company and to the customer. (For statistics to be valid, the expected outcome must be predicted in advance.) The Charter also defines the limitations of the problem and the assumptions and constraints the team members have to work under.

It should include all of the roles and responsibilities of the team members and have a plan as to how the problem will be tackled and how long it will take.

The Technical Stuff

The sigma value

The sigma value relates to the percentage of failures that is acceptable (Fig. 13.4).

A low failure rate (low defects per million opportunities) corresponds to a high sigma. The sigma value can be looked up in a table if you know the DPMO (defects per million opportunities).

Defects per opportunity

The sigma value is the relationship between the actual number of failures and the different possible ways that the product or service can fail.

In the parcel service example the parcel could arrive late, arrive damaged, it could have items missing, it could be delivered to the wrong address, it might not arrive at all or be the wrong parcel. This means it has six possible failure modes or areas where failure is possible. If the letters a, b, c, d, e, and f represent the number of failures in a data analysis of 1000 deliveries, the number of *defects per opportunity* (DPO) is calculated by

$$\frac{a+b+c+d+e+f}{6 \times 1000}$$

Defects per million opportunities

The normal method of reporting defects in Six Sigma is as defects per million opportunities (DPMO); so the number calculated above would be multiplied by 1,000,000.

370 Chapter Thirteen

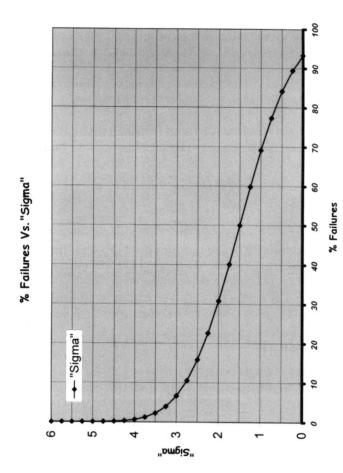

Yield	% Fails	"Sigma"	z-Score
6.68	93.32	0	-1.5
10.56	89.44	0.25	
15.87	84.13	0.5	-1
22.66	77.34	0.75	
30.85	69.15	1	-0.5
40.13	59.87	1.25	
50	50	1.5	0
59.87	40.13	1.75	
69.15	30.85	2	0.5
77.34	22.66	2.25	
84.13	15.87	2.5	1
89.44	10.56	2.75	
93.32	6.68	3	1.5
95.99	4.01	3.25	
97.73	2.27	3.5	2
98.78	1.22	3.75	
99.38	0.62	4	2.5
99.7	0.3	4.25	
99.87	0.13	4.5	3
99.94	0.06	4.75	
99.977	0.023	5	
99.987	0.013	5.25	
99.997	0.003	5.5	
99.99833	0.00167	5.75	
99.99966	0.00034	6	

Figure 13.4 Percentage failures plotted against Six Sigma value.

The Stages of a Six Sigma Analysis

DMAIC	The Process
Define	➢ The team has to identify and clarify the problem. The problem should be defined in words. *It might already be written in the Charter, but it might need to be redefined. If it is, check out that the new definition is agreeable with the management.* ➢ What would be acceptable to the customer as a solution? *If the customer does not see it as a problem, does that mean the problem does not need to be fixed? Remember there is more than one customer. Find out exactly how it affects him, he might not know the answer to the next point....* *How the customer sees the problem might help when seeking a solution or choosing which solution would be most appropriate. The problem might not be simply a choice of one thing or another. Often it is easy to see a problem but it is not so easy to know how to put it right.* ➢ How much does it cost the company? *Use an RCM technique to find out all of the costs associated with the failure.*
Measure	➢ What actually happens when the problem is seen? *Back to RCM again. How do we know we actually have a problem? What visual indications are there? What did the operator or customer see just before it happened? Was there anything unusual? What about the weather: any lightning or was it unusually warm? Is there any recorded process data? What does the problem do to the product?* *If it is an administrative issue, is it related to any non-value-added steps? Is there a way of redesigning the flow of the process?* ➢ Are there any components of the problem that can be measured in units like time, engineering units, defect numbers, quality? *Use TPM and RCM to look for standards.* ➢ Is there a way we can turn the problem into numbers that can be processed?

Check out Chap. 5, the section on quantifying risk assessment.
(For example, Yes = 2, Perhaps = 1, No = 0.)

➢ What "inputs" can be identified that affect the process? If they are complex, where can we collect the data to confirm the relationship graphically?
   Refer to Chap. 11 for a series of problem-solving techniques including brainstorming.

Analyze
➢ Search for the root cause of the problem.
   TPM, RCM, lean, and problem solving.
➢ Teach the team the process.
   Use TPM methodology.
➢ How does the "system" handle its inputs?
➢ Is there a value stream flow or process flow?
➢ What happens if the inputs fluctuate in feed rate or quality or if the room temperature, pressure, or humidity changes?
   Consider all variables when looking for use conditions.
   If the problem is administrative, what happens when there are vacations or sickness? What were the steps leading up to the problem? What were the steps following the problem?
➢ Where have mistakes been made in the past?
   RCM analysis methodology.
➢ Are there any new areas that have been subject to problems recently?
   Are they linked to the current problems?

Improve
➢ Carry out a 5W +1H.
➢ What outcome do we expect to see from the solution?
   (For statistics to be valid the output must be predicted.)
➢ How much will it cost to design and implement the solution?
➢ Who will carry out the work?
➢ How long will it be until the fix is complete and the customer sees the benefits? What if he doesn't notice?

	➢ How can we confirm the fix is correct?
	➢ By changing one thing have we caused any secondary problems?
	➢ Confirm the solution using experiments and maths.
Control	➢ The system needs to become embedded into the normal operation of the section. It cannot remain as an add-on.
	➢ A new operating procedure will need to be written.
	➢ Training on the new procedure will be required.
	➢ Define who will be responsible for collecting and analyzing the monitoring data?
	Note:
	The operatives who will finally be involved in maintaining the system should be kept involved in the team's progress. Their inputs should be sought and, if valuable, be recognized. Their "buy in" is critical for the success of the project.

		Management Actions
13-1	Identify	
	☺	The Six Sigma Controller.
	☺	The Champion/Sponsor and Master Black Belt
	☺	Black Belt(s) and Green Belt(s).
13-2	☺	List the problems to be resolved
		Prioritize them.
13-3	☺	Organize the Six Sigma training.
13-4	☺	Provisionally set up the team(s).

Considerations or Limitations in Using Six Sigma

Something I did not expect happened as I worked my way through writing this chapter. I read a couple of reference books and felt that many of the examples given seemed to have been solved by the lean management technique of value stream analysis. I found I was becoming more discriminating on when I would want to use Six Sigma as a process, even though I completely agree with the advantage of using data as a proof.

Please remember, however, Six Sigma always seems to imply that normal faultfinding techniques "just jump in and fix the problem." This makes a huge assumption that the wrong answer could have been found or that, from all the options considered, the team has just guessed the root cause. (How many times have I promoted Plan-Do-Check-Act in this book?) Have you noticed that Six Sigma also allows for the wrong answer *despite* taking the extra time to prove it first! Sadly, there are lots of people who are not too good at root-cause faultfinding. Even the ones that are good might have to try out more than option if "it could" be the cause. This is one of the reasons for the use of cross-functional teams. It would be interesting to find out if the people who are known to poor faultfinders are also poor at Six Sigma analyses. I think they will be.

I suspect a Six Sigma analysis is more appropriate for processes and systems than for equipment issues. I don't doubt that the system works, but I feel it might be a very slow process and wonder if, in some cases, the time spent seeking statistically valid results might not be better than educated "selective checking" based on normal faultfinding methods. I am probably the example that the Six Sigma consultant uses to prove his point. After all, people judge processes on the basis of their own experiences. I have really not encountered an issue that has taken such a long period to resolve. The one advantage of Six Sigma I can see instantly is that nontechnical people (managers) can usually understand the graphs and data more easily than the actual complexities of the issues.

Faultfinding the cause of a lamp failure

In normal mechanical and electronic problems, the most difficult ones to find are the "intermittent" faults. These are faults that come and go or, even worse, are not visibly present when you are searching for them. The simplest example that springs to mind would be a bad electrical connection in, say, a table lamp. When you notice the lamp is out, you investigate the cause by moving the lamp to a position you can have a look at it. But on testing, you discover the lamp is working. It doesn't matter if you shake it, tap it, swear at it or kiss it, the lamp will not fail. As a best guess, it is most likely (probability) to be a connection problem in the wiring or in the bulb: but where and which one? I can think of five possible methods to find and correct the fault:

1. Use a test meter and measure the electrical resistance at every connection.

However, if the fault has cleared, it is possible the fault would be obscured.

2. Selective replacement: change the bulb and see if the fault returns.
   If it does, it was not the bulb.
3. Check and remake the electrical connections, one at a time, and monitor to see if the fault returns following each check.
   The bulb would also be included in the checks.
4. Tighten all of the connections at the same time and monitor to see if the fault returns.
   If it does, it was not the connections.
5. Tighten all of the connections at the same time and change the bulb.
   The fault would be very unlikely to return.

If the consequences of the failure meant it was essential to know *exactly* what the failure mechanism was, then options 4 and 5 are not appropriate. I guess I see Six Sigma operating like option 3.

		Team Actions
13-5	Define	
	☺ Prepare the charter.	
	☺ Consider if the project is suitable.	
	Evaluate the costs and benefits of a solution.	
13-6	☺ Define the problem and state it in quantifiable terms.	
	☺ Seek input from the customers.	
	☺ Create a process flow diagram.	
	Keep it simple at this point. Too complex will be confusing—details can be added as required.	
	☺ Define what the solution should be.	
13-7	☺ Revise the Charter on the basis of the new information.	
13-8	Measure	
	☺ Collect data on the process to help. Select a range of input parameters that are suspected as influencing the problem. Output data is also required.	
	Set up a way to gather the data needed.	
13-9	☺ Make an initial calculation of the current Sigma value.	

376   Chapter Thirteen

13-10   Analyze
         ☺ Analyze the data and find a root cause or a list of possible causes.
           Use the data to find causal links and validate the cause.
13-11   Improve (and implement)
         ☺ Create a list of possible solutions.
           Evaluate the cost and the benefits of each.
           Identify the timelines needed to implement the fixes.
13-12   Control
         ☺ Set up a system for continuously monitoring the process, ensuring it remains error-free and for alerting the team in the event of problems.
         ☺ Create a procedure.
         ☺ Train the operators.
         ☺ Publicize the team's progress and the final solution. Self-promotion on a continuous improvement activity board.
         ☺ Embed the solution.
13-13   ☺ Make an updated calculation of the new Sigma value.

Possible Limitations with Using Statistics

When looking for data to analyze it is possible to find a link that is not causal. Experiments might need to be carried out to either prove or disprove any theories. Even when we have links that appear to be causal, we need to ensure that the statistics are *statistically significant*. (Statistically significant means an event is unlikely to have happened purely by chance.) This is one of the limitations of statistics. It is a mathematical requirement that finding a link by chance that was not predicted beforehand is not a proof. It is merely the discovery of a new question to be considered. The experiment would have to be repeated with the particular objectives clearly stated for the link to be accepted.

Consider mad cow disease as a case in point. It began with cows developing an illness that had symptoms which were previously seen only in sheep. The sheep disease was called scrapie. There was a series of debates because the disease does not usually jump species. Eventually, the disease did ***not*** jump species again. This time, it did not jump from cows to people. A link with the human equivalent, CJD, was suspected. It looked like humans had developed the same brain disease as cows were suffering. "This is impossible. It just cannot happen...," we were told. Possible causes were discussed early on, but there was no *statistically significant* proof. So, we were told that we should not panic and should wait until the data has been studied and analyzed. Part of the problem was that the proof of the disease was only available after a postmortem examination.

The details from the early "discussions" were considered to be enough proof for many ordinary people. After all, where else could the disease be coming from they asked. So, they made a leap of faith and decided to stop eating meat immediately: Option 5 of the lamp failure example. They simply did not want to take the risk on the basis of the existing evidence, even knowing that it was *statistically* unfounded. As far as they could see, the worst-case scenario was they would miss eating meat for a while. The best case was they might avoid getting the disease.

There was a potential complication in arriving at a solution. One of the consequences of a wrong decision would be a slump in the meat industry, and so absolute proof was sought. In effect, the government had a "lose-lose" situation. On the one hand the meat industry would be affected and on the other people would possibly die. The government fell back on the old chestnut of absolute proof. However, scientific data took time to become statistically significant, but we waited and waited and waited. Theories were thrown into the air from all corners of the medical profession. Members of Parliament bought hamburgers for their children on camera. As seems to be normal for statistics, some studies pointed to *this* cause and some pointed to *that* cause and nothing was conclusive. But statistics being statistics means that even in the best-run studies, you can still get the wrong answer despite doing everything correctly. Eventually, to the enormous surprise of very few, the link was proved and millions of cattle were destroyed.

Mad cow disease is not the only example of the time taken to find an absolute proof. What about cigarettes? Here we have an argument that ran for decades. Then there is the occasional suspected or alleged link between medicines and disease; for example, autism and MMR vaccine. This will be very difficult to verify, but just as in the situation with mad cow disease, many parents have opted for single injections, despite the medical advice from the government. We are also bombarded with new statistics alleging a relationship between how many portions of vegetables we need to eat to avoid cancer, how much alcohol is bad for us, how high-cholesterol foods make us more susceptible to heart disease, the potential health issues for women who use birth control tablets or HRT treatments, and so on. Statistics are often used to prove both sides of an argument at the same time. One current major debate concerns global warming. I am still unsure if it is real despite the barrage of statistics. I keep wondering if this changing weather pattern has not always been around. After all, when I look at the ancient ruins in Egypt, like the Sphinx and the pyramids, I wonder who would want to build such fabulous monuments in the desert. It seems it was not always a desert. Quite the opposite in fact.

Which all brings us back to the point. How much proof do we need to take positive, preventive, action? As I have said before, I have used graphical analysis to prove mathematical links between a given action and a corresponding event happening at the output of the machine. It usually manifests as a variation in the specification of the product. Not having enough Six Sigma experience, my colleagues tell me that most issues are resolved without the need for complex links. However, Six Sigma is a tool that is available for us to use. We also have access to all of the other faultfinding methods in our toolbox to consider, should it not prove to be reaching an appropriate conclusion.

Having said that, I find it difficult to see how it can fail. We have already used all of our toolbox worth of ideas to find our list of possible causes. The only difference is how we choose *the order* of the solutions we try. In Six Sigma we look for mathematical links before and after we test the fix, and in faultfinding we look for the fault going away and not coming back. Every other stage seems pretty much the same....

I need more experience.

Index

Page numbers followed by f indicate figure and page numbers followed by t indicate table.

A

activity boards, 327–329
  layout of activity board, example, 33f
  training records and, 108f
AM (autonomous maintenance). See also AM (autonomous maintenance), creating standards/preparation for
  action list, 60
  cleaning map, 50–52
  discovery of serious fault during cleaning, 58
  initial clean/inspect and F-tagging, 45–50
    initial clean process flowchart, 48, 49f
  as pillar of TPM, 9
  recording funguai with F-tags, 52–57
    F-tag log sheet, 53–55
    F-tag log sheet, example, 54f
    red F-tag, 53
    task certification sheet, 55
    task certification sheet, example, 56f
    white F-tag, 53
  tracking progress of initial cleans, 58–59
    graph, example, 59f
AM (autonomous maintenance), creating standards/preparation for
  AM team, 216
    responsibilities of, 92
    tasks of, 85, 86f
  PM/Zero Fails teams, 97–101
    master failure list, 99–100f
    PM tasks, 85–86f
    PM team responsibilities, 98–99
    Zero Fails tasks, 85, 86f, 87
  pre-AM safety checks, 88
  task transfer, 89–94
    F-tag embedding/responsibility spreadsheet, 89, 90f, 95–96
    reviewing hard-to-access areas, 89, 91

AM teams, 85, 86f, 87, 92, 216
area map, 127–129
  for Nova implanter, example, 128f

B

basic condition neglect, 71–72
bell curve. See binomial distribution
binomial distribution, 99, 364f
bottleneck tools, 307

C

categorization log sheet, 64
cause and effect diagram. See fishbone diagram
CEDAC chart (Cause and Effect Diagram with Additional Cards), 322
changeover. See SMED
changeover parts, 16
cleaning. See AM (autonomous maintenance)
cleaning map, 63
  alternative style, 176f
cost of ownership, 2
costs spreadsheet, 64

D

decision diagram, 228–244
  column sequence in tabular format, 229f
  example, 236f
  failure-finding decision blocks, 234f
  missing sign, 242f
  multiple failure frequencies, 239f
  on-condition monitoring decision blocks for hidden failures, 230f
  recording process on decision worksheet, 249
    completed decision worksheet using previous examples, 257f

379

decision diagram (*contd.*)
    decision worksheet layout, 250*f*
    right side, 231*f*
    scheduled restoration/discard decision blocks for hidden failures, 234*f*
    sequence in tabular format, 237*t*
    three rows of operational decision blocks, 238*f*
defect chart, 63
    for categories of causes, 76*f*
    showing historical fails, 76*f*
defect map, 63, 82*f*–83
Deming, W. Edwards, 12
design weakness, 73
deterioration
    forced, 16, 32, 33
        use conditions role in, 34–37
    natural, 32–33
    reasons for, 38
    unchecked, 69–70
discard, scheduled, 226
DIY (do-it-yourself) repairs, 214
double redundancy, 253
drawings record sheet, 65
dummy wafers, 261

E

education/training pillars, TPM, 7–9. *See also* safety pillars, TPM
    introduction to, 103–105
    TPM education/training pillar, 106–126
        assessing skill level of team members, 120–126
            skill log for transferring tasks to operators, example, 123*f*
        equipment training, 113–116
        equipment training sequence, 116–120
        management actions for education/training, 113
        one-point lesson, 110
        prerequisite training for teams, 107
        tool-specific training record, 109–110
        tool-specific training record, example, 109*f*
        training record, 107
        training record on "activity board," example, 108*f*
        Zero Fails team, 107–108, 110
            composition-membership, 111*f*
            composition-membership, overlapping management, 111*f*
efficiency, overall equipment, 24, 26–32
    availability, 29–30
    performance, 31
    quality, 31–32*t*
equipment training, 113–116
    sequence for, 116–120

F

failure. *See also* RCM PM analysis
    functional, 268
    hidden, 234*f*
    non-time-based, 297–298
        infant mortality, 298
    partial, 15, 271, 284–285
    time-based, 296*f*
    total, 15, 271, 284
Failure Analysis Sheet, 38, 41, 63
    example of, 39*f*
failure data, analyzing/categorizing
    data sources
        F-tags, 63
        machine history log, 63–65
        minor stops/unrecorded losses, categorizing, 65–71
            actions for, 68–71
        area photographs/drawings record sheet, 65
        impact of minor stops, 65–68
        minor stops log sheets, 65
    defect map, 82*f*–83
    modified defect chart showing historical fails, 76*f*
    Pareto charts for, 80*t*–82, 81*f*
    TPM causes for F-tags, 71–79
        basic condition neglect, 71–72
        data as percentage of fails/percentage hours of repair time, 79*t*
        design weakness, 73
        historical fails data sorted as repair time, 78*f*
        inadequate skill level, 73
        modified defect chart showing historical fails, 77*f*
        operating standards not followed, 72
        possible F-tag category spreadsheet example, 75*f*
        sample data as number of fails/hours of repair time, 79*t*

standard defect chart for categories of causes, 76f
unchecked deterioration, 72
unknown, 73–74
failure-finding interval (FFI), 253–255
failure modes, 268, 285–288, 286f
fault analysis
  fault tree diagram, 323–324
  fault tree diagram, example, 323f
  fishbone diagram, 320–323, 321f, 322f
    alarm fishbone diagram, example, 322f
    general fishbone diagram, example, 321f
  5 why's, 317–320
    "why-why" analysis sheet, example, 320f
  operating standards not followed, 319
  out-of-control action plans, 324–326
5S
  benefits of, 154–156
  decision to implement, 156–157
    summary of stages, 158f
  initial management implementation, 157–162
    audit sheets, 159–160
    implementation plan, example, 158f
    red tag holding area, 160–162
      holding area log sheet, example, 160–161
  meaning of 5S, 154
    currently used "S" equivalents, 154t
  schematic illustration of 5S process, 157f
  step 1: seiri-sort, 162–164
    red tag details, 162–164
    red tag, example, 163f
  step 2: seiton-set in order, 164–174
    floor markings, 172–173
    preferred storage locations, example, 166f
    shadow boards, 167
    spaghetti map, 170–171
    spaghetti map, before/after, example, 170f, 171f
    standard safety warnings, 173f
    storage labeling, example, 165f
    storing by first-in first-out sequence, 168–169
  step 3: seiso-shine, 174–179
    alternative style cleaning map, 176f
    clean and inspect checklist, example, 177f
    5S area audit sheet, 178f
    Plan-Do-Check Act cycle, 174f
    5S cleaning map or assignment map, 175–179
  step 4: seiketsu-standardization, 179–181
  step 5: shitsuke-self-discipline, 181–182
5 why's, 317–320
flip charts, 201t, 275
forced deterioration, 16, 32, 33
  use conditions role in, 34–37
Ford, Henry, 2
F-tags
  for initial clean/inspect, 45–50
  recording funguai with, 52–57
    F-tag log sheet, 53–55
    F-tag log sheet, example, 54f
    red F-tag, 53
    task certification sheet, 55
    task certification sheet, example, 56f
    white F-tag, 53
  record sheets for
    cleaning map, 63
    defect chart, 63
    defect map, 63
    failure analysis sheet, 63
    F-tag category spreadsheet, 63
    F-tag log sheet, 63
    task certification sheet, 63
  TPM causes for, 71–79
    basic condition neglect, 71–72
    data as percentage of fails/percentage hours of repair time, 79t
    design weakness, 73
    historical fails data sorted as repair time, 78f
    inadequate skill level, 73
    modified defect chart showing historical fails, 77f
    operating standards not followed, 72
    possible F-tag category spreadsheet example, 75f
    sample data as number of fails/hours of repair time, 79t
    standard defect chart for categories of causes, 76f
    unchecked deterioration, 72
    unknown, 73–74

fuel gauge system of skill log/task transfer sheet, 89
functions. *See* RCM

H

hard-to-access areas, 89, 91, 145
hidden failures, 234*f*
hidden functions, 275–276
    examples, 276*f*

I

inadequate skill level, 69, 73
infant mortality, 298

J

JIPM (Japanese Institute of Plant Maintenance), 4–5
Just in Time (JIT), 348

L

lean manufacturing, 6, 12, 340–357
  defects, 342–343
  equipment, 353–355
    right-sized tools, 355
  overprocessing, 347–348
  overproduction, 343–344
  pull, 355–357
  7-wastes as defined by Ohno, 342*f*
  transporting, 344–347
    production villages *vs.* cells, 346*f*
  unnecessary inventory, 348–349
  unnecessary operator movement, 349–350
  value, 350–353
    balancing costs, example, 350*f*
    simple value stream flow, example, 353*f*
  waiting, 344
lock out tag out (LOTO), 85, 131, 135, 145

M

machine history log, 63–65
  record sheets for
    categorization log sheet, 64
    costs spreadsheet, 64
    malfunction map, 64

maintenance/production friction, reasons for, 305–311
  introduction of new problem found after tool returned, 310–311
  preventing passing of post-maintenance test run, 310
  task times not established, 305
  turnaround parts not used, 305–308
  unexpected problems, 309–310
  when starting from scratch, 311–313
maintenance strategy, deciding on. *See* RCM PM analysis; TPM PM analysis
maintenance systems, history of development, 1–3
malfunction map, 64
Material Safety Data Sheet (MSDS), 51
Mean Time Between Failures (MTBF), 65, 225
Mean Time To Fail (MTTF), 35
Mean Time To Repair (MTTR), 65
memory prompter, 272*f*
minor stops/unrecorded losses, categorizing, 65–71
  actions for, 68–71
  area photographs/drawings record sheet, 65
  impact of minor stops, 65–68
  minor stops log sheets, 65
modular system, 128–129
Moubray, John, 272, 293

N

natural deterioration, 32–33
normal distribution. *See* binomial distribution
North, Marshall, 263

O

Observation Sheet, 194, 195*f*, 199
OCAPs (out-of-control action plans), 324–326
Ohno, Taiichi, 12, 183, 184, 185, 343
on-condition maintenance, 302–304, 303*f*
one-point lesson, 110
operating standards not followed, 72, 319

Index

Overall Equipment Efficiency (OEE), 10, 93–94
   formula for, 31–32

P

Pareto charts, 80t–82, 81f
partial failure, 15, 271, 284–285
persuading tool, 17
P-F curve, 99, 303f
photographs, 65, 194
pillars of TPM, 7–12. *See also* education/training pillars, TPM; safety pillars, TPM
   autonomous maintenance, 9
   education and training, 7–9
   focused improvement, 11
   health & safety, 7
   initial phase management, 11–12
   overview of, 8f
   planned maintenance, 9–10
   quality maintenance, 10–11
   support systems, 11
PM teams, 85–86f, 87, 97–101
Post-it Notes, 193–194, 199, 273, 322
Predictive Maintenance (PdM), 300
primary function, 269
process wafers, 261
protective clothing (PPE), 51

Q

Quick Changeover methodology. *See* SMED (single minute exchange of die)

R

RCM PM analysis, 227–255
   decision diagram, 228
      column sequence in tabular format, 229f
      example, 236f
      failure-finding decision blocks, 234f
      missing sign, 242f
      multiple failure frequencies, 239f
      on-condition monitoring decision blocks for hidden failures, 231f
      scheduled restoration/discard decision blocks for hidden failures, 234f
      sequence in tabular format, 231t
      three rows of operational decision blocks, 238f
   failure finding/calculating acceptable risk, 252–255
      failure finding, 253–255
      most common FFI values as percentage of MTBF, 255t
      proving testing frequency formula, 254f
   failure unacceptable/redesign system, 243
   failure of thermocouple measuring temperature of heating plate, 244f
   vacuum pump exhaust line blocks with dust, 246f
   recording process on decision worksheet, 249–252
      decision worksheet layout, 250f
RCM (Reliability Centered Maintenance)
   development of, 5–6
   equipment defined as functions, 267–280
      defining functions, 272–275
      duplicate function, 277
      duplicate function elimination flowchart, 278f
      ESCAPES, as memory prompter, 270f
      extra "numbers" columns on Information Worksheet, 281f
      fail safe device, 276
      failure modes, 268
      features that can be turned into functions for use in analysis, 275
      features *vs.* functions, 268t
      functional failure, 268
      hidden function examples, 276
      hidden functions, 275–277
      information worksheet, 273–274f
      modified Information Worksheet, 279f
      partial failure, 271
      performance standards for, 272
      primary functions, 269, 272
      protective covers, 277
      secondary functions, 269–270t
      total failure, 271
   flexibility of, 6
   functional failures/failure effects, 282

RCM (*contd.*)
  causes of failures, 287
  failure effect, 289–292
    costs of, 292
  failure modes, 285–288
    summary of, 288
  functions/functional failures/failure modes, example, 286*f*
  Information Worksheets, examples, 293*f*, 294*f*, 295*f*
  partial failures, 284–285
  serious consequences of failure, 288
  total failures, 284
  ideal team composition, 260*t*
  identifying functions/labeling, 280, 282–83
    defining functions, flowchart, 282*f*
    simplifying finding functions by subdividing areas, 283*f*
  non-time-based failures, 297–298, 297*f*
  operating context of analysis, 257–261
    contents of operating context, example, 259*f*
    process flowchart, 258*f*
  origin of, 292
  time-based failures, 296*f*
  tool analysis level, example, 261
    paddle drive and positioning, 264*f*
    TMX furnace photograph/schematic, 262*f*
    use of overlapping diagrams to show linkage, 265
    wafers/boats/paddle, 262*f*
reactive maintenance, 1–2, 269
risk assessment. *See* safety pillars, TPM
root-cause analysis, 73, 74, 112

S

safety issues. *See also* safety pillars, TPM
  authority for working in specific machine areas, 334–335
    area responsibility/certification table, example, 336*t*
    responsibility map, 335*f*
  in cleaning, 51
  standard safety warnings, 173*f*
safety pillars, TPM, 126–151
  area map, 127–129
  area map for Nova implanter, example, 128*f*

countermeasures, 141–145
  eliminating risk, 141
  enclosure, 141–142
  equipment condition, 143–145
  substitution, 141
  training, 142
  using personal protection equipment, 142–143
hazard map, 129–131
hazard map, example, 130*f*
person who creates risk assessment, 145–146
quantifying risk, 139–141
risk assessment, 131–137
  Level 1 risk assessment, example, 134*f*
  Level 2 risk assessment, example, 135*f*
  Level 3 risk assessment, example, 136*f*
  natural levels of, 132*t*–136, 133*f*
risk assessment categories, 137–139
  biological, 139
  chemical, 137–138
  electrical, 137
  gas, 138
  height, 138
  magnetism, 139
  manual handling, 138
  mechanical, 137
  moving vehicles, 139
  radiation, 139
  slips/trips, 138
  stored energy, 138
  temperature, 138
risk variation in initial assessment, 137*f*
safe working procedures as standards, 146–151
  "step-by-step" safe working procedure, examples, 147*f*, 148*f*
scheduled discard, 226
secondary function, 271–272*t*
second-sourced parts, 36–37
self-discipline, 181–182
shadow boards, 167
Six Sigma
  analysis stages, 371–373
  analyze, 372
  control, 373
  define, 371
  improve, 372–373
  measure, 371–372

champion or sponsor, 368
considerations/limitations in, 373–375
  faultfinding cause of lamp failure,
    374–376
  graphs/their use in, 361
    average/standard deviation, 362–364
      average wages, 363f
      data/histogram of wage
        distribution, 363t
      binomial distribution, 362f
      standard deviation, 362
      standard deviation/z scores, 364
    x-y graph, 365–366
      simple x-y graph, 366f
  main terms of Six Sigma, 367–370
    customer, 367
    teams/leaders, 369
  possible limitations with using
    statistics, 376–378
  rules/expectations, 368–369
  Six Sigma controller, 368
  technical stuff, 369–373
    defects per million opportunities, 369
    defects per opportunity, 369
    sigma value, 369
      percentage failures plotted against,
        370f
Six Sigma Charter, 368–369
slip, 263
SMED (single minute exchange of die), 6
  analysis sequence for changeover, 186f
    step 1: creating SMED team, 186–188
      specific functions for analysis,
        example, 187t
      team members/responsibilities of,
        186–188
    step 2: select tool, 188–189
    step 3: document every step of
      changeover, 189–196
      Observation Sheet, 194, 195f
      parallel task allocation, 193f
      photographs, 194
      Post-it Notes/notebook, 193–194
      video, 194, 196
    step 4: viewing changeover as bar
      graph, 196–197
      Observation Sheet, 195f, 196
    step 5: define target time for
      changeover, 197–198
      reduction plan, example, 198f
    step 6: analysis of elements, 198–204
      analysis sheet, 200f
      flip chart analysis, 201t
    step 7: repeating exercise, 211
  applying to maintenance/use of
    turnaround parts, 211–212
  goal of, 191
  origin of, 183–185
    pros/cons of large batch production,
      185t
  SMED analysis, 204–210
    analysis chart with improvements
      and ideas, 209f
    cost vs. improvement impact chart,
      205f
    create new procedure, 208–210
    implementing ideas, 208
spaghetti map, 170–171
spaghetti map, before/after, example, 170f,
    171f
spare parts trolley, 309

T

task certification sheet, 63
team
  AM, 85, 86f, 87, 92, 216
  ideal composition for RCM, 260t
  PM, 85–86f, 87, 97–101
  Six Sigma, 367
  SMED, 186–188
  Zero Fails, 10, 85, 86f, 87, 107–108,
    110
    composition-membership, 111f
    composition-membership,
      overlapping management, 111f
team objectives and activity boards
  activity boards, 327–328
    layout of activity board, example,
      33f
  authority for working in specific
    machine areas, 334–335
    area responsibility/certification table,
      example, 336t
    responsibility map, 335f
  monitoring progress, 331
    calculating failure rate/targeting
      improvement, 331
    initial improvement sheet, example,
      332f
    what to monitor, 331
  results of real RCM analysis, 335
    summary of boatloader analysis,
      337–340
      boatloader assembly, 338f
  team goals, 329–330

throughput, 26, 31
time- and condition-based maintenance
   introduction to on-condition
      maintenance, 302–304
      costs of, 304
      P-F curve used for, example, 303*f*
   maintenance/production friction,
      reasons for, 304
      introduction of new problem
         found after tool returned, 310–311
         preventing passing of
            post-maintenance test run, 310
      task times not established, 305
      turnaround parts not used, 305
      turnaround parts not used, example, 306*f*
      unexpected problems, 309–310
TM. *See* TPM
tool-specific training record, 109–110
   example of, 109*f*
Toyota Production System. *See* lean manufacturing
TPM PM analysis, 214–227
   interpreting PM maps, 220–223
      map example, 222*f*
   malfunction/PM maps, 215–220
      analysis sheet, 218*t*
      flowchart, 219*t*
      green dot allocation table, 217*t*
      malfunction map, 215*f*
      map of mechanical "electrode" assembly, 217*f*
   scheduled maintenance or scheduled restoration, 223–226
      statistical pattern of failures of part over time, 225*f*
   scheduled replacement or scheduled discard, 226–227
   visual guide of main TPM steps, 224*f*
TPM (Total Productive Maintenance), overview of. *See also* AM; AM (autonomous maintenance), creating standards/preparation for; education/training pillars, TPM; failure data, analyzing/categorizing; safety pillars, TPM
   basic condition, 21–24
   fault development, 15–21
      alternative, 16
      examples of, 16–21
      partial failures, 15
      total failures, 15
   ideal condition, 37
   improvement methodology, 37–41
   natural and forced deterioration, 32–33
   overall equipment efficiency, 24, 26–32
      availability, 29–30
      performance, 31
      quality, 31–32*t*
   restoring basic condition, 41–42
   technical standards, 24–26
      examples of, 25–26
   use conditions, 34–37
training. *See* education/training pillars, TPM
training record, 107
   on "activity board," example, 108*f*
triple redundancy, 255
turnaround parts, 211–212, 227, 305, 306*f*

U

unchecked deterioration, 69–70

V

value, 350–353
   balancing costs, example, 350*f*
   simple value stream flow, example, 353*f*
vibration monitoring, 303–304
voice of customer (VOC), 367

W

"why-why" analysis sheet, example, 320*f*

X

x-y graph, 365–366
   simple x-y graph, 366*f*

Z

Zero Fails (ZF) team, 10, 85, 86*f*, 87, 107–108, 110
   composition-membership, 111*f*
   composition-membership, overlapping management, 111*f*

ABOUT THE AUTHOR

Steven Borris has 33 years of maintenance and training experience; 20 of which were with U.S. companies. While a senior engineer at Varian's European Implanter Group, he oversaw all operating issues, supported customers, advised on improvement initiatives, and provided training for operation/maintenance and electronics. He resides in Glasgow, Scotland.